D0990141

John T. Brosnan, D. Phil.

MAMMALIAN PROTEASES:
A Glossary and Bibliography

MAMMALIAN PROTEASES:
A Glossary and Bibliography
Volume 2
Exopeptidases

J. Ken McDonald
Department of Biochemistry
Medical University of South Carolina
Charleston, South Carolina, U.S.A.

Alan J. Barrett
Department of Biochemistry
Strangeways Research Laboratory
Worts Causeway, Cambridge, U.K.

1986

ACADEMIC PRESS
Harcourt Brace Jovanovich, Publishers

London Orlando San Diego New York Austin
Montreal Sydney Tokyo Toronto

ACADEMIC PRESS INC. (LONDON) LTD
24/28 Oval Road
London NW1

United States Edition published by
ACADEMIC PRESS INC.
Orlando, Florida 32887

Copyright © 1986 by
ACADEMIC PRESS INC. (LONDON) LTD

All Right Reserved
No part of this book may be reproduced in any form by photostat,
microfilm, or any other means, without written permission
from the publishers

British Library Cataloguing in Publication Data
Mammalian proteases: a glossary and bibliography.
Vol. 2
1. Proteinase — Handbooks, manuals, etc.
2. Proteinase — Bibliography
I. Title   II. Barrett, A.J.
599.01'9256   Q609.P75
ISBN 0-12-079502-7

Typeset by Communitype, The Grange, Highfield Drive,
Wigston, Leicester.
Printed by Whitstable Litho Ltd., Whitstable, Kent.

PREFACE

Many of the remarks contained in the Preface to Volume 1, which focused on the endopeptidases, apply equally well to the present volume. The field of proteolytic enzymes continues to be the scene of vigorous research in many countries, and the enzyme systems responsible for protein breakdown are generally viewed as being vitally important for the normal health and development of living organisms. Like the endopeptidases, the exopeptidases are the subject of growing interest, for both basic and utilitarian reasons, to chemists, biochemists, physiologists, pharmacologists and pathologists.

It has been our contention that review articles and books written and printed in conventional ways are inadequate to the bench-top needs of research workers, especially to those working in research areas that are peripheral to the subject covered. Accordingly, Volume 2, like its predecessor, was generated as a possible remedy, using the same computer-aided system of codification and presentation of information. Our objective has been to provide a concise summary of the properties of each enzyme, together with a bibliography that includes titles, and is sufficiently complete to allow the scientist to find whatever additional information he may need with the minimum of difficulty. The data are stored in a computer-accessible form, and we plan to keep the text under constant revision for use in future editions.

Whereas Volume 1 was concerned with the proteases (endopeptidases) that act, in general, to initiate protein degradation by cleaving internal peptide bonds within long polypeptide chains, Volume 2 is concerned with proteases (exopeptidases) that act at one or the other terminus of a polypeptide chain to remove one, two or three amino acids at a time, action sometimes being restricted to dipeptides or tripeptides. It would appear that the exopeptidases evolved as part of a synergistic, proteolytic mechanism that is especially effective in catalyzing the release for subsequent reuse of essential amino acids from dietary proteins, and from intracellular proteins that are degraded as part of the dynamic process of protein turnover. For example, in the gastro-intestinal tract, chymotrypsin and trypsin act on proteins to release fragments possessing C-terminal aromatic or basic residues, respectively. The specificities of these endopeptidases are perfectly complemented by those of the exopeptidases, carboxypeptidases A and B. The former acts to release the exposed C-terminal aromatic amino acids such as phenylalanine and tryptophan, and the latter releases the exposed arginine and lysine residues. Since these amino acids include several that are essential dietary components, one can readily recognize the survival value of the complementary specificities of endopeptidases and exopeptidases.

Specific exopeptidases of the blood and tissues are responsible also for the production and inactivation of poly peptide hormones and other biologically-active peptides that exhibit a wide range of activities, including the regulation of fluid and electrolyte balance, blood pressure, smooth muscle contraction, vascular permeability, chemotaxis and neuro transmission. Because of their well-defined specificities, certain exopeptidases are used by protein chemists as valuable tools for protein sequencing, end-group analysis and structural modifications.

The present volume provides a ready reference on exopeptidase properties and specificities. It also serves as an effective key to the literature for the study of the exopeptidases and their contribution to intracellular protein degradation and to physiological regulation. In addition, it provides useful information for those wishing to utilize purified exopeptidases as reagents.

J.K.McD.

A.J.B.

ACKNOWLEDGEMENTS

We gratefully acknowledge the valuable help and advice given by friends and colleagues including C. A. Arce, R.-M. Bålöw, J. Butterworth, B. J. Campbell, D. W. Cushman, K. Docherty, J. C. Hutton, G. Kalnitsky, H. Kirschke, J. F. Lenney, K. K. Mäkinen, M. T. McQuillan, G. O'Cuinn, M. Orlowski, S. Tsunasawa, E. Shaw, W. H. Simmons, R. E. Smith, S. H. Snyder, E. Söderling, W. Stauber and S. Wilk.

A special expression of gratitude goes to Ellenor (wife of JKM) for her valuable and thoughtful assistance with bibliographic searches, cataloging, and proofreading. We also take pleasure in thanking Mrs Joan Eynon (Charleston) and Mrs Ann Hall (Cambridge) for their skilled secretarial assistance and word-processor wizardry. Mr Neil Rawlings (Cambridge) skillfully compiled the index. Also, a "well-done" commendation goes to the Postal Services of the United States and the United Kingdom for not having lost a single piece of mail during the unfolding of this tale of two cities!

J. K. McD.

A. J. B.

CONTENTS

Section 18
Omega Peptidases

CONTENTS OF VOLUME 1

Section 2
Cysteine Proteinases

Section 3
Aspartic Proteinases

Section 4
Metallo-Proteinases

Section 5
Unclassified Proteinases

ABBREVIATIONS AND CONVENTIONS

Generally, the names of amino-acids and their derivatives have been abbreviated according to common practice (IUPAC-IUB Commission on Biochemical Nomenclature. Symbols for amino-acid derivatives and peptides. Recommendations 1972. J. Biol. Chem. **247**, 977-983, 1972). Amino-acids may be assumed to be in the L-configuration unless D- or DL- is specified.

Less common amino-acid residues

Orn	L-Ornithine
Sar	N-Methylglycine (sarcosine)
γ-Glu	γ-(Linked)-glutamic acid
PCA	Pyrrolid-2-one-5-carboxylic acid

Substituents at the amino group

Ac-	Acetyl
Boc-	t-Butyloxycarbonyl-
Bz-	Benzoyl-
Hip-	Benzoylglycyl- (hippuryl-)
HCO-	Formyl-
FA-	Furanacryloyl-
Z-	Benzyloxycarbonyl-

Substituents at the carboxyl group

$-CH_2Cl$	Chloromethane
$-CHN_2$	Diazomethane
-NH2	Amide
-NMec	7-(4-methyl)coumarylamide

-NNap	2-Naphthylamide
-NNapOMe	2-(4-methoxy)naphthylamide
-NPhNO$_2$	4-Nitroanilide
-OEt	Ethyl ester
-OMe	Methyl ester

Other abbreviations

Dip-F	Diisopropyl fluorophosphate
IAcNH$_2$	Iodoacetamide
IAcOH	Iodoacetate
fMet	Formylmethionine
Hyp	4-hydroxyproline
MalNEt	N-Ethylmaleimide
NNapOMe	4-Methoxy-2-naphthylamide
Pms-F	Phenylmethanesulfonyl fluoride
SDS	Sodium dodecyl sulfate
Argininic acid	2,L-Hydroxy-5-guanidino-valeric acid

Conventions

Antipapain, chymostatin, elastatinal, leupeptin, pepstatin and phosphoramidon are microbial proteinase inhibitors (see Umezawa, H. & Aoyagi, T. Activities of proteinase inhibitors of microbial origin. In: *Proteinases in Mammalian Cells and Tissues* (Barrett, A. J. ed.), pp. 637-662, North-Holland Publishing Co., Amsterdam, 1977).

INTRODUCTION

Organization

The present volume, like its predecessor, has been compiled with the aim of providing a convenient and concise source of information on the mammalian proteolytic enzymes. Whereas Volume 1 dealt with the **endopeptidases**, this volume focuses on the **exopeptidases** - a term that was introduced by Bergmann and Ross to describe proteases that degrade peptide chains from their ends (see J. Biol. Chem. *114*: 717-726, 1936). The text of this work was assembled in Cambridge by a computer-assisted technique, and our intention is to continue to up-date and revise the information in subsequent publications.

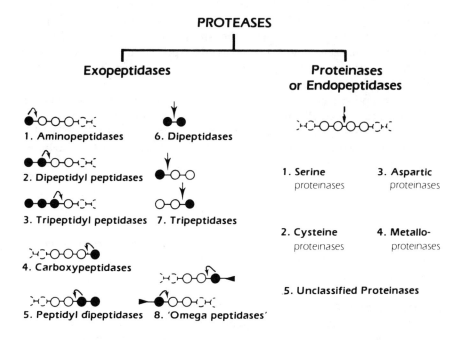

Figure 1

Sections 1 through 10 were reserved for the **endopeptidases** and Sections beginning with 11 were reserved for the **exopeptidases**. Thus, Volume 1 begins with **Entry 1.01**, the first endopeptidase (in the **serine proteinase** class) to be considered. Volume 2 begins with **Entry 11.01**, the first exopeptidase (in the **aminopeptidase** class) to be considered. The **Tables of Contents** for both volumes are included here.

Classification

For many years exopeptidase classification has been unsettled, to say the least. To a large degree, this confusing state of affairs has reflected our incomplete and changing views concerning the identities and specificities of the exopeptidases. However, the last few years have witnessed a greatly increased understanding of this diverse group of proteases, and, with some reservation, it is probably safe to say that a rational system of classification (and nomenclature) is emerging that is based on a more complete and more accurate knowledge of exopeptidase properties and specificities.

In contrast to the endopeptidases, which are classified in Volume 1 according to their catalytic mechanisms (as **serine, cysteine, aspartic** and **metallo**), the exopeptidases are herein classified (as illustrated in Fig. 1) according to their substrate specificity, and given **trivial** class names that indicate the site (terminus) of attack and the size of the liberated fragment. In Figure 1, the solid circles identify the liberated moiety, which may be an amino acid, a dipeptide, or a tripeptide, and in the case of the **omega peptidases**, a modified or derivatized amino acid. The names assigned to specific enzymes within a class usually indicate the preferred or required terminal (or penultimate) residue, either specifically or by class.

The exopeptidase classes described in this volume, and listed in Figure 1, include the (**aminoacyl**) **aminopeptidases** (α-aminoacylpeptide hydrolases, EC 3.4.11), **dipeptidyl peptidases** (dipeptidylpeptide hydrolases EC 3.4.14), **tripeptidyl peptidases** (tripeptidylpeptide hydrolases, EC 3.4.-), **carboxypeptidases** (peptidylamino acid hydrolases, EC 3.14.16 - 18), **peptidyl dipeptidases** (peptidyldipeptide hydrolases, EC 3.4.15), **dipeptidases** (EC 3.4.13), **tripeptidases** (tripeptide aminopeptidase, EC 3.4.11.4), and **"omega peptidases"** (EC 3.4.19). In the case of two classes, the dipeptidases and tripeptidases, the names are based simply on substrate size requirements, in terms of the required number of unsubstituted amino acids. Trivial names (as opposed to the systematic names shown within the parentheses) are used throughout Volume 2.

Nomenclature

As illustrated in Figure 1, the term **protease**, herein considered to be a general term equivalent to **peptide hydrolase**, is applied equally to both exopeptidases and endopeptidases. On the other hand, the term **proteinase** is applied only to proteases that exhibit endopeptidase activity, and the term **peptidase** to proteases that exhibit only exopeptidase activity on small peptides. A guideline used in both Volumes 1 and 2 requires that a protease be classified as a proteinase if it exhibits a significant degree of endopeptidase activity, regardless of whether it exhibits exopeptidase activity.

We have proposed that the term **omega peptidase** be used to designate a new class of exopeptidases with specificities that do not conform to classical definitions for aminopeptidases and carboxypeptidases. The Enzyme Nomenclature Committee has recently adopted the **omega peptidases** as a new class (3.4.19) of exopeptidases capable of removing terminal residues that (a) lack a free α-amino or α-carboxyl group (i.e. the pyroglutamyl and aminoacylamide groups), or (b) are linked through a scissile bond that involves a carboxyl or amino group that is not attached to an -carbon (i.e. a linkage involving the ω-carboxyl group of aspartic or glutamic acid, or the ϵ-amino group of lysine). Such bonds are generally refered to as **isopeptide bonds**. By comparison, the classical exopeptidases are restricted to hydrolyzing α-peptide linkages (which involve only α-amino and α-carboxyl groups) to release amino acids, dipeptides or tripeptides from unsubstituted N- or C-termini. The sequential release of these fragments from oligopeptide substrates can result in a substantial degree of degradation.

Names applied here to many of the (aminoacyl) amino peptidases are based on their preferences or requirements for a particular N-terminal amino acid. However, because so few kinetic data have been published, preferences used to name these enzymes could not be based on V_{max} values or catalytic coefficients, k_{cat}/K_m. Thus, in keeping with tradition, a name that reflects a preference for a particular terminal amino acid is generally based on relative rates of hydrolysis determined under optimal conditions and at substrate concentrations that were usually well in excess of K_m values. For example, an enzyme that shows its highest rate of hydrolysis on N-terminal alanyl bonds is named **alanyl aminopeptidase** (not alanine amino peptidase), and is abbreviated **AAP**. Further, in order to avoid ambiguous abbreviations, the one-letter-code was adopted to identify the preferred or required amino acid. Thus, the abbreviation for **arginyl aminopeptidase** is RAP. In an effort to discourage the use of letters to name aminopeptidases, names such as aminopeptidase A, aminopeptidase B, aminopeptidase M, aminopeptidase N, and aminopeptidase P have been replaced with others that reflect substrate specificity and, if needed to avoid ambiguity, subcellular localization.

Similarly, names applied to carboxypeptidases that release single amino acids serve to identify the required or preferred C-terminal residue. An enzyme that removes C-terminal arginine faster than any other amino acid is called an **arginine carboxypeptidase** (not arginyl carboxypeptidase). In contrast to **arginyl** aminopeptidase, the carboxypeptidase does not actually cleave an arginyl linkage. Accordingly, an arginyl residue would not be expected to participate in the acyl-enzyme intermediate. This nomenclature distinction becomes especially discriminating, as illustrated in Figure 2, when it is applied to the naming of enzymes such as **prolyl carboxypeptidase** and **proline carboxypeptidase**. The former term identifies a carboxypeptidase that recognizes a prolyl residue in the **penultimate position**, with little regard for the identity of the C-terminal amino acid, and specifically hydrolyzes the prolyl linkage, resulting in the release of a C-terminal amino acid. The latter term, proline carboxypeptidase, identifies a carboxypeptidase that preferentially removes C-terminal proline residues. As illustrated in Figure 2, the same principles apply to the naming of **prolyl aminopeptidase, proline aminopeptidase, proline dipeptidase ("prolidase"), and** prolyl dipeptidase **("prolinase")**.

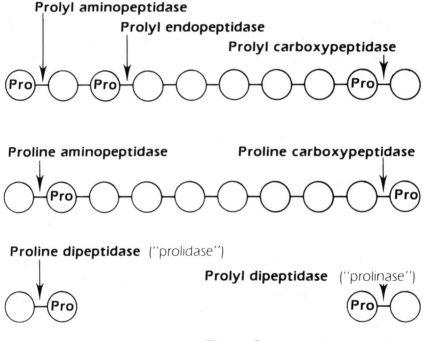

Figure 2

As illustrated in **Figure 3, the dipeptidyl peptidases** and the **peptidyl dipeptidases** are aminopeptidases and carboxy peptidases, respectively, that catalyze the (sequential) release of dipeptides from oligopeptides. The dipeptidyl peptidases were previously known as **dipeptidyl aminopeptidases (DAP's)** and Roman numerals were used to identify the (four) known examples. This numbering convention has been retained. However, in the case of peptidyl dipeptidases, which are carboxypeptidases (in the sense that their action requires the presence of an unsubstituted α-carboxyl group), it is recommended that capital letters be used to designate various members of this group. Such a convention would be consistent with the long-standing practice of naming carboxypeptidases alphabetically. As a broader generalization, it may prove helpful to use Roman numerals whenever it becomes necessary to distinguish between exopeptidases acting at the N-terminus, and to use capital letters when distinguishing between those acting at the C-terminus. This convention should help to distinguish between "tongue-twisters" like those in Figure 3!

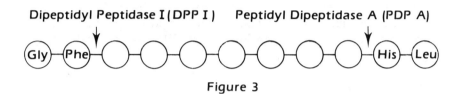

Figure 3

The two exopeptidase classes represented in Figure 3 exhibit very different mechanisms of action. The dipeptidyl peptidases are believed to form an active dipeptidyl-enzyme intermediate that mediates the transfer of the dipeptidyl moiety. In fact, DPP I, exhibits considerable dipeptidyl transferase and polymerase activity. In such systems, reactants other than water can act as acceptors. These include hydroxylamine and other unprotonated amino groups. Such a reaction mechanism is not exhibited by peptidyl dipeptidase A (angiotensin converting enzyme), and for this reason one of its earlier names, "dipeptidyl carboxypeptidase", was never really appropriate.

Subcellular localization and tissue distribution
Because some exopeptidases have similar or identical specificities, it has proved useful to give them names that distinguish them by their different subcellular localizations. Thus, some exopeptidases were given names that include prefixes such as **soluble, lysosomal,** or **microsomal**. These operational (working) terms serve to designate subcellular fractions that are usually obtained by differential centrifugation. For example, **microsomal alanyl aminopeptidase** (mAAP) is found primarily in vesicularized membrane fractions sedimented by high-speed ultracentrifugaton. These findings provide only indirect evidence for a true membrane-bound localization within intact cells. **Soluble alanyl aminopeptidase** (sAAP) by way of comparison, is believed to be freely soluble (cytosolic) in cells on the basis of its presence in high-speed supernatants obtained during differential centrifugation. Evidence for the

lysosomal localization of some exopeptidases is usually derived from sedimentation and latency studies, but in many instances has been confirmed by cytochemistry.

Tissue distributon is also used in the naming of exopep tidases. For example, three different carboxypeptidases with B-type specificity (that preferentially remove C-terminal basic amino acids, arginine and lysine) have been identified. One occurs in the pancreas, one in the blood, and one in the lysosomes of most cells. The pancreatic enzyme has for many years been referred to simply as **carboxypeptidase B**, abbreviated **CPB**. More recently an enzyme with the same properties (and given the same EC number) has been detected in tissues other than the pancreas. Consequently, this enzyme (regardless of tissue source) is herein referred to as **tissue carboxypeptidase B**, abbreviated **tCPB**. The two remaining enzymes have their own identities, and are herein referred to as **plasma carboxypeptidase B (pCPB) and** lysosomal carboxypeptidase B (*l*CPB).

Generalizations

Most of the exopeptidases seem to be metallo-proteins with high Mr values, often reaching 300,000. They are generally found to be glycoproteins, and to have complex subunit structures. Thus, they differ from the endopeptidases, which commonly have relatively simple molecular structure, and accordingly the exopeptidases tend to be more difficult to prepare and to handle.

The specificities of the exopeptidases often allow for a high degree of synergism with the endopeptidases, as well as with one another. It appears that the specificities have evolved as part of a proteolytic mechanism that provides for the efficient retrieval of essential amino acids (such as lysine, arginine, leucine and the aromatic amino acids) from dietary proteins, and from intracellular proteins that are degraded as part of the dynamic process of protein turnover.

A BRIEF HISTORY OF
THE STUDY OF
MAMMALIAN EXOPEPTIDASES

by J. K. McDonald

*In memory of Peter Dehm, Ph.D., and his contributions
to the exopeptidases, born in Munich, July 8, 1941; died
in Charleston, October 25, 1980.*

Exopeptidases were among the first proteases to be discovered in mammalian tissues, and the contributions of many of those who worked in the field have been described in articles by Waldschmidt-Leitz (1), Johnson & Berger (2), Smith (3, 4), Smith & Hill (5), Hanson (6), McDonald *et al.* (7), DeLange & Smith (8), and McDonald & Schwabe (9).

Over half a century has passed since "dipeptidase" was detected in extracts of hog intestinal mucosa by Josephson & von Euler (10). These extracts, which were referred to as intestinal "erepsin" preparations, and were the original source of many exopeptidases, proved to contain a complex mixture of enzymes of diverse specificity (11). As early as 1929, it was recognized in the laboratories of Linderstrøm-Lang (12) and Waldschmidt-Leitz (13) that the "dipeptidase" activity of intestinal extracts was attributable to more than one enzyme. Linderstrøm-Lang (12) described an exopeptidase that was distinct from the dipeptidase and aminopolypeptidase of Waldschmidt-Leitz (13). This enzyme, which Linderstrøm-Lang referred to as "leucyl peptidase", was destined to become known as leucine aminopeptidase, the prime example of an N-terminal exopeptidase (11).

Also in 1929, Waldschmidt-Leitz & Purr (14) discovered in the beef pancreas a protease that cleaved C-terminal amino acids from acylated peptides. They called this "carboxypolypeptidase" to distinguish it from the aminopolypeptidase of intestinal origin. Eight years later, when the carboxypeptidase was crystallized by Anson (15), it became the first of the proteases active on simple substrates of known structure to be obtained in crystalline form. It was often referred to as "Anson's enzyme". In 1931, Waldschmidt-Leitz and co-workers

(16) described another pancreatic protease that released arginine from various acetylated protamines, but not from protamine esters. About 25 years later, Folk (17) showed that the activity of "protaminase" was attributable to a second carboxypeptidase which he named "basic carboxypeptidase" because of its almost absolute specificity for liberation of the basic amino acids, lysine and arginine. Folk & Gladner (18) then proposed that this protease be called "carboxypeptidase B", and that the classical carboxypeptidase of Waldschmidt-Leitz be called "carboxypeptidase A". By way of contrast, the latter exhibited a distinct preference for C-terminal aromatic and branched-chain aliphatic amino acids. Both carboxypeptidases were shown to be secreted as zymogens (procarboxypeptidases) activatable by trypsin.

After about 1950, mammalian tisues other than those of the gastrointestinal tract were increasingly studied in the search for new exopeptidases, as well as for richer sources of the known enzymes. Because these tissue proteases were apparently committed to intracellular (rather than extracellular) protein degradation, there was a growing interest in their intracellular localization. In the remainder of this brief review, a descriptive comment will be included for each of the most familiar and well-characterized exopeptidases, and an attempt will be made to cite the original (or at least a primary) reference for its discovery. The exopeptidases will be grouped according to their subcellular localization, i.e., cytosolic, membrane-bound, or lysosomal. Within each group they will be cited primarily in their order of discovery. The names that will be used are the most (historically) familiar ones, but, wherever appropriate, the names that we have adopted for the entries in the present volume will be given too (in italics). A group designated the "omega peptidases" has been introduced. Peptidases assigned to this group will include those whose actions are restricted to one end or the other of a polypeptide chain, but whose specificity requirements do not include unsubstituted (N-terminal) α-amino or (C-terminal) α-carboxyl groups. Generally, the soluble and membrane-bound proteases can be assumed to have pH optima in the neutral-to-alkaline range, whereas the lysosomal ones have acidic pH optima.

Cytosolic exopeptidases

Linderstrøm-Lang's leucine aminopeptidase (*leucyl aminopeptidase*) was the first of the cytosolic aminopeptidases to be characterized. In 1955, Spackman *et al.* (19) reported that hog kidney was a more stable and plentiful source than intestinal mucosa. They obtained the enzyme in nearly homogeneous form, as did Hanson *et al.* (20) and Schwabe (21) who found beef eye lenses and dental pulp, respectively, to be rich sources. In 1964, a cytosolic chloride- and sulfhydryl-dependent aminopeptidase was discovered in rat liver extracts (22). It was called "aminopeptidase B" (*arginyl aminopeptidase*) because of its strict specificity for the removal of N-terminal Arg- and Lys- residues from 2-

naphthylamide derivatives and peptides. Aminopeptidase B appears to act as the N-terminal counterpart of carboxypeptidase B. In 1966, Behal *et al.* (23) described an alanine aminopeptidase (*soluble alanyl aminopeptidase*) in human liver that was Co^{2+}-dependent and puromycin-sensitive. Although alanine was the preferred residue, many other N-terminal residues were cleaved. In contrast to this broad specificity, Cheung & Cushman (24) identified a Mn^{2+}-stimulated exopeptidase in dog kidney that was highly specific for N-terminal Asp- and Glu-residues of 2-naphthylamides and polypeptides; the rate of Asp-cleavage was about 3-fold that for Glu-, hence the name "aspartate aminopeptidase" (*aspartyl aminopeptidase*).

In 1967, Ellis & Nuenke (25) characterized a dipeptidyl arylamidase from beef pituitary extracts that was later renamed (7) "dipeptidyl aminopeptidase III" (*dipeptidyl peptidase III*). This enzyme, which could be specifically assayed with Arg-Arg-2-naphthylamide (at pH 9.0), catalyzed the removal of dipeptidyl moieties sequentially from the N-terminus of certain unsubstituted polypeptides.

The first example of a cytosolic "tripeptidyl aminopeptidase" (*tripeptidyl peptidase II*) was reported in 1983 by Bålöw *et al.* (25a). The enzyme purified from rat liver catalyzes the consecutive release of tripeptides at pH 7.5 from the N-termini of oligopeptides derived from rat liver pyruvate kinase.

In 1973, Chenoweth *et al.* (26) purified an exopeptidase from hog kidney that released N-terminal amino acids from tripeptides only. This enzyme, known as "tripeptidase" (*tripeptide aminopeptidase*) was a metal- and sulfhydryl-dependent exopeptidase that acted only on tripeptides having unsubstituted N- and C-termini. Tripeptides with proline or hydroxyproline in the central position were not attacked. Also in 1973, Nordwig & Mayer (27) isolated "proline imino-peptidase" (*prolyl aminopeptidase*) from hog kidney. They showed that it specifically removed N-terminal prolyl residues from polypeptides.

Examples of cytosolic carboxypeptidases could not be found, but some of the best known dipeptidases are cytosolic. In 1932, Grassmann *et al.* (28) postulated the existence of a distinct dipeptidase in hog intestinal mucosa that hydrolyzed proline-containing dipeptides such as Pro-Gly. They ascribed this activity to a "prolinase" (*prolyl dipeptidase*). Five years later, Bergmann & Fruton (29) found that the tissue contained another dipeptidase, which they termed "prolidase" (*proline dipeptidase*), that hydrolyzed Gly-Pro. During the ensuing thirty years, prolinase and prolidase remained the only two mammalian proteases known that specifically hydrolyzed peptide bonds involving proline and hydroxyproline. The literature abounds with reports of other dipeptidases; these reports are too numerous to mention individually and are often difficult to interpret. Many peptidase activities have been incorrectly attributed to dipeptidases as a consequence of the uncritical use of certain substrates. Nevertheless, it appears that Josephson & von Euler (10) correctly attributed the hydrolysis of Gly-Gly to a specific dipeptidase (*glycylglycine dipeptidase*) that is strongly activated by Co^{2+} (30, 31). Similarly, the Mn^{2+}-activated, intestinal dipeptidase (*glycylleucine dipeptidase*) first reported by Smith (32) also appears to be a true cytosolic dipeptidase. Other cytosolic dipeptidases are the "carnosinase" (*ß-alanylhistidine dipeptidase*) described by Hanson & Smith (33), the "homocarnosinase" (*γ-homoalanylhistidine dipeptidase*) of Lenney (34), and the "cysteinylglycinase" (*cysteinylglycine dipeptidase*) of Binkley (35).

Membrane-bound exopeptidases

Exopeptidases of almost every class have been found in membranes, and usually appear in the microsomal fraction. Over the past twenty years, four membrane-bound, N-terminal exopeptidases have been discovered in kidney extracts. In 1962, Glenner *et al.* (36) detected "aminopeptidase A" (*glutamyl aminopeptidase*) in rat kidney and characterized it as a Ca^{2+}-activated enzyme that was relatively specific for the removal of N-terminal dicarboxylic amino acids. The rate of Glu-removal exceeded that for Asp- by about six-fold. In the following year, Pfleiderer & Celliers (37) discovered "aminopeptidase M" (*microsomal alanyl aminopeptidase*), which has subsequently become well known. They showed that it had a broad range of action on N-terminal, neutral amino acids residues, and was a Zn^{2+}-metallo-protein.

Hopsu-Havu & Glenner (38) discovered another renal enzyme that released intact Gly-Pro from its 2-naphthylamide derivative. Its ability to catalyze the removal of Gly-Pro from oligopeptide substrates was confirmed and extended by McDonald *et al.* (6), and on the basis of its ability to release dipeptides sequentially it was named "dipeptidyl aminopeptidase IV" (*dipeptidyl peptidase IV*). In 1970, Dehm & Nordwig (39) uncovered a renal X-Pro-aminopeptidase (*proline aminopeptidase*) that was capable of catalyzing the cleavage of the bond between an N-terminal amino acid residue and a penultimate prolyl residue. Good rates were obtained with Gly-Pro-Hyp.

The only well-characterized membrane-bound dipeptidase was discovered in hog kidney by Robinson *et al.* (40). This "renal dipeptidase" (*microsomal dipeptidase*) cleaved Ala-Gly faster than Gly-Gly, and was Zn^{2+}-activated. In contrast, the cytosolic glycylglycine dipeptidase mentioned above was Co^{2+}-activated, and hydrolyzed only Gly-Gly.

Two C-terminal exopeptidases are known to be membrane-bound, and both have been extensively characterized. The first to be discovered was "angiotensin converting enzyme" (*peptidyl dipeptidase A*). It was detected by Skeggs *et al.* (41) as an enzyme in horse plasma that removed the C-terminal dipeptide His-Leu from angiotensin I. The specificity of this Zn^{2+}-metallo-"carboxypeptidase" is now recognized as being broader than was at first realized. The second of the enzymes was "carboxypeptidase P" (*microsomal prolyl carboxypeptidase*). Dehm & Nordwig (42) isolated this enzyme from hog kidney microsomes and showed that it catalyzed a Mn^{2+}-activated release, primarily of neutral C-terminal amino acids. The rates of release were greatest when a prolyl residue was in the penultimate position. Relative to the rate obtained on Z-Pro-Ala (the best assay substrate), ten percent of "prolinase-type" activity was observed on Pro-Ala.

Lysosomal exopeptidases

As early as 1929, Willstatter & Bamann (43) had observed that aqueous extracts of animal tissues were capable of digesting protein substrates at acidic pH, and they termed the responsible enzyme "Kathepsin" from the Greek "to digest". Fruton and his collaborators (44) later detected at least four distinct .PA proteases in cathepsin preparations. In 1952, the cathepsin nomenclature was revized (45), and cathepsins A, B and C became widely known as the intracellular counterparts of pepsin, trypsin and chymotrypsin, respectively. Because the cathepsins, by definition, comprised tissue proteases that exhibited optimal activity at acidic pH, it was not surprising that all of these enzymes were eventually found to be localized in lysosomes - the intracellular organelles that were first shown by de Duve and his associates (46) to contain the bulk of the acid hydrolases of the cell.

In more recent years, several of the cathepsins have been shown to be exopeptidases and not endopeptidases (proteinases). In 1966, McDonald et al. (47) identified a chloride- and sulfhydryl-activated N-terminal exopeptidase in beef pituitary extracts that catalyzed, at acidic pH, the release of Ser-Tyr from its 2-naphthylamide. This was termed "dipeptidyl arylamidase I", and was subsequently renamed "dipeptidyl aminopeptidase I" because of the broad specificity it exhibited on a range of (hormonal) polypeptides (7). It represented the first to be discovered in a new class of exopeptidases now known as the "dipeptidylpeptide hydrolases". The demonstration of a similar chloride requirement for beef spleen cathepsin C (48) led to the demonstration by McDonald and co-workers that the activities of cathepsin C (49) and "glucagon-degrading enzyme" (50) were attributable to one lysosomal (51) N-terminal exopeptidase, and the enzyme has now been designated *dipeptidyl peptidase I* (51a). In the course of work with dipeptidyl peptidase I, a second dipeptidyl peptidase was discovered in the beef pituitary (51, 52). It, too, was lysosomal (7, 51), but its specificity was very different from, and somewhat complementary to, that of dipeptidyl peptidase I. *Dipeptidyl peptidase II* exhibited no activator requirements, and was recently shown to be active in the cleavage of prolyl bonds, and to be a relatively rare example of a lysosomal serine protease (53).

In 1978, Doebber et al. (53a) discovered a protease in the beef pituitary gland that released tripeptides from the N-terminus of beef growth hormone. This "tripeptidyl aminopeptidase" (*tripeptidyl peptidase I*), or an enzyme with similar properties, was subsequently demonstrated in the hog ovary by McDonald et al. (53b). At pH 5.0, this lysosomal serine protease catalyzed the depolymerization of poly-Gly-Pro-Ala, a model collagen chain.

For many years, evidence for the existence of a lysosomal (aminoacyl) aminopeptidase was weak. In 1941, Fruton *et al.* (44) described a sulfhydryl-dependent peptidase activity from beef spleen that hydrolyzed, at pH 5.0, Leu-NH$_2$ and Leu-Gly-Gly, but not Z-Leu-Gly-Gly. This enzyme, which they named "cathepsin III", was virtually forgotten in the years that followed. In 1968, Sylvén & Snellman (54) employed Leu-2-naphthylamide as a substrate for cathepsin B, but McDonald *et al.* (55) showed that the activities in cathepsin B preparations that hydrolyzed Leu-2-naphthylamide and Lys-2-naphthylamide were distinct and separable activities that could be eluted ahead of cathepsin B from a column of DEAE-cellulose. In 1977, Kirschke *et al.* (56) used a similar technique to separate such activities from cathepsin B of rat liver lysosomes. They showed that their purified enzyme, termed "cathepsin H", was a cysteine proteinase that hydrolyzed proteins as well as aminoacyl arylamides, and for this reason it was described as an "endoaminopeptidase". It therefore appears that cathepsin H is the illusive lysosomal protease responsible for the earlier accounts of a sulfhydryl-dependent lysosomal aminopeptidase assayable on Leu-2-naphthylamide.

As was previously mentioned, some of the enzymes originally called cathepsins have proved to be lysosomal exopeptidases. "Cathepsin A" is another of these. It was originally detected in beef spleen by Fruton & Bergmann (57) on the basis of its ability to hydrolyze Z-Glu-Tyr at acidic pH. The enzyme showed no activator requirements, and was believed to be pepsin-like. Years later, in 1967, Iodice (58) reported that highly purified cathepsin A from beef spleen showed only carboxypeptidase activity on glucagon. Its specificity has now been shown to resemble that of pancreatic carboxypeptidase A, and for this reason it was suggested (9) that cathepsin A be named "carboxypeptidase A, lys" (here, *lysosomal carboxypeptidase A*).

The activity of a second carboxypeptidase was detected in beef spleen by Fruton & Bergmann (57). It, too, hydrolyzed Z-Glu-Tyr at pH 5.5, but was readily distinguished from cathepsin A by its sulfhydryl requirement (44). Because it represented the fourth proteolytic component of the cathepsin complex, it was named "cathepsin IV". In 1962, Greenbaum & Sherman (59) showed that the best available preparations of beef cathepsin B contained a sulfhydryl-activated "catheptic carboxypeptidase" that was active on both Bz-Gly-Arg and Z-Glu-Tyr at pH 4.0. Several years later, when McDonald *et al.* (55) used a beef spleen preparation to prepare cathepsins B' (B1) and B (B2) according to Otto's procedure (60), it was observed (but not reported) that a sulfhydryl-dependent carboxypeptidase activity assayed on Bz-Gly-Arg at pH 5.0, was present in Otto's cathepsin B2. Attempts to separate activities toward Bz-Arg-NH$_2$ and Bz-Gly-Arg were unsuccessful (9), and it was tentatively concluded that these activities were attributable to a single enzyme, termed "carboxypeptidase B, lys" (here, *lysosomal carboxypeptidase B*). A purified preparation of beef spleen lysosomal carboxypeptidase B (which contained both activities) was used by McDonald & Ellis (61) to characterize its carboxypeptidase specificity. A preparation of cathepsin B2 purified from rat liver lysosomes by Ninjoor *et al*. (62) also was found to contain an apparently-inseparable carboxypeptidase activity which the authors believed to be responsible for the "histone hydrolase" activity they had previously encountered in their cathepsin B2 preparations (63). Since it had not been possible to demonstrate endopeptidase activity of cathepsin B2 (61, 62), it no longer seemed appropriate to classify this enzyme as a cathepsin. Use of the term "cathepsin B2" was discontinued in favor of *lysosomal carboxypeptidase B*, and it followed that cathepsin B1 (an established cysteine proteinase) needed only to be referred to as "cathepsin B".

A third, and quite different, lysosomal carboxypeptidase was discovered in hog kidney by Yang *et al*. (64). They called the enzyme "angiotensinase C". (Two aminopeptidases in blood had previously been named angiotensinases A and B.) Matsunaga (65) described a similar enzyme from rat kidney and named it "lysosomal angiotensinase", but it was subsequently demonstrated by McDonald *et al*. (66) that dipeptidyl peptidase I was probably responsible for the bulk of the acid angiotensinase activity in rat liver. Yang & Erdos (67) later renamed their enzyme "prolylcarboxypeptidase" to reflect its more general ability to cleave (penultimate) prolyl bonds (at acidic pH) to release C-terminal amino acids. McDonald *et al*. (68) characterized a similar exopeptidase that had been detected in preparations of rat liver dipeptidyl peptidase I as a carboxypeptidase sensitive to diisopropyl fluorophosphate. The enzyme was termed "catheptic carboxypeptidase C". Subsequently, however, both activities were attributed to the same enzyme, herein referred to as *lysosomal prolyl carboxypeptidase*.

As already noted, most dipeptidases occur in the cytosol and are active at mildly alkaline pH. However, one metal-dependent dipeptidase was detected by McDonald *et al*. (68) in rat liver lysosomes. It cleaved a wide range of dipeptides at pH 5.5, and required free N- and C-termini. The enzyme was provisionally termed "Ser-Met dipeptidase", and later renamed lysosomal dipeptidase (9) (here, *lysosomal dipeptidase I*).

Omega peptidases

The last group of exopeptidases to be considered is comprised of the omega peptidases. Representatives are found in all three subcellular locations. γ-Glutamyl transpeptidase is a membrane-bound omega peptidase that has been studied extensively. It was discovered and named by Hanes *et al*. (69) in 1952, and shown to catalyze both the hydrolysis and the transfer of N-terminal, γ-glutamyl residues. Glutathione served as a substrate for the enzyme, but much more convenient was the chromogenic substrate developed in conjunction with the purification of the enzyme from hog kidney by Orlowski & Meister (70), γ-glutamyl-p-nitroanilide, used at pH 9.0.

Another well-known omega peptidase, whose action at the amino-terminus does not require a free α-amino group, is "pyrrolidone-carboxylyl peptidase" (*pyroglutamyl peptidase*). The enzyme was first discovered in *Pseudomonas fluorescens* (71), but was found in rat liver soon after (72). Regardless of source, this peptidase required sulfhydryl activation and was capable of releasing pyroglutamyl residues from the N-termini of polypeptides at pH 7.5 without cleaving peptide bonds elsewhere in the chain.

An omega peptidase that removes C-terminal γ-glutamyl residues (as well as other amino acids held in γ-glutamyl linkage) is present in the intestinal mucosa. It was known for many years as "folate conjugase", and was recognized as being responsible for the release of unconjugated forms of folate (pteroyl monoglutamates) from the folate polyglutamates in the diet. Hoffbrand & Peters (73) showed that the enzyme, which they called *γ-glutamyl carboxypeptidase* (a name which continues to be appropriate), was concentrated in the lysosomes of guinea pig intestinal mucosa, and acted optimally at pH 4.6. It is now known to have a wide species and tissue distribution, and to be lysosomal in rat liver (74). Still another omega peptidase has been isolated from rat liver; this catalyzed the release of acylamino acids from the N-termini of peptides and proteins (75); the activity of this "acylamino acid-releasing enzyme" (*acylaminoacyl peptidase*) was greatest at about pH 7.2, and the rates depended on the nature of the N-terminal acyl group.

An omega peptidase present in mammalian tissues was originally identified in toad bladder by Campbell and co-workers (76). This "carboxamidopeptidase" (*peptidyl aminoacylamidase*) released substituted C-terminal amino acids from peptides, and was recognized for its ability to catalyze the release, at neutral pH, of Gly-NH$_2$ from the C-terminus of vasopressin (76). The last omega peptidase to be included in this group, about which a limited amount of information is available, is ß-*aspartyl peptidase* (77). This enzyme releases N-terminal aspartic acid from peptides held in ß-linkage.

References

1. Waldschmidt-Leitz, E. The mode of action and differentiation of proteolytic enzymes. Physiol. Reviews **11**: 358-370, 1931.
2. Johnson, M. L. & Berger. J. The enzymatic properties of peptidases. Adv. Enzymol. **2**: 69-92, 1942.
3. Smith, E. L. Proteolytic enzymes. Annu. Rev. Biochem. **18**: 35-58, 1949.
4. Smith, E. L. The specificity of certain peptidases. Adv. Enzymol. **12**: 191-257, 1951.
5. Smith, E. L. & Hill, R. L. Leucine aminopeptidase. In: *The Enzymes* (Boyer, P. D., Lardy, H. & Myrbäck, K., eds), 2nd edn, vol. 4, part A, pp. 37-62, Academic Press, New York, 1960.
6. Hanson, H. Hydrolasen: Peptidasen (Exopeptidasen). In: *Hoppe-Seyler/*

Thierfelder-Handbuch der physiologisch- und pathologisch-chemischen Analyse (Lang, K., Lehnartz, E., Hoffmann-Ostenhof, O. & Siebert, G., eds.), 10th edn, vol. 6, part C, pp. 1-229, Springer-Verlag, Berlin, 1966.

7. McDonald, J. K., Callahan, P. X., Ellis, S. & Smith, R. E. Polypeptide degradation by dipeptidyl aminopeptidase I (cathepsin C) and related peptidases. In: *Tissue Proteinases* (Barrett, A. J. & Dingle, J. T., eds), pp. 69-107, North-Holland Publishing Co., Amsterdam, 1971.

8. DeLange, R. J. & Smith, E. L. Leucine aminopeptidase and other N-terminal exopeptidases. In: *The Enzymes* (Boyer, P. D. ed.), 3rd edn, vol. 3, pp.81-118, Academic Press, New York, 1971.

9. McDonald, J. K. & Schwabe, C. Intracellular exopeptidases. In: *Proteinases in Mammalian Cells and Tissues* (Barrett, A. J., ed.), pp. 311-391, North-Holland Publishing Co., Amsterdam, 1977.

10. Josephson, K. & von Euler, H. Enzymatische Spaltung von Dipeptiden. IV. Uber die Wirkungsweise des Darmerepsins (Darmpeptidase). Hoppe-Seyler's Z. Physiol. Chem. **162**: 85-94, 1926.

11. Smith, E. L. & Bergmann, M. The peptidases of intestinal mucosa. J. Biol. Chem. **153**: 627-651, 1944.

12. Linderstrøm-Lang, K. Uber Darmerepsin. Hoppe-Seyler's Z. Physiol. Chem. **182**: 151-174, 1929.

13. Waldschmidt-Leitz, E., Balls, A. K. & Waldschmidt-Graser, J. Uber Dipeptidase und Polypeptidase aus Darm-Schleimhaut. (XVI. Mitteilung zur Spezifität tierischer Proteasen). Berichte **62B**, 2217-2226, 1929.

14. Waldschmidt-Leitz, E. & Purr, A. Uber Proteinase und Carboxy-Polypeptidase aus Pankreas. (XVII. Mitteilung zur Spezifität tierischer Proteasen). Berichte **62B**, 956-962, 1929.

15. Anson, M. L. Carboxypeptidase. I. The preparation of crystalline carboxypeptidase. J. Gen. Physiol. **20**: 663-669, 1937.

16. Waldschmidt-Leitz, E., Ziegler, F., Schäffner, A. & Weil, L. Uber die Struktur der Protamine. I. Protaminase und die Produkte ihrer Einwirkung auf Clupein und Salmin. Z. Physiol. Chem. **197**: 219-236, 1931.

17. Folk, J. E. A new pancreatic carboxypeptidase. J. Am. Chem. Soc. **78**: 3541-3542, 1956.

18. Folk, J. E. & Gladner, J. A. Carboxypeptidase B. I. Purification of the zymogen and specificity of the enzyme. J. Biol. Chem. **231**: 379-391, 1958.

19. Spackman, D. H., Smith, E. L. & Brown, D. M. Leucine aminopeptidase. IV. Isolation and properties of the enzyme from swine kidney. J. Biol. Chem. **212**: 255-269, 1955.

20. Hanson, H., Glässer, D. & Kirschke, H. Leucinaminopeptidase aus Rinderaugenlinsen. Kristallisation, Eigenschaften und optimale Wirkungsbedingungen des Enzyms. Z. Physiol. Chem. **340**: 107-125, 1965.

21. Schwabe, C. Peptide hydrolases in mammalian connective tissue. II.

Leucine aminopeptidase. Purification and evidence for subunit structure. Biochemistry **8**: 783-794, 1969.

22. Hopsu, V. K., Kantonen, U.-M. & Glenner, G. G. A peptidase from rat tissue selectively hydrolyzing N-terminal arginine and lysine residues. Life Sci. **3**: 1449-1453, 1964.

23. Behal, F. J., Klein, R. A. & Dawson, F. B. Separation and characterization of aminopeptidase and arylamidase components of human liver. Arch. Biochem. Biophys. **115**: 545-554, 1966.

24. Cheung, H. S. & Cushman, D. W. A soluble aspartate aminopeptidase from dog kidney, Biochim. Biophys. Acta **242**: 190-193, 1971.

25. Ellis, S. & Nuenke, J. M. Dipeptidyl arylamidase III of the pituitary. Purification and characterization. J. Biol. Chem. **242**: 4623-4629, 1967.

25a. Bålöw, R. M., Ragnarsson, U. & Zetterqvist, D. Tripeptidyl aminopeptidase in the extralysosomal fraction of rat liver. J. Biol. Chem. **258**: 11622-11628, 1983.

26. Chenoweth, D., Mitchel, R. E. J. & Smith, E. L. Aminotripeptidase of swine kidney. I. Isolation and characterization of three different forms; utility of the enzyme in sequence work. J. Biol. Chem. **248**: 1672-1683, 1973.

27. Nordwig, A. & Mayer, H. The cleavage of prolyl peptides by kidney peptidases. Detection of a new peptidase capable of removing N-terminal proline. Hoppe-Seyler's Z. Physiol. Chem. **354**: 380-383, 1973.

28. Grassmann, W., von Schoenebeck, O. & Auerbach, G. Uber die enzymatische Spaltbarkeit der Prolinpeptide. II. Z. Physiol. Chem. **210**: 1-14, 1932.

29. Bergmann, M. & Fruton, J. S. On proteolytic enzymes. XII. Regarding the specificity of aminopeptidase and carboxypeptidase. A new type of enzyme in the intestinal tract. J. Biol. Chem. **117**: 189-202, 1937.

30. Maschmann, E. Zur Kenntnis tierischer Peptidasen (V). Biochem. Z. **310**: 28-41, 1941.

31. Smith, E. L. The glycylglycine dipeptidases of skeletal muscle and human uterus. J. Biol. Chem. **173**: 571-583, 1948.

32. Smith, E. L. Studies on dipeptidases. II. Some properties of the glycyl-L-leucine dipeptidases of animal tissues. J. Biol. Chem. **176**: 9-19, 1948.

33. Hanson, H. T. & Smith, E. L. Carnosinase: an enzyme in swine kidney. J. Biol. Chem. **179**: 789-801, 1949.

34. Lenny, J. F., Kan, S.-C., Siu, K. & Sugiyama, G. H. Homocarnosinase: a hog kidney dipeptidase with a broader specificity than carnosinase. Arch. Biochem. Biophys. **184**: 257-266, 1977.

35. Binkley, F. & Nakamura, K. Metabolism of glutathione. I. Hydrolysis by tissues of the rat. J. Biol. Chem. **173**: 411-421, 1948.

36. Glenner, G. G., McMillan, P. J. & Folk, J. E. A mammalian peptidase specific for the hydrolysis of N-terminal α-L-glutamyl and aspartyl residues. Nature **194**: 867 only, 1962.

37. Pfleiderer, G. & Celliers, P. G. Isolierung einer Aminopeptidase aus Nierenpartikeln. Biochem. Z. **339**: 186-189, 1963.

38. Hopsu-Havu, V. K. & Glenner, G. G. A new dipeptide naphthylamidase hydrolyzing glycyl-prolyl-ß-naphthylamide. Histochemie **7**: 197-201, 1966.

39. Dehm, P. & Nordwig, A. The cleavage of prolyl peptides by kidney peptidases. Partial purification of a "X-prolyl-aminopeptidase" from swine kidney microsomes. Eur. J. Biochem. **17**: 364-371, 1970.

40. Robinson, D. S., Birnbaum, S. M. & Greenstein, J. P. Purification and properties of an aminopeptidase from kidney cellular particulates. J. Biol. Chem. **207**: 1-26, 1953.

41. Skeggs, L. T., Jr., Kahn, J. R. & Shumway, N. P. The preparation and function of the hypertensin-converting enzyme. J. Exp. Med. **103**: 295-299, 1956.

42. Dehm, P. & Nordwig, A. The cleavage of prolyl peptides by kidney peptidases. Isolation of a microsomal carboxypeptidase from swine kidney. Eur. J. Biochem. **17**: 372-377, 1970.

43. Willstätter, R. & Bamann, E. Uber die Proteasen der Magenschleimhaut Erste Abhandlung über die Enzyme der Leukocyten. Hoppe-Seyler's Z. Physiol. Chem. **180**: 127-143, 1929.

44. Fruton, J. S., Irving, G. W., Jr. & Bergmann, M. On the proteolytic enzymes of animal tissues. III. The proteolytic enzymes of beef spleen, beef kidney and swine kidney. Classification of the cathepsins. J. Biol. Chem. **141**: 763-774, 1941.

45. Tallan, H. H., Jones, M. E. & Fruton, J. S. On the proteolytic enzymes of animal tissues. X. Beef spleen cathepsin C. J. Biol. Chem. **194**: 793-805, 1952. 46. de Duve, C., Pressman, B. C., Gianetto, R., Wattiaux, R. & Appelmans, F. Tissue fractionation studies. 6. Intracellular distribution patterns of enzymes in rat-liver tissue. Biochem. J. **60**: 604-617, 1955.

47. McDonald, J. K., Ellis, S. & Reilly, T. J. Properties of dipeptidyl arylamidase I of the pituitary. Chloride and sulfhydryl activation of seryltyrosyl-ß-naphthylamide hydrolysis. J. Biol. Chem. **241**: 1494-1501, 1966.

48. McDonald, J. K., Reilly, T. J., Zeitman, B. B. & Ellis, S. Cathepsin C: a chloride-requiring enzyme. Biochem. Biophys. Res. Commun. **24**: 771-775, 1966.

49. McDonald, J. K., Zeitman, B. B., Reilly, T. J. & Ellis, S. New observations on the substrate specificity of cathepsin C (dipeptidyl aminopeptidase I). Including the degradation of γ-corticotropin and other peptide hormones. J. Biol. Chem. **244**: 2693-2709, 1969.

50. McDonald, J. K., Callahan, P. X., Zeitman, B. B. & Ellis, S. Inactivation and degradation of glucagon by dipeptidyl aminopeptidase I (cathepsin C) of rat liver. Including a comparative study of secretin degradation. J. Biol. Chem. **244**: 6199-6208, 1969.

51. McDonald, J. K., Reilly, T. J., Zeitman, B. B. & Ellis, S. Dipeptidyl arylamidase II of the pituitary; properties of lysylalanyl-ß-naphthylamide hydrolysis: inhibition by cations, distribution in tissues and subcellular localization. J. Biol. Chem. 243: 2028-2037, 1968.

51a. Nomenclature Committee of the International Union of Biochemistry (1979). *Enzyme Nomenclature 1978* (Adademic Press, New York), and supplement (1980) in Eur. J. Biochem. 104: 1-4.

52. McDonald, J. K., Leibach, F. H., Grindeland, R. E. & Ellis, S. Purification of dipeptidyl aminopeptidase II (dipeptidyl arylamidase II) of the anterior pituitary gland; peptidase and dipeptide esterase activities. J. Biol. Chem. 243: 4143-4150, 1968.

53. McDonald, J. K. & Schwabe, C. Dipeptidyl peptidase II of beef dental pulp. Initial demonstration and characterization as a fibroblastic, lysosomal peptidase of the serine class active on collagen-related peptides. Biochim. Biophys. Acta 616: 68-81, 1980.

53a. Doebber, T. W., Divor, A. R. & Ellis, S. Identification of a tripeptidyl aminopeptidase in the anterior pituitary gland: effect on the chemical and biological properties of rat and bovine growth hormone. Endocrinology 103: 1794-1804, 1978.

53b. McDonald, J. K., Hoisington, A. R. & Eisenhauer, D. A. Partial purification and characterization of an ovarian tripeptidyl peptidase: a lysosomal exopeptidase that sequentially releases collagen-related (Gly-Pro-X) triplets. Biochem. Biophys. Res. Commun. 126: 63-71, 1985.

54. Sylvén, B. & Snellman, O. Studies on the histochemical "leucine aminopeptidase" reaction. V. Cathepsin B as a potent effector of LNA hydrolysis. Histochemie 12: 240-243, 1968.

55. McDonald, J. K., Zeitman, B. B. & Ellis, S. Leucine naphthylamide: an inappriopriate substrate for the histochemical detection of cathepsins B and B'. Nature 225: 1048-1049, 1970. (See Nature 226; 90 for erratum correcting a typographic error in the original title).

56. Kirschke, H., Langner, J., Wiederanders, B., Ansorge, S., Bohley, P. & Hanson, H. Cathepsin H: an endoaminopeptidase from rat liver lysosomes. Acta Biol. Med. Ger. 36: 185-199, 1977.

57. Fruton, J. S. & Bergmann, M. On the proteolytic enzymes of animal tissues. I. Beef spleen. J. Biol. Chem. 130: 19-27, 1939.

58. Iodice, A. A. The carboxypeptidase nature of cathepsin A. Arch. Biochem. Biophys. 121: 241-242, 1967.

59. Greenbaum, L. M. & Sherman, R. Studies on catheptic carboxypeptidase. J. Biol. Chem. 237: 1082-1085, 1962.

60. Otto, K. Uber ein neues Kathepsin. Reinigung aus Rindermilz, Eigenschaften, sowie Vergleich mit Kathepsin B. Hoppe-Seyler's Z. Physiol. Chem. 348: 1449-1460, 1967.

61. McDonald, J. K. & Ellis, S. On the substrate specificity of cathepsins B1

and B2 including a new fluorogenic substrate for cathepsin B1. Life Sci. **17**: 1269-1276, 1975.

62. Ninjoor, V., Taylor, S. L. & Tappel, A. L. Purification and characterization of rat liver lysosomal cathepsin B2. Biochim. Biophys. Acta **370**: 308-321, 1974.

63. DeLumen, B. O. & Tappel, A. L. Histone hydrolase activity of rat liver lysosomal cathepsin B2. Biochim. Biophys. Acta **293**: 217-225, 1973.

64. Yang, H. Y. T., Erdös, E. G. & Chiang, T. S. New enzymatic route for the inactivation of angiotensin. Nature **218**: 1224-1226, 1968.

65. Matsunaga, M. Nature of lysosomal angiotensinase activity. Jpn. Circ. J. **35**: 333-338, 1971.

66. McDonald, J. K., Zeitman, B. B., Callahan, P. X. & Ellis, S. Angiotensinase activity of dipeptidyl aminopeptidase I (cathepsin C) of rat liver. J. Biol. Chem. **249**: 234-240, 1974.

67. Yang, H. Y. T. & Erdös, E. G. Prolylcarboxypeptidase: a recently described lysosomal enzyme. In: *Immunopathology of Inflammation* (Forscher, B. K. & Houck, J. C., eds.), pp. 146-148, 1971.

68. McDonald, J. K., Zeitman, B. B. & Ellis, S. Detection of a lysosomal carboxypeptidase and a lysosomal dipeptidase in highly-purified dipeptidyl aminopeptidase I (cathepsin C) and the elimination of their activities from preparations used to sequence peptides. Biochem. Biophys. Res. Commun. **46**: 62-70, 1972.

69. Hanes, C. S., Hird, F. J. R. & Isherwood, F. A. Enzymic transpeptidation reactions involving γ-glutamyl peptides and α-amino-acyl peptides. Biochem. J. **51**: 25-35, 1952.

70. Orlowski, M. & Meister, A. Isolation of γ-glutamyl transpeptidase from pig kidney. J. Biol. Chem. **240**: 338-347, 1965.

71. Doolittle, R. F. & Armentrout, R. W. Pyrrolidonyl peptidase. An enzyme for selective removal of pyrrolidonecarboxylic acid residues from polypeptides. Biochemistry **7**: 516-521, 1968.

72. Armentrout, R. W. Pyrrolidonecarboxylyl peptidase from rat liver. Biochim. Biophys. Acta **191**: 756-759, 1969.

73. Hoffbrand, A. V. & Peters, T. J. The subcellular localization of pteroyl polyglutamate hydrolase and folate in guinea pig intestinal mucosa. Biochim. Biophys. Acta **192**: 479-485, 1969.

74. Silink, M. & Rowe, P. B. The localization of glutamate carboxypeptidase in rat liver lysosomes. Biochim. Biophys. Acta **381**: 28-36, 1975.

75. Tsunasawa, S., Narita, K. & Ogata, K. Purification and properties of acylamino acid-releasing enzyme from rat liver. J. Biochem. **77**: 89-102, 1975.

76. Campbell, B. J., Thysen, B. & Chu, F. S. Peptidase catalyzed hydrolysis of antidiuretic hormone in toad bladder. Life Sci. **4**: 2129-2140,1965.

77. Haley, E. E. ß-Aspartyl peptidase from rat liver. Methods Enzymol. **19**: 737-741, 1970.

Section 11
AMINOPEPTIDASES

Leucyl Aminopeptidase

Summary

EC Number: 3.4.11.1

Earlier names: Leucyl peptidase, aminoleucyl peptidase, leucine-aminoexopeptidase, leucine aminopeptidase, "alcaline leucinamidase", cytosol aminopeptidase I.

Abbreviation: LAP.

Distribution: Enzymes with similar or identical specificity are present in the cytosol of virtually all mammalian cells, as well as in the cells of plants and microorganisms.

Source: Hog intestinal mucosa, kidney, pancreas, and muscle. Beef eye lens and connective tissue (dental pulp) fibroblasts. Most work has been done with LAP preparations from hog kidney [1], beef lens [2] and beef dental pulp [3].

Action: Catalyzes the removal of unsubstituted N-terminal L-amino acids (including proline) that are linked through an unsubstituted nitrogen atom to an amino acid (other than proline) of L-configuration. Amino acid amides, arylamides and methyl esters also serve as substrates. LAP exhibits a preference for large hydrophobic residues, in particular leucine.

Requirements: pH 9.0-9.5 at 37°C for Leu-NH$_2$, with some activity exhibited over the range pH 6-10. Activation with Mn^{2+} or Mg^{2+} (at 1 mM) is required for full activity. Compared to the Mg^{2+}-activated enzyme, the Mn^{2+}-activated enzyme is two to three-fold more active on most substrates, but the former is more stable.

Substrate, usual: Leu-NH$_2$ [4].

Substrate, special: Leu-NNap and Leu-NPhNO$_2$ are useful for the assay of enriched and purified preparations of LAP. These arylamide substrates are hydrolyzed at rates that are much less than 1% of that on Leu-NH$_2$ [5,6], but their fluorogenic and chromogenic properties offer convenience and sensitivity. On the other hand, they should not be used for the assay of LAP in crude extracts (see Comment).

Inhibitors: The native (Zn-metallo-) enzyme is inhibited by 1,10-phenanthroline, dipyridyl, cupferron, sodium diethyldithiocarbamate, sodium sulfide and sodium cyanide. Inhibition is reversible with Mn^{2+} or Mg^{2+}. EDTA does not inhibit native LAP (as in crude extracts), but it does reversibly inhibit the Mn^{2+}- or Mg^{2+}-activated enzyme. Tertiary butyl threonyl peptides such as Thr(OBu^t)-Phe-Pro [7], chloromethane derivatives such as Leu-CH$_2$Cl and Phe-CH$_2$Cl [8], and N-(leucyl)-o-aminobenzenesulfonate [9], D-leucine hydroxamate [10], and L-leucinal [11] are all potent competitive inhibitors.

Bestatin, with a K_i of 0.02 μM [12], is also a competitive inhibitor. Amastatin [13] and 2,3-dihydroxybenzyl-Ala-Thr (Bu-2743E) [14] are other microbial metabolites that strongly inhibit LAP. Amastatin also inhibits glutamyl aminopeptidase ("aminopeptidase A", Entry 11.07), but has no effect on soluble arginyl aminopeptidase ("aminopeptidase B") [13]. Puromycin has no effect on LAP [15]. Hydrophobic compounds such as long-chain alcohols and fatty acids are also competitive inhibitors. Reactive thiol groups are present, but do not appear to be essential for activity. Dip-F has no effect.

Molecular properties: LAP is a Zn-containing metalloprotein belonging to the class of "two-metal proteases". M_r about 326,000; pI 4.0-5.0. No carbohydrate. Consists of six identical bilobal subunits (M_r about 54,000) containing two zinc atoms each and a single polypeptide chain. One zinc atom occurs at the catalytic site (the "slow-exchange site") and the other at the activation site (the "fast-exchange site"). Activation is typically associated with the replacement of Zn^{2+} at the latter site with Mg^{2+} (or Mn^{2+}) to yield a Zn^{2+}-Mg^{2+} LAP [16]. K_m values are significantly affected by the substitution of metals (Mg^{2+} or Co^{2+}) at either site [17]. Although the native Zn-containing enzyme is activated and stabilized by Mg^{2+} (2 mM), a greater degree of activation can be obtained with Mn^{2+} (resulting in a lowered K_m and an elevated V_{max} on Leu-NH$_2$; however, the Mn^{2+}-activated enzyme is unstable.

The native oligomeric molecule exhibits the shape of a hollow cylinder, 67 Å x 107 Å. Thr N-terminus. $A_{280, 1\%}$ 10. Amino acid composition of beef lens LAP closely resembles that of the hog kidney enzyme [6]. The entire sequence of he 478-amino acid subunit has been reported [18] for the former. No sequence homology with any other published primary structure was detected.

Comment: Purified LAP hydrolyzes Leu-NH$_2$ far more rapidly than the corresponding arylamide derivatives [5,6]. For the hog kidney enzyme the rate is 122 times faster than Leu-NNap and 265 times faster than Leu-NPhNO$_2$; for the beef lens enzyme the rate is 47 times faster than Leu-NNap and 117 times faster than Leu-NPhNO$_2$ [6]. Although the arylamide substrates are useful for the assay of purified LAP, their use for the assay of LAP in crude extracts can be grossly misleading because most tissues contain other aminopeptidases that act preferentially on the arylamide derivatives, as indicated by the observation that extracts of hog kidney hydrolyze Leu-NNap faster than Leu-NH$_2$ [6], and that leucine is not the preferred N-terminal residue, in contrast to the results of a survey of naphthylamides [15] conducted with the purified hog kidney enzyme.

Highly purified preparations of LAP have been found to contain traces of contaminating endopeptidases that could be inactivated by treatment with Dip-F and IAcOH [19]. This may be necessary when the enzyme is used as a reagent for N-terminal and sequence analyses.

Purifed LAP is stable indefinitely when stored frozen between pH 7 and 9 in the presence of MgCl$_2$, or as a complex with calcium phosphate gel [20], but it is rapidly inactivated below pH 5 and above pH 11.

Levels of LAP assayed on Leu-Gly, which are normally very low in human serum, are typically greatly elevated during various liver diseases such as obstructive jaundice and infectious hepatitis [21], often exceeding transaminase (GOT and GPT) activities [22].

References
[1] Himmelhoch *Methods Enzymol.* **19**: 508-513, 1970.
[2] Hanson & Frohne *Methods Enzymol.* **45**: 504-521, 1976.
[3] Schwabe *Biochemistry* **8**: 783-794, 1969.
[4] Smith *Methods Enzymol.* **2**: 88-93, 1955.
[5] Patterson *et al. J. Biol. Chem.* **238**: 3611-3620, 1963.
[6] Hanson *et al. Hoppe-Seyler's Z. Physiol. Chem.* **348**: 689-704, 1967.
[7] Jost *et al. FEBS Lett.* **23**: 211-214, 1972.
[8] Fittkau *et al. Eur. J. Biochem.* **44**: 523-528, 1974.
[9] Taylor *et al. J. Biol. Chem.* **257**: 11571-11576, 1982.
[10] Wilkes & Prescott *J. Biol. Chem.* **258**: 13517-13521, 1983.
[11] Andersson *et al. Biochemistry* **21**: 4177-4180, 1982.
[12] Suda *et al. Arch. Biochem. Biophys.* **177**: 196-200, 1976.
[13] Aoyagi *et al. J. Antibiot.* **31**: 636-638, 1978.
[14] Kobaru *et al. J. Antibiot.* **36**: 1396-1398, 1983.
[15] McDonald *et al. Biochem. Biophys. Res. Commun.* **16**: 135-140, 1964.
[16] Carpenter & Vahl *J. Biol. Chem.* **248**: 294-304, 1973.
[17] Allen *et al. Biochemistry* **22**: 3778-3783, 1983.
[18] Cuypers *et al. J. Biol. Chem.* **257**: 7086-7091, 1982.
[19] Frater *et al. J. Biol. Chem.* **240**: 253-257, 1965.
[20] Schwabe *Biochemistry* **8**: 795-802, 1969.
[21] Fleisher *et al. Clin. Chim. Acta* **9**: 254-258, 1964.
[22] Plaquet *et al. Clin. Chim. Acta* **46**: 91-103, 1973.

Bibliography

1928

Linderstrøm-Lang, K. & Sato, M.   Die Spaltung von Glycylglycin, Alanylglycin und Leucylglycin durch Darm- und Malzpeptidasen. *Hoppe-Seyler's Z. Physiol. Chem.* **184**: 83-92.

1929

Linderstrøm-Lang, K.   Uber Darmerepsin. *Hoppe-Seyler's Z. Physiol. Chem.* **182**: 151-174.

1930

Linderstrøm-Lang, K.   Uber die Einheitlichkeit der Darmdipeptidase. *Hoppe-Seyler's Z. Physiol. Chem.* **188**: 48-68.

1936

Johnson, M. J., Johnson, G. H. & Peterson, W. H.   The magnesium-activated leucyl peptidase of animal erepsin. *J. Biol. Chem.* **116**: 515-526.

1939

Berger, J. & Johnson, M. J.   Metal activation of peptidases. *J. Biol. Chem.* **130**: 641-654.

1941

Maschmann, E.   Zur Kenntnis, tierischer Peptidasen (V.). *Biochem. Z.* **310**: 28-41.

Maschmann, E.   Zur Kenntris tierischer Peptidasen. IV. *Biochem. Z.* **309**: 179-189.

Smith, E. L. & Bergmann, M.   The activation of intestinal peptidases by manganese. *J. Biol. Chem.* **138**: 789-790.

1942

Johnson, M. J. & Berger, J.   The enzymatic properties of peptidases. *Adv. Enzymol.* **2**: 69-92 (see pp. 79-80).

1944

Smith, E. L. & Bergmann, M.   The peptidases of intestinal mucosa. *J. Biol. Chem.* **153**: 627-651.

1949

Smith, E. L.   Catalytic action of the metal peptidases. *Fed. Proc. Fed. Am. Soc. Exp. Biol.* **8**: 581-588.

Smith, E. L.   Proteolytic enzymes. *Annu. Rev. Biochem.* **18**: 35-58 (see pp. 38-39.

Smith, E. L. & Polglase, W. J.   The specificity of leucine aminopeptidase.

II. Optical and side chain specificity. *J. Biol. Chem.* **180**: 1209-1223.

1951

Smith, E. L.   The specificity of certain peptidases. *Adv. Enzymol.* **12**: 191-257 (see pp. 196-206).

1952

Binkley, F.   Evidence for the polynucleotide nature of cysteinylglycinase. *Exp. Cell. Res. Suppl.* **2**: 145-157.

Smith, E. L., Spackmann, D. H. & Polglase, W. J.   The specificity of leucine aminopeptidase. III. Action on diastereoisomers. *J. Biol. Chem.* **199**: 801-817.

1955

Green, M. N., Tsou, K., Bressler, R. & Seligman, A. M.   The colorimetric determination of leucine aminopeptidase activity with L-leucyl-β-naphthylamide hydrochloride. *Arch. Biochem.* **57**: 458-474.

Smith, E. L.   Aminopeptidases. B. Leucine aminopeptidase. *Methods Enzymol.* **2**: 88-93.

Smith, E. L. & Spackman, D. H.   Leucine aminopeptidase. V. Activation, specificity and mechanism of action. *J. Biol. Chem.* **212**: 271-299.

Spackman, D. H., Smith, E. L. & Brown, D. M.   Leucine aminopeptidase. IV. Isolation and properties of the enzyme from swine sidney. *J. Biol. Chem.* **212**: 255-269.

1956

Hill, R. L. & Smith, E. L.   Hydrolysis of amino acid amide derivatives of 2-aminofluorene, 4-amino-biphenyl and 4,4'-diaminobiphenyl by leucine aminopeptidase. *Proc. Soc. Exp. Biol. Med.* **92**: 500-503.

1957

Waldschmidt-Leitz, E. & Keller, L.   Zur frage der existenz einer leucyl-peptidase. *Hoppe-Seyler's Z. Physiol. Chem.* **309**: 228-238.

1958

Hanson, H. & Methfessel, J.   Zur Spezifität der Peptidaseaktivität der Rinderlinse. *Acta Biol. Med. Ger.* **1**: 414-421.

Hill, R. L., Spackman, D. H., Brown, D. M. & Smith, E. L.   Leucine aminopeptidase. *Biochem. Prep.* **6**: 35-48.

1959

Folk, J. E., Gladner, J. A. & Viswanatha, T.   A simplified chromatographic purification of leucine aminopeptidase. *Biochim. Biophys. Acta* **36**: 256-257.

Patterson, E. K.   Leucine aminopeptidase fragments from an ascites tumor. *J. Biol. Chem.* **234**: 2327-2337.

1960

Binkley, F. Resistant peptidases of swine kidney. *J. Am. Chem. Soc.* **82**: 987-990.

Binkley, F. & Torres, C. Spectrophotometric assay of peptidase activity. *Arch. Biochem. Biophys.* **86**: 201-203.

Fittkau, S., Glässer, D. & Hanson, H. Zur Aktivität und Spezifität der Leucinaminopeptidase in Augenlinsen. Aminosäure- und Dipeptidanilide als Substrate. *Hoppe-Seyler's Z. Physiol. Chem.* **322**: 101-111.

Nachlas, M. M., Monis, B., Rosenblatt, D. & Seligman, A. M. Improvement in the histochemical localization of leucine aminopeptidase with a new substrate, L-leucyl-4-methoxy-2-naphthylamide. *J. Biophys. Biochem. Cytol.* **7**: 261-264.

Smith, E. L. & Hill, R. L. Leucine aminopeptidase. In: *The Enzymes* (Boyer, P. D., Lardy, H. & Myrbäck, K. eds), 2nd edn, vol. 4, part A, pp. 37-62, Academic Press, New York.

Talanti, S. & Hopsu, V. K. The influence of methyl-thiouracil on the leucine aminopeptidase activity in the thyroid gland of the rat. *Acta Endocrinol.* **35**: 481-484.

1961

Talanti, S. & Hopsu, V. K. Leucine aminopeptidase in the mammary gland of the cow and rat. *Nature* **191**: 86-87.

1962

Behal, F. J., Kanavage, C. B. & Hamilton, R. D. Separation of leucylaminopeptidase components in normal human sera by DEAE cellulose column chromatography. *Proc. Soc. Exp. Biol. Med.* **109**: 411-412.

Fasold, H., Linhart, P. & Turba, F. Einfache darstellung hochgereinigter leucinaminopeptidase durch präparative Elektrophorese und Chromatographie. *Biochem. Z.* **336**: 182-190.

Glässer, D. & Hanson, H. Einfaches Verfahren zur präparativen Gewinnung hoch-gereinigter Leucinaminopeptidase aus Rinderaugenlinsen. *Hoppe-Seyler's Z. Physiol. Chem.* **329**: 249-256.

Hanson, H., Glässer, D. & Kleine, R. Uber Aminosäureesterase-Wirkung der Leucinaminopeptidase der Rinderaugenlinsen. *Hoppe-Seyler's. Z. Physiol. Chem.* **329**: 257-263.

Hopsu, V. K., Riekkinen, P. & Luostarinen, E. Hormonal influences on leucine aminopeptidase in the accessory reproductive tracts of the rat. *Experientia* **18**: 15-17.

Spector, A. A study of peptidase and esterase activity in calf lens. *Exp. Eye Res.* **1**: 330-335.

1963

Glässer, D. & Hanson, H. Kristallisierte Leucinaminopeptidase aus Rinderaugenlinsen. *Naturwissenschafter* **50**: 595-596.

Matheson, A. T., Bjerre, S. & Hanes, C. S. A conjugated form of aminopeptidase from autolyzed extracts of kidney tissue. *Can. J. Biochem. Physiol.* **41**: 1741-1770.

Patterson, E. K., Hsiao, S. H. & Keppel, A. Studies on dipeptidases and aminopeptidases. I. Distinction between leucine aminopeptidase and enzymes that hydrolyze L-leucyl-β-naphthylamide. *J. Biol. Chem.* **238**: 3611-3620.

Roth, M. Une ultramicro-méthode fluorimétrique pour le dosage de la leucine-aminopeptidase dans les liquides biologiques. *Clin. Chim. Acta* **9**: 448-453.

1964

Bryce, G. F. & Rabin, B. R. The assay and reaction kinetics of leucine aminopeptidase from swine kidney. *Biochem. J.* **90**: 509-512.

Bryce, G. F. & Rabin, B. R. The function of the metal ion in leucine aminopeptidase and the mechanism of action of the enzyme. *Biochem. J.* **90**: 513-518.

Fleisher, G. A., Pankow, M. & Warmka, C. Leucine aminopeptidase in human serum: an ultramicromethod for the determination of the rates of hydrolysis of L-leucylglycine. Clin. Chim. Acta **9**: 254-258.

Glässer, D. & Hanson, H. Ein weiteres Verfahren zur Gewinnung kristallisierter Leucinaminopeptidase aus Rinderaugenlinsen. *Naturwissenschaften.* **51**: 110-111.

McDonald, J. K., Reilly, T. J. & Ellis, S. The hydrolysis of amino acyl-β-naphthylamides by plasma aminopeptidases. *Biochem. Biophys. Res. Commun.* **16**: 135-140.

Sylvén, B. & Snellman, O. Studies on the histochemical "leucine aminopeptidase" reaction. III. On the different LNA-splitting enzymes from spleen. *Histochemie* **3**: 484-486.

1965

Frater, R., Light, A. & Smith, E. L. Chemical and enzymic studies on the amino-terminal sequence of papain. *J. Biol. Chem.* **240**: 253-257.

Hanson, H., Glässer, D. & Kirschke, H. Leucinaminopeptidase aus Rinderaugenlinsen. Kristallisation, Eigenschaften und optimale Wirkungsbedingungen des Enzyms. *Hoppe-Seyler's Z. Physiol. Chem.* **340**: 107-125.

Kretschmer, K. & Hanson, H. Kristallisierte Leucinaminopeptidase aus Rinderaugenlinsen. Physikalische Konstanten I. *Hoppe-Seyler's. Z. Physiol. Chem.* **340**: 126-137.

Uriel, J. & Avrameas, S. Systematic fractionation of swine pancreatic hydrolases. I. Fractionation of enzymes soluble in ammonium sulfate solution at 0.40 saturation. *Biochemistry* **4**: 1740-1749.

1966

Felgenhauer, K. & Glenner, G. G. Quantitation of tissue-bound renal aminopeptidase by a microdensitometric technique. *J. Histochem.*

Cytochem. **14**: 53-63.

Hanson, H. Hydrolasen: Peptidasen (Exopeptidasen). In: *Hoppe-Seyler/Thierfelder - Handbuch der physiologisch- und pathologisch-chemischen Analyse* (Lang, K., Lehnartz, E., Hoffmann-Ostenhof, O. & Siebert, G. eds.), 10th edn., vol. **6**, part C, pp. 1-229, Springer-Verlag, Berlin.

Hanson, H. & Hütter, H.-J. Zur Darstellung von Leucinaminopeptidase aus Schweinenieren. *Hoppe-Seyler's Z. Physiol. Chem.* **347**: 118-126.

Joseph, R. L. & Sanders, W. J. Leucine aminopeptidase in extracts of swine muscle. *Biochem. J.* **100**: 827-832.

Kleiner, H. & Schram, E. Séparation des isozymes hydrolysant la L-leucyl-β-naphthylamide par électrophorèse verticale en gel d'acrylamide. *Clin. Chim. Acta* **14**: 377-385.

Meirovich, C.-I. Synthése et libération de leucine-amino-peptidase par les glandes parathyroïdiennes de rat in vitro. *Ann. Biol. Clin. (Paris).* **24**: 831-838.

Rybák, M., Mansfeld, V., Petáková, M. & Simonianová, E. The breakdown of L-leucyl-p-nitroanilide by serum and tissue hormogenates. *Physiol. Bohemoslov.* **15**: 276-280.

Schechter, I. & Berger, A. The hydrolysis of diastereoisomers of alanine peptides by carboxypeptidase A and leucine aminopeptidase. *Biochemistry* **5**: 3371-3375.

Wachsmuth, E. D., Fritze, I. & Pfleiderer, G. An aminopeptidase occurring in pig kidney. II. A study on the mechanism of the hydrolysis. *Biochemistry* **5**: 175-182.

1967

Hanson, H. & Lasch, J. Enzymatische Transamidierungsreaktion mit kristalliner Rinderaugenlinsen-Leucinaminopeptidase. *Hoppe-Seyler's Z. Physiol. Chem.* **348**: 1525-1539.

Hanson, H., Glässer, D., Ludewig, M., Mannsfeldt, H.-G., John, M. & Nesvadba, H. Struktur- und Wirkungsidentität der Leucinaminopeptidase aus Schweinenieren und Rinderaugenlinsen und Vergleich mit der Partikelaminopeptidase aus Schweinenieren. *Hoppe-Seyler's Z. Physiol. Chem.* **348**: 689-704.

Hanson, H., Hütter, H.-J., Mannsfeldt, H.-G., Kretschmer, K. & Sohr, C. Zur Darstellung und Substratspezifität einer von der Leucinaminopeptidase unterscheidbaren Aminopeptidase aus Nierenpartikeln. *Hoppe-Seyler's Z. Physiol. Chem.* **348**: 680-688.

Light, A. Leucine aminopeptidase (LAP). *Methods Enzymol.* **11**: 426-436.

Moseley, M. H. & Melius, P. Barbital stabilization of a purified leucine aminopeptidase against EDTA inactivation. *Can. J. Biochem.* **45**: 1641-1644.

1968

Himmelhoch, S. R. & Peterson, E. A. Preparation of leucine aminopeptidase free of endopeptidase activity. *Biochemistry* **7**: 2085-2093.

Hütter, H.-J., Barth, T. & Rychlik, I.   Zur Darstellung hochgereinigter Schweinenieren-Leucinaminopeptidase mittels der Methode der Gelfiltration an Sephadex G-200. *Hoppe-Seyler's Z. Physiol. Chem.* **349**: 113-114.

Knight, J. A. & Hunter, D. T.   A continuous spectrophotometric method for the determination of leucine aminopeptidase. *Clin. Chem.* **14**: 555-564.

Kretschmer, K.   Kristallisierte Leucinaminopeptidase aus Rinderaugenlinsen. Bestimmung der Molekelgestalt und der Molmasse durch elektronenoptische Untersuchungen. *Hoppe-Seyler's Z. Physiol. Chem.* **349**: 715-718.

Kretschmer, K. & Hanson, H.   Kristallisierte Leucinaminopeptidase aus Rinderaugenlinsen. Untersuchungen über die Untereinheiten: Physikalische Konstanten. *Hoppe-Seyler's Z. Physiol. Chem.* **349**: 831-845.

Marks, N., Datta, R. K. & Lajtha, A.   Partial resolution of arylamidases and aminopeptidases. *J. Biol. Chem.* **243**: 2882-2889.

1969

Casey, A. E., Gilbert, F. & Downey, E. L.   Direct association of serum levels of beta globulin and leucine aminopeptidase. *Ala. J. Med. Sci.* **6**: 285-286.

Deftos, L. J. & Potts, J. T.   Contamination of DFP-treated leucine aminopeptidase with multiple endopeptidases. *Biochim. Biophys. Acta* **171**: 121-127.

Frohne, M. & Hanson, H.   Untersuchungen zur Reaktion der kristallinen Leucinaminopeptidase aus Rinderaugenlinsen mit SH-reagentien. *Hoppe-Seyler's Z. Physiol. Chem.* **350**: 213-222.

Himmelhoch, S. R.   Leucine aminopeptidase: a zinc metalloenzyme. *Arch. Biochem. Biophys.* **134**: 597-602.

Iwig, M., Glässer, D. & Hanson, H.   Eignung der klassischen Leucinaminopeptidase (LAP) als Indikator-Enzym bei Hepatitis epidemica acuta. Differenzierung zwischen Leucinaminopeptidase und Arylamidase bei hepatobiliären Erkrankungen. *Z. Klin. Chem. Klin. Biochem.* **7**: 225-238.

Kretschmer, K. & Kollin, G.   Kristallisierte Leucinaminopeptidase aus Rinderaugenlinsen. Bestimmung der Gestalt und der Hydratation durch Röntgen-Kleinwinkelstreuung. *Hoppe-Seyler's Z. Physiol. Chem.* **350**: 431-438.

Schwabe, C.   Leucine aminopeptidase. A calcium phosphate gel complex. *Biochemistry* **8**: 795-802.

Schwabe, C.   Peptide hydrolases in mammalian connective tissue. II. Leucine aminopeptidase. Purification and evidence for subunit structure. *Biochemistry* **8**: 783-794.

1970

Glässer, D., John, M. & Hanson, H.   Zur Organ- und Speziesspezifität der Leucin-Aminopeptidase aus Rinderaugenlinsen. *Hoppe-Seyler's Z. Physiol. Chem.* **351**: 1337-1343.

Himmelhoch, S. R.   Leucine Aminopeptidase from swine kidney. *Methods*

Enzymol. **19**: 508-513.

Kettmann, U. & Hanson, H. Zur Bedeutung des Zinks in der Leucinaminopeptidase aus Rinderaugenlinsen. *FEBS Lett.* **10**: 17-20.

Lisowski, J., Rajkumar, T. V., Wolf, D. P. & Stein, E. A. Evidence for tightly-bound zinc in leucine aminopeptidase from pig kidney. *Acta Biochim. Pol.* **17**: 311-324.

Melius, P., Moseley, M. H. & Brown, D. M. Characterization of the subunits of swine kidney leucine aminopeptidase. *Biochim. Biophys. Acta* **221**: 62-68.

1971

Delange, R. J. & Smith, E. L. Leucine aminopeptidase and other N-terminal exopeptidases. In: *The Enzymes* (Boyer, P. D. ed.), 3rd. edn., vol. 3, pp. 81-118, Academic Press, New York.

Melbye, S. W. & Carpenter, F. H. Leucine aminopeptidases (bovine lens). Stability and size of subunits. *J. Biol. Chem.* **246**: 2459-2463.

Sandström, B. Electron-microscope demonstration of leucyl-naphthylamidase activity. *Histochemie* **26**: 40-43.

1972

Carpenter, F. H. & Harrington, K. T. Intermolecular cross-linking of monomeric proteins and cross-linking of oligomeric proteins as a probe of quaternary structure. Application to leucine aminpeptidase (bovine lens). *J. Biol. Chem.* **247**: 5580-5586.

Fittkau, S. Synthese und Eigenschaften von 1-Chlor-3-amino-4-phenylbutanon-2, einem Inhibitor der Leuzinaminopeptidase aus Rinderaugenlinsen. *Acta Biol. Med. Ger.* **28**: 259-267.

Jost, R., Masson, A. & Zuber, H. A strong competitive peptide-inhibitor for leucine aminopeptidase. *FEBS Lett.* **23**: 211-214.

Metrione, R. M. The thiolesterase activity of leucine aminopeptidase. *Biochim. Biophys. Acta* **268**: 518-522.

Murakami, M., Shimasue, A., Tsubokura, T. & Nagayama, T. Studies on the isozymes of leucine aminopeptidase. LAP isozymes of the normal human serum. *Hiroshima J. Med. Sci.* **21**: 91-99.

1973

Carpenter, F. H. & Vahl, J. M. Leucine aminopeptidase (bovine lens). Mechanism of activation by Mg^{2+} and Mn^{2+} of the zinc metalloenzyme, amino acid composition, and sulfhydryl content. *J. Biol. Chem.* **248**: 294-304.

Dey, S. K., Gupta, J. S. & Deb, C. Histochemical studies on the Leydig-cell-leucine aminopeptidase activity in the guinea-pig testis. *J. Reprod. Fert.* **34**: 475-479.

Fittkau, S. α-Aminochlormethylketone aus Aminosäuren und Peptiden als substratanaloge Inhibitoren der Leucinaminopeptidase. *J. Prakt. Chem.* **315**: 1037-1044.

Gupta, J. S., Dey, S. K. & Deb, C. Histochemical studies on leucine aminopeptidase activity in the rat uterus. *J. Reprod. Fertil.* **34**: 467-473.

Jost, R. Competitive peptide-inhibitors of leucine aminopeptidase: specific interaction of threonine (tertiary butyl)-peptides with the enzyme from swine kidney and from bovine eye lens. *FEBS Lett.* **29**: 7-9.

Lasch, J., Kudernatsch, W. & Hanson, H. Bovine-lens leucine aminopeptidase. Kinetic studies with activators and competitive inhibitors. *Eur. J. Biochem.* **34**: 53-57.

Plaquet, R., Ledeme, N., Vincent-Fiquet, O. & Biserte, G. [Automatic determination of leucine aminopeptidase in human serum.] Clin. Chim. Acta **46**: 91-103.

Royer, G. P. & Andrews, J. P. Immobilized derivatives of leucine aminopeptidase and aminopeptidase M. *J. Biol. Chem.* **248**: 1807-1812.

Wiederanders, B., Lasch, J., Kirschke, H., Bohley, P., Ansorge, S. & Hanson, H. The suitability of bovine-lens leucine aminopeptidase for sequence analysis and limited hydrolysis of polypeptides. *Eur. J. Biochem.* **36**: 504-508.

1974

Fittkau, S., Förster, U., Pascual, C. & Schunck, W.-H. Bovine-lens leucine aminopeptidase. Kinetic studies with substrates and substrate-like inhibitors. *Eur. J. Biochem.* **44**: 523-528.

Phillips, R. W. & Manildi, E. R. Abnormal serum isoenzyme of leucine aminopeptidase (LAP) in malignant neoplastic disease. *Cancer* **34**: 350-356.

Wacker, H., Lehky, P. & Vanderhaeghe, F. On the subunit structure of particulate aminopeptidase from pig kidney. *Experientia* **30**: 693 only.

1975

Ludewig, M., Frohne, M., Marquardt, I. & Hanson, H. Photoinactivation and carbethoxylation of leucine aminopeptidase. *Eur. J. Biochem.* **54**: 155-162.

Misselwitz, R., Zirwer, D., Frohne, M. & Hanson, H. Bovine lens aminopeptidase. Number and state of tryptophyl residues. *FEBS Lett.* **55**: 233-236.

1976

Fittkau, S., Schunck, W.-H. & Mootsi, S. Versuche zur Affinitätsmarkierung der Leuzinaminopeptidase mit neuen substratanalogen Inhibitoren. *Acta Biol. Med. Ger.* **35**: 365-378.

Hanson, H. & Frohne, M. Crystalline leucine aminopeptidase from lens (α-aminoacyl-peptide hydrolase; EC 3.4.11.1). *Methods Enzymol.* **45**: 504-521.

Kleine, R. & Lehmann, J. Das Verhalten der Leuzinaminopeptidase aus Rinderaugenlinsen in Guanidinhydrochlorid. Dissoziations- und Reassoziationsversuche. *Acta Biol. Med. Ger.* **35**: 331-341.

Suda, H., Aoyagi, T., Takeuchi, T. & Umezawa, H. Inhibition of aminopeptidase B and leucine aminopeptidase by bestatin and its stereoisomer. *Arch. Biochem. Biophys.* **177**: 196-200.

Thompson, G. A. & Carpenter, F. H. Leucine aminopeptidase (bovine lens). Effect of pH on the relative binding of Zn^{2+} and Mg^{2+} on the activation of the enzyme. *J. Biol. Chem.* **251**: 53-60.

Thompson, G. A. & Carpenter, F. H. Leucine aminopeptidase (bovine lens). The relative binding of cobalt and zinc to leucine aminopeptidase and the effect of cobalt substitution on specific activity. *J. Biol. Chem.* **251**: 1618-1624.

1977

Blum, M. & Sirota, P. Serum cystine aminopeptidase and leucine aminopeptidase activity in women with benign and malignant uterine and ovarian tumors. *Israel J. Med. Sci.* **13**: 875-880.

Carmel, A., Kessler, E. & Yaron, A. Intramolecularly-quenched fluorescent peptides as fluorogenic substrates of leucine aminopeptidase and inhibitors of clostridial aminopeptidase. *Eur. J. Biochem.* **73**: 617-625.

Jurnak, F., Rich, A., van Loon-Klaassen, L., Bloemendal, H., Taylor, A. & Carpenter, F. H. Preliminary X-ray study of leucine aminopeptidase (bovine lens), an oligomeric metalloenzyme. *J. Mol. Biol.* **112**: 149-153.

Kenny, A. J. Proteinases associated with cell membranes. In: *Proteinases in Mammalian Cells and Tissues* (Barrett, A. J. ed.), pp. 393-444 (see pp. 409-410), North-Holland Publishing Co., Amsterdam.

Kiselev, N. A., Stel'mashchuk, V. Y., Tsuprun, V. L., Ludewig, M. & Hanson, H. Electron microscopy of leucine aminopeptidase. *J. Mol. Biol.* **115**: 33-43.

Lassman, G., Damerau, W., Schwarz, D., Kleine, R. & Frohne, M. Zur stabilität der raumstruktur von leucinaminopeptidase bei guanidin.HCl-einwirkung. ESR-spektroskopische und biochemische untersuchungen. *Studia Biophysica* **63**: 149-162.

Ludewig, M., Hanson, H., Kiselev, N. A., Stel'Mashchuk, V. Y. & Tsuprun, V. L. On the quaternary structure of leucine aminopeptidase. *Acta Biol. Med. Ger.* **36**: 157-165.

McDonald, J. K. & Schwabe, C. Intracellular exopeptidases. In: *Proteinases in Mammalian Cells and Tissues* (Barrett, A. J. ed.), pp. 311-391 (see pp. 349-355), North-Holland Publishing Co., Amsterdam.

Shen. C.-C. & Melius, P. Leucine aminopeptidase from swine kidney : purification, molecular weight, subunit and amino acid composition. *Prep. Biochem.* **7**: 243-256.

Spector, G. J. Leucine and alanine aminopeptidase activities in experimentally induced intradermal granulomas and late stages of wound healing in the rat. *Lab. Invest.* **36**: 1-7.

1978

Aoyagi, T., Tobe, H., Kojima, F., Hamada, M., Takeuchi, T. & Umezawa, H.

Amastatin, an inhibitor of aminopeptidase A, produced by actinomycetes. *J. Antibiot.* **31**: 636-638.

Friedrich, M., Uhlig, J. & Noack, R. Hemmung der aminopeptidatischen Aktivitäten des Rattendünndarmes durch das Tripeptid H-Thr-(O-tert. butyl)-Phe-Pro-OH. *Acta Biol. Med. Ger.* **37**: 1513-1522.

Saifuku, K., Sekine, T., Namihisa, T., Takahashi, T. & Kanaoka, Y. A novel fluorometric ultramicro determination of serum leucine aminopeptidase using a coumarine derivative. *Clin. Chim. Acta* **84**: 85-91.

Taylor, A. & Carpenter, F. H. Kinetic parameters for the hydrolysis of various substrates by leucine aminopeptidase (bovine lens) [LAP]. *Fed. Proc. Fed. Am. Soc. Exp. Biol.* **37**: 2352 only.

1979

Fittkau, S., Kämmerer, G. & Damerau, W. The binding center of leucine aminopeptidase. Investigations with substrate analogs and spin-labelled inhibitors. *Ophthalmic Res.* **11**: 381-385.

Klante, K.-P., Mqotsi, S. & Fittkau, S. Leucine aminopeptidase from bovine eye lens. Affinity labelling of the enzyme with diazopeptide inhibitors. *Ophthalmic Res.* **11**: 386-388.

Lasch, J. Kinetic properties of bovine lens leucine aminopeptidase. *Ophthalmic Res.* **11**: 372-376.

Minato, S. A new colorimetric method for the determination of serum enzyme, γ-glutamyl transpeptidase, cystine aminopeptidase, and leucine aminopeptidase. *Clin. Chim. Acta* **92**: 249-255.

Mizutani, S., Yoshino, M., Oya, M., Noto, H., Inamoto, Y., Sakura, H. & Kawashima, Y. A comparison of angiotensinase and placental leucine aminopeptidase during normal pregnancy. *Clin. Biochem.* **12**: 50-51.

Müller-Frohne, M. Leucine aminopeptidase from bovine eye lens. Studies on the active centre by chemical modification techniques. *Ophthalmic Res.* **11**: 377-380.

Taylor, A., Carpenter, F. H. & Wlodawer, A. Leucine aminopeptidase (bovine lens): an electron microscopic study. *J. Ultrastruct. Res.* **68**: 92-100.

van Loon-Klaassen, L., Cuypers, H. T. & Bloemendal, H. Limited tryptic digestion of bovine eye lens leucine aminopeptidase. *FEBS Lett.* **107**: 366-370.

1980

Fittkau, S. & Damerau, W. Spinmarkierte Oligopeptide als substratanaloge Inhibitoren der Leucinaminopeptidase. *J. Prakt. Chem.* **322**: 1032-1038.

Klante, K.-P. & Fittkau, S. Affinitätsmarkierung und differenzierte Markierung der Leuzinaminopeptidase mit Diazo-Peptid-Inhibitoren. *Acta Biol. Med. Ger.* **39**: 781-790.

Nagaoka, I. & Yamashita, T. Leucine aminopeptidase as an ecto-enzyme of polymorphonuclear neutrophils. *Biochim. Biophys. Acta* **598**: 169-172.

van Loon-Klaassen, L. A. H., Cuypers, H. T., van Westreenen, H., de Jong,

W. W. & Bloemendal, H. The primary structure of bovine lens leucine aminopeptidase. Complete amino acid sequence of the N-terminal cyanogen bromide fragment and site of limited tryptic digestion. *Biochem. Biophys. Res. Commun.* **95**: 334-341.

1981

Taylor, A., Tisdell, F. E. & Carpenter, F. H. Leucine aminopeptidase (bovine lens): synthesis and kinetic properties of ortho-, meta-, and para-substituted leucyl-anilides. *Arch. Biochem. Biophys.* **210**: 90-97.

Van Wart, H. E. & Lin, S. H. Metal binding stoichiometry and mechanism of metal ion modulation of the activity of porcine kidney leucine aminopeptidase. *Biochemistry* **20**: 5682-5689.

1982

Andersson, L., Isley, T. C. & Wolfenden, R. α-Aminoaldehydes: transition state analogue inhibitors of leucine aminopeptidase. *Biochemistry* **21**: 4177-4180.

Cuypers, H. T., van Loon-Klaassen, L. A. H., Vree Egberts, W. T. M., de Jong, W. W. & Bloemendal, H. Sulfhydryl content of bovine eye lens leucine aminopeptidase. Determination of the reactivity of the sulfhydryl groups of the zinc metalloenzyme, of the enzyme activated by Mg^{2+}, Mn^{2+}, and Co^{2+}, and of the metal-free apoenzyme. J. Biol. Chem. **257**: 7086-7091.

Cuypers, H. T., van Loon-Klaassen, L. A. H., Vree Egberts, W. T. M., de Jong, W. W. & Bloemendal, H. The primary structure of leucine aminopeptidase from bovine eye lens. *J. Biol. Chem.* **257**: 7077-7085.

Lin, S. H. & Van Wart, H. E. Effect of cryosolvents and subzero temperatures on the hydrolysis of L-leucine-p-nitroanilide by porcine kidney leucine aminopeptidase. *Biochemistry* **21**: 5528-5533.

Taylor, A., Sawan, S. & James, T. L. Structural aspects of the inhibitor complex formed by N-(leucyl)-o-aminobenzenesulfonate and manganese with Zn^{2+}-Mn^{2+} leucine aminopeptidase (EC 3.4.11.1). *J. Biol. Chem.* **257**: 11571-11576.

1983

Allen, M. P., Yamada, A. H. & Carpenter, F. H. Kinetic parameters of metal-substituted leucine aminopeptidase from bovine lens. *Biochemistry* **22**: 3778-3783.

Antonucci, A. & Solinas, S. P. Leucine aminopeptidase from human urine. *Biochem. Int.* **6**: 617-625.

Kobaru, S., Tsunakawa, M., Hanada, M., Konishi, M., Tomita, K. & Kawaguchi, H. Bu-2743E, a leucine aminopeptidase inhibitor produced by *Bacillus circulans. J. Antibiot.* **36**: 1396-1398.

Ledeme, N., Vincent-Fiquet, O., Hennon, G. & Plaquet, R. Human liver L-leucine aminopeptidase: evidence for two forms compared to pig liver enzyme. *Biochimie* **65**: 397-404.

Taylor, A., Brown, M. J., Daims, M. A. & Cohen, J. Localization of leucine aminopeptidase in normal hog lens by immunofluorescence and activity assays. *Invest. Ophthalmol. Visual Sci.* **24**: 1172-1180.

Taylor, A., Daims, M., Lee, J. & Surgenor, T. Identification and quantification of leucine aminopeptidase in aged normal and cataractous human lenses and ability of bovine lens LAP to cleave bovine crystallins. Current Eye Res. **2**: 47-56.

Wilkes, S. H. & Prescott, J. M. Stereospecificity of amino acid hydroxamate inhibition of aminopeptidases. *J. Biol. Chem.* **258**: 13517-13521.

1984

Taylor, A., Volz, K. W., Lipscomb, W. N. & Takemoto, L. J. Leucine aminopeptidase from bovine lens and hog kidney. Comparison using immunological techniques, electron microscopy, and X-ray diffraction. *J. Biol. Chem.* **259**: 14757-14761.

Aspartyl Aminopeptidase

Summary

EC Number: 3.4.11.-

Earlier names: Aspartate aminopeptidase, Mg^{2+}-activated aspartate aminopeptidase, aspartyl naphthylamidase, angiotensinase A_2, cytosol aminopeptidase II.

Abbreviation: DAP (D, aspartic acid).

Distribution: Cytosol of kidney.

Source: Dog kidney.

Action: Catalyzes the removal of N-terminal, dicarboxylic amino acids from peptides and arylamide derivatives. α-Asp-NNap is hydrolyzed three times faster than α-Glu-NNap. There is no detected hydrolysis of 2-naphthylamides of other amino acids. DAP rapidly removes N-terminal aspartic acid from [Asp^1,Ile^5]-angiotensin I, but not N-terminal asparagine from [Asn^1,Val^5]-angiotensin II.

Requirements: pH 7.5 at 37°C, with preincubation in 50 mM $MnCl_2$.

Substrate, usual: α-Asp-NNap.

Substrate, special: None reported, but α-Asp-NMec would probably offer greater sensitivity.

Inhibitors: None reported. EDTA is not inhibitory, but in view of the activation obtained with Mn^{2+}, other chelating agents may prove to be inhibitory.

Molecular properties: None reported. The fact that the enzyme was "only slightly retarded" during gel permeation chromatography on BioGel P-300 suggests that DAP has M_r in the range 250,000-300,000.

Comment: Unlike glutamyl aminopeptidase, aspartyl aminopeptidase (DAP) is neither stimulated by Ca^{2+} nor inhibited by EDTA. By way of further contrast, DAP is activated by Mn^{2+}, and appears to be freely in solution in the cytoplasm. This new aminopeptidase was discovered by Cheung & Cushman during their study of glutamyl aminopeptidase present in dog kidney extracts. Although DAP was only partially (33-fold) purified from this source, it does appear that these workers adequately distinguished the enzyme from other aminopeptidases, and for this reason DAP has been included here as a unique aminopeptidase, although thus far only demonstrated in dog kidney. Its rapid action on the N-terminal aspartic acid residues of angiotensin analogs points to a possible physiological role in angiotensin degradation and in the conversion of angiotensin II to angiotensin III, des-Asp[1]-angiotensin II.

Bibliography

1971

Cheung, H. S. & Cushman, D. W. A soluble aspartate aminopeptidase from dog kidney. *Biochim. Biophys. Acta* **242**: 190-193.

Soluble Alanyl Aminopeptidase

Summary

EC Number: 3.4.11.14

Earlier names: Arylamidase, neutral arylamidase, alanyl arylamidase, naphthylamidase, amino acid naphthylamidase, thiol-activated aminopeptidase, human liver aminopeptidase (HLA), puromycin-sensitive aminopeptidase, aminopolypeptidase, cytosol aminopeptidase III.

Abbreviation: sAAP (s, soluble).

Distribution: sAAP has a wide distribution among the tissues and fluids of the body. It is especially abundant in the pancreas and organs and tissues of the hepatobiliary tract. Several organ-specific isozymes occur in the serum. Most of the circulating activity is believed to arise from the liver, and that in the urine from the kidney.

Source: Liver [1-3], small intestine [4,5], kidney [6-7], and pancreas [8], primarily of human origin; brain from rats [9] and monkeys [10], and beef heart [11].

Action: Catalyzes the release of unsubstituted, N-terminal amino acids from dipeptides, tripeptides, oligopeptides, and the amide and arylamide derivatives of amino acids, with a preference for derivatives of alanine, methionine, and phenylalanine over leucine [1]. The puromycin-sensitive activity in human plasma exhibits the same order of preference [12].

　　As reported for human liver sAAP [3], the relative rates of hydrolysis (at pH 6.8) of amino acid 2-naphthylamides are Ala- 100, Phe- 63, Met- 58, Leu- 36, Arg- 28, Trp- 19, Gly- 14, Lys- 13, Ser- 6, Thr- 6, Glu- 4, Val- 3, and Ile- 3. Whereas sAAP from human kidney is reported to be inactive on Pro-NNap [6], the enzyme from monkey brain is reportedly active, although its K_m is high [10]. Partially-purified sAAP from hog liver exhibits a V_{max} (at pH 6.5) on Phe-NNap that is slightly higher than that for Ala-NNap [12a].

　　The relative rates at which dipeptides are hydrolyzed (at pH 6.8) by the human liver enzyme [3] are as follows: Ala-Trp 100, Ala-Val 47, Ala-Ala 45, Ala-Phe 43, Leu-Leu 36, Ala-Leu 27, Leu-Ala 26, Ala-Ser 23, Phe-Gly 19, Ala-Glu 19, Ala-Gly 18, Ala-Asp 11 and Leu-Gly 5. It is also reported [1] that the human liver enzyme has no activity on Leu-Gly or Leu-NH$_2$, substrates for LAP.

Affinities are higher for the oligopeptide substrates as compared to di- and tri-peptides [13], and several physiologically-active peptides and hormones are susceptible to N-terminal attack, as shown for the enzyme from monkey brain [13] and human pancreas [14]. Although sAAP has kinin-converting activity on Met-Lys-bradykinin and Lys-bradykinin, it has no kininase activity on bradykinin or substance P, wherein the N-terminal residues are adjacent to proline. It does, however, exhibit pronounced angiotensinase activity [14].

The K$_m$ values for Ala-, Leu-, Arg- and Met-NNap are, respectively, 125, 69, 60 and 36 μM; K$_{cat}$ values are 2.5, 0.82, 0.38 and 1.6 x 10^4 s^{-1}. Ala-NNap has the highest catalytic coefficient (k$_{cat}$/K$_m$).

Requirements: pH 6.8 at 37°C in 0.1 M phosphate buffer containing 0.5 mM dithiothreitol for activation [13] and stabilization [9]; cysteine and 2-mecaptoethanol are less effective. Rates may be enhanced up to 2.5-fold [3] by the incorporation of Co^{2+} at 2 mM [2,15].

Substrate, usual: Ala-NNap [3].

Substrate, special: Ala-NMec [16].

Inhibitors: sAAP is extremely sensitive to puromycin [5,9,12,13]. The inhibition is competitive, with K$_i$ values reported to be 2 μM (at pH 7.0 and 37°C) for rat brain sAAP [9], 0.5 μM for monkey brain sAAP [13], and 12.5 μM for the human intestinal enzyme. Puromycin has no effect on most other aminopeptidases, including leucyl aminopeptidase [9,12]. sAAP is also inhibited competitively by bestatin [13], but so, too, are leucyl aminopeptidase and arginyl aminopeptidase. Penicillin and methicillin exhibit noncompetitive inhibition [5]. Inhibition of human liver sAAP by such β-lactam antibiotics has been reported for methicillin (K$_i$ 2.4 mM), oxacillin (K$_i$ 1.6 mM), and cloxacillin (K$_i$ 0.51 mM) [16a].

Chelating agents, especially 1,10-phenanthroline and 8-hydroxy-quinoline, are potent inhibitors [5] by virtue of their ability to bind to the zinc; however, the actual removal of the zinc requires drastic treatment. Sulfhydryl blocking reagents are reported to be effective inhibitors of sAAP from brain [9,10] and to a lesser degree of the human liver enzyme [1].

Other inhibitors include unsubstituted amino acids with hydrophobic side chains [17]. Inhibition by phenylalanine, for example, is noncompetitive, with Leu-Leu it is mixed, and with Ala-Ala-Ala it is competitive. These different mechanisms are attributed to the binding of the amino acid to the S$_2'$ subsite on the enzyme, which is the active site residue that binds to the substrate residue located third from the N-terminus - the P$_2'$ residue, as defined by Schechter & Berger [18]. The S$_2'$ binding site preferentially binds large hydrophobic residues as indicated by a much lower K$_i$ value for Ala-Ala-Phe compared to Ala-Ala-Ala as inhibitors of Ala-NNap hydrolysis [17]. Additional inhibitors include carboxylic acids and aliphatic amines [3].

Molecular properties: A Zn-sialo-glycoprotein containing about 21%
carbohydrate, M_r 235,000 [19]. The native enzyme consists of two subunits
(M_r 118,000), each containing a firmly-bound Zn^{2+} atom [19]. A
temperature-dependent equilibrium appears to exist between the monomeric
and dimeric forms [14]. The enzyme is rich in tryptophan, with 31
residues in each monomer, the majority of which are apparently located
in a hydrophobic environment [20]. Cysteine and cystine, on the other
hand, are present in very low amounts. The carbohydrate moiety is
comprised of 9.0% hexoses, 4.3% glucosamine and 4.1% sialic acid. Several
electrophoretically-distinct isozymes of sAAP, all with common antigenic
determinants [21], occur in the blood, and are very likely attributable
to varying amounts of sialic acid. One isozyme arises from the liver,
one from the kidney, one from the duodenum and two from the pancreas. K_m
and k_{cat} values for a particular substrate are statistically identical
for the various isozymes [8]. The average extinction coefficient
($A_{280,1\%}$) is 16.0 [21]. The enzyme undergoes thermal denaturation at
60°C [13], and the rate is greatly accelerated in the presence of EDTA
[22].

Comment: The various isozymes of sAAP that occur in the blood are
potentially useful tissue/organ specific enzyme markers for particular
diseases. They have pI values ranging from 3.6 to 4.7 [23]. Their serum
levels appear to be elevated in certain pathological conditions, e.g.
adenocarcinoma of the colon and pancreas, and nephrotic syndrome, as
well as in pregnancy and following surgery.

In the earlier literature, sAAP (EC 3.4.11.14) was frequently confused
with LAP (EC 3.4.11.1), primarily because of the use of Leu-NNap as an
assay substrate for LAP. However, it is now well established that these
exopeptidases are separate, and can readily be distinguished on the
basis of their substrate specificities alone. For example, whereas sAAP
hydrolyzes Leu-NNap more rapidly than Leu-Gly, the opposite is true for
LAP. In addition, sAAP is distinguished by its sensitivity to puromycin
and to a degree by its thiol dependence. These identifying
characteristics were first reported for an arylamidase activity in
bovine pituitary extracts [24].

Although sAAP and LAP are now less frequently confused, a study of
the current literature shows that sAAP (EC 3.4.11.14) is now commonly
confused with membrane-bound alanyl aminopeptidase (mAAP, EC 3.4.11.2).
Papers [23,25] dealing with one or the other of these enzymes often make
little if any distinction between them; in fact, they are sometimes
discussed as though they had a common identity. Although their substrate
specificity and pH optima are similar, the puromycin sensitivity of sAAP
and heat stability of mAAP should still serve as distinguishing
characteristics. Furthermore, whereas mAAP is characteristically found
in the microsome fraction during subcellular fractionation and requires
treatment with proteases or detergents for solubilization, sAAP has been

variously located in the "soluble" fraction [9-11,15], the lysosome fraction [26], and in intermediate fractions that are difficult to characterize, except possibly as protein aggregates [1]. It may be that sAAP in certain tissues tends to associate with other components in homogenates, thus explaining why deoxycholate treatment [1] or autolysis [2] often facilitate solubilization.

It can now be stated with a degree of certainty that the thiol-dependent leucyl arylamidase present in lysosomal fractions is attributable to cathepsin H (see lysosomal arginyl aminopeptidase, Entry 11.10). This unusual cysteine proteinase exhibits pronounced exopeptidase activities, at its pH 6.8 optimum, on both Leu-NNap and Arg-NNap. As points of distinction, the rate of Leu-NNap hydrolysis is typically about half that for Arg-NNap, and the enzyme is unaffected by puromycin [27,28]. Additionally, it has a low (28,000) M_r.

Numerous publications describe arylamidase activities in various body fluids, tissue extracts, and impure preparations. If confusion is to be minimized, and the responsible enzymes correctly identified, a more rigorous discrimination will be required in future studies involving these activities. Probably, new members of the (aminoacyl) aminopeptidase class remain unrecognized among the "arylamidases".

References

[1] Smith *et al. J. Biol. Chem.* **240**: 1718-1721, 1965.
[2] Behal *et al. Biochim. Biophys. Acta* **178**: 118-127, 1969.
[3] Little *et al. Methods Enzymol.* **45**: 495-503, 1976.
[4] Behal & Little *Clin. Chim. Acta* **21**: 347-355, 1968.
[5] McClellan & Garner *Biochim. Biophys. Acta* **613**: 160-167, 1980.
[6] Behal & Story *Arch. Biochem. Biophys.* **131**: 74-82, 1969.
[7] Kao *et al. Biochemistry* **17**: 2990-2994, 1978.
[8] Sidorowicz *et al. Clin. Chim. Acta* **104**: 169-179, 1980.
[9] Marks *et al. J. Biol. Chem.* **243**: 2882-2889, 1968.
[10] Hayashi & Oshima *J. Biochem.* **81**: 631-639, 1977.
[11] Brecher *et al. Biochim. Biophys. Acta* **191**: 472-475, 1969.
[12] McDonald *et al. Biochem. Biophys. Res. Commun.* **16**: 135-140, 1964.
[12a]Kawata *et al. J. Biochem.* **92**: 1093-1101, 1982.
[13] Hayashi *J. Biochem.* **84**: 1363-1372, 1978.
[14] Sidorowicz *et al. Clin. Chim. Acta* **111**: 69-79, 1981.
[15] Behal *et al. Arch. Biochem. Biophys.* **115**: 545-554, 1966.
[16] Parsons *et al. Int. J. Biochem.* **10**: 217-221, 1979.
[16a]Starnes *et al. Eur. J. Biochem.* **124**: 363-370, 1982.
[17] Garner & Behal *Biochemistry* **14**: 3208-3212, 1975.
[18] Schechter & Berger *Biochem. Biophys. Res. Commun.* **27**: 157-162, 1967.
[19] Starnes & Behal *Biochemistry* **13**: 3221-3227, 1974.
[20] Garner & Behal *Physiol. Chem. Phy.* **9**: 47-54, 1977.
[21] Sidorowicz *et al. Clin. Chim. Acta* **107**, 245-256, 1980.

[22] Garner & Behal *Biochemistry* **13**: 3227-3233, 1974.
[23] Böhme *et al. Enzyme* **21**: 464-470, 1976.
[24] Ellis & Perry *Biochem. Biophys. Res. Commun.* **15**: 502-505, 1964.
[25] Hiwada *et al. Clin. Chim. Acta* **93**: 113-117, 1979.
[26] Mahadevan & Tappel *J. Biol. Chem.* **242**: 2369-2374, 1967.
[27] Kirschke *Acta Biol. Med. Ger.* **36**: 1547-1548, 1977
[28] Barrett *et al. Biochem. J.* **187**: 909-912, 1980.

Bibliography

1963

Ellis, S. A thiol-activated aminopeptidase of the pituitary. *Biochem. Biophys. Res. Commun.* **12**: 452-456.
Patterson, E. K., Hsiao, S.-H. & Keppel, A. Studies on dipeptidases and aminopeptidases. I. Distinction between leucine aminopeptidase and enzymes that hydrolyze L-leucyl-β-naphthylamide. *J. Biol. Chem.* **238**: 3611-3620.

1964

Behal, F. J., Hamilton, R. D., Dawson, F. B. & Terrell, L. C. A study of the activity of human plasma aminopeptidase components on dipeptides and chromogenic substrates. *Arch. Biochem. Biophys.* **108**: 207-214.
Ellis, S. & Perry, M. Inhibition of thiol-activated aminopeptidase by puromycin. *Biochem. Biophys. Res. Commun.* **15**: 502-505.
McDonald, J. K., Reilly, T. J. & Ellis, S. The hydrolysis of amino acyl-β-naphthylamides by plasma aminopeptidases. *Biochem. Biophys. Res. Commun.* **16**: 135-140.

1965

Behal, F. J., Asserson, B., Dawson, F. & Hardman, J. A study of human tissue aminopeptidase components. *Arch. Biochem. Biophys.* **111**: 335-344.
Patterson, E. K., Hsiao, S-I., Keppel, A. & Sorof, S. Studies on dipeptidases and aminopeptidases. II. Zonal electrophoretic separation of rat liver peptidases. *J. Biol. Chem.* **240**: 710-716.
Smith, E. E., Kaufman, J. T. & Rutenburg, A. M. The partial purification of an amino acid naphthylamidase from human liver. *J. Biol. Chem.* **240**: 1718-1721.

1966

Behal, F. J., Klein, R. A. & Dawson, F. B. Separation and characterization of aminopeptidase and arylamidase components of human liver. *Arch. Biochem. Biophys.* **115**: 545-554.
Ellis, S. & Perry, M. Pituitary arylamidases and peptidases. *J. Biol. Chem.* **241**: 3679-3686.

Panveliwalla, D. K. & Moss, D. W. A comparison of aminoacyl-β-naphthylamide hydrolases in extracts of human tissues. *Biochem. J.* **99**: 501-506.

1967

Mahadevan, S. & Tappel, A. L. Arylamidases of rat liver and kidney. *J. Biol. Chem.* **242**: 2369-2374.

1968

Behal, F. J. & Little, G. H. Arylamidase of human duodenum. *Clin. Chim. Acta* **21**: 347-355.

Marks, N., Datta, R. K. & Lajtha, A. Partial resolution of brain arylamidases and aminopeptidases. *J. Biol. Chem.* **243**: 2882-2889.

Nagatsu, I., Nagatsu, T. & Glenner, G. G. Species differences of serum amino acid β-naphthylamidases. *Enzymologia* **34**: 73-76.

Peters, J. E., Rehfeld, N., Beier, L. & Haschen, R. J. Immunologische charakterisierung der isoenzyme der alanin-aminopeptidase. *Clin. Chim. Acta* **19**: 277-286.

1969

Behal, F. J. & Story, M. N. Arylamidase of human kidney. *Arch. Biochem. Biophys.* **131**: 74-82.

Behal, F. J., Little, G. H. & Klein, R. A. Arylamidase of human liver. *Biochim. Biophys. Acta* **178**: 118-127.

Brecher, A. S. & Suszkiw, J. B. Brain arylamidase. Purification and characterization of the soluble bovine enzyme. *Biochem. J.* **112**: 335-342.

Brecher, A. S., König, M. & Barefoot, S. W. Partial purification and some properties of bovine heart arylamidase. *Biochim. Biophys. Acta* **191**: 472-475.

1971

Little, G. H. & Behal, F. J. Human liver arylamidase: molecular weight and subunit structure. *Biochim. Biophys. Acta* **243**: 312-319.

Little, G. H. & Behal, F. J. Hydrolysis of di- and oligopeptides by human liver arylamides. *Proc. Soc. Exp. Biol. Med.* **136**: 954-957.

1972

Camargo, A. C. M., Ramalho-Pinto, F. J. & Greene, L. J. Brain peptidases: conversion and inactivation of kinin hormones. *J. Neurochem.* **19**: 37-49.

1974

Garner, C. W. & Behal, F. J. Human liver aminopeptidase. Role of metal ions in mechanism of action. *Biochemistry* **13**: 3227-3233.

Starnes, W. L. & Behal, F. J. A human liver aminopeptidase. The amino acid and carbohydrate content, and some physical properties of a sialic acid containing glycoprotein. *Biochemistry* **13**: 3221-3227.

1975

Bornschein, W. Isoenzyme der alaninarylamidase (AAP, EC 3.4.11.2) und gamma-glutamyltranspeptidase (GGTP, EC 2.3.2.2.) bei chronischer pankreatitis und pankreasneoplasma. *Clin. Chim. Acta* **61**: 325-333.

Garner, C. W. & Behal, F. J. Effect of pH on substrate and inhibitor kinetic constants of human liver alanine aminopeptidase. Evidence for two ionizable active center groups. *Biochemistry* **14**: 5084-5088.

Garner, C. W. & Behal, F. J. Human liver alanine aminopeptidase. Inhibition by amino acids. *Biochemistry* **14**: 3208-3212.

1976

Böhme, I., Hütter, H. J., Gerlach, W. & Haschen, R. J. Nature of the multiple forms of alanine aminopeptidase. *Enzyme* **21**: 464-470.

Herrmann, W. P. & Uhlenbruck, G. Carbohydrate moieties of human seminal plasma arylamidases. *Andrologia* **8**: 308-312.

Little, G. H., Starnes, W. L. & Behal, F. J. Human liver aminopeptidase. *Methods Enzymol.* **45**: 495-503.

Oya, M., Wakabayashi, T., Yoshino, M. & Mizutani, S. Subcellular distribution and electrophoretic behaviour of aminopeptidase in human placenta. *Physiol. Chem. Phys.* **8**: 327-335.

1977

Garner, C. W. & Behal, F. J. Fluorescence of human liver alanine aminopeptidase. *Physiol. Chem. Phys.* **9**: 47-53.

Garner, C. W. & Behal, F. J. Hydrophobic binding sites of human liver alanine aminopeptidase. *Arch. Biochem. Biophys.* **182**: 667-673.

Hayashi, M. & Oshima, K. Purification and characterization of arylamidase from monkey brain. *J. Biochem.* **81**: 631-639.

Herrmann, W. P. Incomplete arylamidase in psoriasis scales. *Arch. Dermatol. Res.* **255**: 231-236.

Hiwada, K. & Kokubu, T. Comparison of soluble and membrane- bound neutral arylamidases from renal cell carcinoma. *Clin. Chim. Acta* **80**: 395-401.

Spector, G. J. Leucine and alanine aminopeptidase activities in experimentally induced intradermal granulomas and late stages of wound healing in the rat. *Lab. Invest.* **36**: 1-7.

1978

Behal, F. J. Human kidney alanine aminopeptidase. The physical and kinetic properties of a sialic acid containing glycoprotein. *Fed. Proc. Fed. Am. Soc. Exp. Biol.* **37**: 1438 only.

Hayashi, M. Monkey brain arylamidase. II. Further characterization and studies on mode of hydrolysis of physiologically active peptides. *J. Biochem.* **84**: 1363-1372.

Kao, Y. J., Starnes, W. L. & Behal, F. J. Human kidney alanine aminopeptidase: physical and kinetic properties of a sialic acid containing glycoprotein. *Biochemistry* **17**: 2990-2994.

1979

Freitas, J. O., Jr., Guimaràes, J. A., Borges, D. R. & Prado, J. L. Two arylamidases from human liver and their kinin-converting activity. *Int. J. Biochem.* **10**: 81-89.

Niinobe, M., Tamura, Y., Arima, T. & Fujii, S. Immunological properties of and neuraminidase action on aminopeptidases and arylamidases in human normal and cancer tissues. *Cancer Res.* **39**: 4212-4217.

Parsons, M. E., Godwin, K. O. & Pennington, R. J. T. Further studies on aminopeptidases of rat muscle. *Int. J. Biochem.* **10**: 217-221.

Sidorowicz, W. & Behal, F. J. Human pancreas alanine aminopeptidase. *Fed. Proc. Fed. Am. Soc. Exp. Biol.* **38**: 841 only.

1980

McClellan, J. B., Jr. & Garner, C. W. Purification and properties of human intestine alanine aminopeptidase. *Biochim. Biophys. Acta* **613**: 160-167.

Sidorowicz, W., Hsia, W.-C., Maslej-Zownir, M. & Behal, F. J. Multiple molecular forms of human alanine aminopeptidase: immunochemical properties. *Clin. Chim. Acta* **107**: 245-256.

Sidorowicz, W., Jackson, G. C. & Behal, F. J. Multiple molecular forms of human pancreas alanine aminopeptidase. *Clin. Chim. Acta* **104**: 169-179.

1981

Diener, U., Knoll, E., Langer, B., Rautenstrauch, H., Ratge, D. & Wisser, H. Urinary excretion of N-acetyl-β-D-glucosaminidase and alanine aminopeptidase in patients receiving amikacin or *cis*-platinum. *Clin. Chim. Acta* **112**: 149-157.

Sidorowicz, W., Zownir, O. & Behal, F. J. Action of human pancreas alanine aminopeptidase on biologically active peptides: kinin converting activity. *Clin. Chim. Acta* **111**: 69-79.

1982

Kawata, S., Imamura, T., Ninomiya, K. & Makisumi, S. Purification and characterization of an aminopeptidase from porcine liver. *J. Biochem.* **92**: 1093-1101.

Starnes, W. L., Szechinski, J. & Behal, F. J. Human liver alanine aminopeptidase. A kinin-converting enzyme sensitive to β-lactam antibiotics. *Eur. J. Biochem.* **124**: 363-370.

1983

Imamura, T., Kawata, S., Ninomiya, K. & Makisumi, S. Porcine liver aminopeptidase. Further characterization of its sulfhydryl groups. *J. Biochem.* **94**: 267-273.

Soluble Arginyl Aminopeptidase

Summary

EC Number: 3.4.11.6

Earlier names: Aminopeptidase B (APB), arginyl arylamidase, arylamidase B, arginine aminopeptidase, chloride-activated arginine aminopeptidase, cytosol aminopeptidase IV.

Abbreviation: sRAP (s, soluble; R, arginine).

Distribution: Widely distributed in mammalian tissues and cells, i.e. liver, muscle, erythrocytes, leukocytes, macrophages, synovial fluid, gingival fluid, and inflammatory exudates. Cytosolic.

Source: Rat liver [1], hog liver [2], human liver [3], and human erythrocytes [4].

Action: Catalyzes specifically the removal of unsubstituted, N-terminal Arg- and Lys- residues from peptides and 2-naphthylamide derivatives. The rate of arginine release is typically about twice that for lysine. Action on peptides has been demonstrated for dipeptides and tripeptides possessing an unsubstituted, N-terminal arginyl or lysyl residue. Dipeptides are preferred over tripeptides, with no action occurring on oligopeptides. K_m values for the best substrates increase in the following order: Arg-Phe, Arg-Trp, Arg-Lys $<$ Arg-Ala, Arg-Gly, Arg-NNap $<$ Arg-Gly-Gly. Highest k_{cat} values are exhibited on Arg-NNap, followed by Arg-Phe and Arg-Trp, for sRAP purified from rat liver and inflammatory exudate, and human erythrocytes [4a]. Except for some relatively weak endopeptidase (kininase) activity associated with the red cell enzyme [5], sRAP from other sources shows no kinin-converting activity [3] or action on endopeptidase assay substrate such as Bz-Arg-NNap and Bz-Arg-OEt.

Requirements: pH 7.0 at 30°C on Arg-NNap, with some activity exhibited over the range pH 6-8 [1]. Characteristically, a 3- to 4-fold activation is attained when Cl⁻ is incorporated at 0.15 M into (0.25 M) phosphate buffer [4,6].

Substrate, usual: Arg-NNap [1,7].

Substrate, special: Arg-NMec.

Inhibitors: Non-competitive inhibitors include thiol-blocking reagents, in particular 4-chloromercuribenzoate (K_i 1.8 x 10^{-6} M) and divalent cations such as Hg^{2+}, Pb^{2+}, Cd^{2+}, Cu^{2+} and Zn^{2+}. Metal chelating agents such as 1,10-phenanthroline and EDTA are also inhibitory [6,7]. Bestatin is a potent competitive inhibitor of both the rat liver [7] and hog liver [8] enzymes, with K_i values of 0.06 μM and 0.014 μM, respectively. Unlike bestatin, which also inhibits other aminopeptidases, arphamenine B is a specific inhibitor of sRAP that shows potent competitive inhibition, K_i 0.8 nM [8a]. Several amino acids exhibit competitive inhibition, especially basic ones such as arginine (K_i 0.3 mM) and homoarginine (K_i 0.23 mM). Competitive inhibition is also shown by Bz-Arg-NPhNO$_2$ (K_i 0.17 mM), an endopeptidase substrate. Non-competitive inhibition is seen with D-arginine (K_i 0.77 mM).

Molecular properties: M_r 95,000 for freshly-purified sRAP from rat liver [1] and human fetal liver [9]. High-molecular weight aggregates of (rat liver) sRAP occur when solutions are stored at 4°C; however, these aggregates dissociate to monomers in the presence of substrate [10]. Values of 43,000, 20,000 and 66,000 have been reported for sRAP from human liver [3], hog liver [2] and rat muscle [10], respectively. All are markedly activated by Cl^-, and have pI values close to 5.0.

Inhibition by 4-chloromercuribenzoate and heavy metal ions suggests a functional role for cysteine at or near the active center. Only the enzyme from human erythrocytes (M_r 95,000) exhibits subunit structure. This tetrameric form of sRAP generates active dimers (M_r 52,000) during purification, and shows little sensitivity to 4-chloromercuri-benzoate [4]. However, titration of hog liver and human red cell sRAP with 5,5'-dithiobis-(2-nitrobenzoic acid) (Ellman's reagent) reveals one essential thiol group per active enzyme unit. At least one essential imidazole group is also present near the active site. Neither enzyme contains essential carboxyl or arginyl groups, nor do they contain zinc [12].

The human liver and erythrocyte forms of the enzyme are immunologically related [5]. Optimum stability is at pH 6-9. The enzyme is rather thermolabile and is unstable to freezing at -20°C unless stored in 40% (v/v) glycerol.

Comment: Arginyl aminopeptidases, regardless of source, are characteristically markedly stimulated by Cl^- and are sensitive to sulfhydryl reagents, and metal chelators. No significant activation is exhibited by sulfhydryl compounds or metals. In contrast to soluble alanyl aminopeptidase (sAAP), which also hydrolyzes Arg-NNap, sRAP is not sensitive to puromycin [8]. And, in contrast to glutamyl aminopeptidase ("aminopeptidase A") and leucyl aminopeptidase, sRAP is unaffected by amastatin [13].

The use of Arg-NNap to assay or to localize sRAP in tissue preparations is further complicated by a significant level of arginyl aminopeptidase activity exhibited by cathepsin H, a lysosomal enzyme that exhibits a pH 6.8 optimum on Arg-NNap (see Entry 11.10). Possible useful points of distinction include the following. Whereas sRAP is inhibited by (0.1 mM) bestatin, cathepsin H is not; the opposite effect is obtained with E-64, L-3-carboxy-2,3-*trans*-epoxypropionyl-leucyl-amido(4-guanidino)butane [14]. In addition, sRAP shows negligible activity on Leu-NNap, whereas cathepsin H hydrolyzes this substrate at about 50% of the rate exhibited on Arg-NNap.

Conflicting reports concerning inhibition by Pms-F and Tos-Phe-CH₂Cl have been taken as evidence for a functional histidyl residue. There is also evidence [15] for the existence in leukocytes and macrophages of a lysosomal form of arginyl aminopeptidase that is typically Cl⁻-activated and SH-dependent. The enzyme present in inflammatory exudates may be lysosomal, but should not be confused with the lysosomal arginyl aminopeptidase activity exhibited by cathepsin H. As a useful point of distinction, it was recently noted [14] that cathepsin H is unaffected by bestatin at 0.1 mM (see Entry 11.10).

Although the sensitivity of purified sRAP to chelating agents suggests that the enzyme is a metalloprotein, this has yet to be unequivocally established. As regards the restoration of activity with added metals, both negative [5,6] and positive [2,7] reports exist. Like LAP, rat liver sRAP is reported to be inhibited by 1,10-phenanthroline and reactivated by Zn^{2+} [7]. On the other hand, unlike LAP, metal analyses have failed to show the presence of zinc in either the typical (rat liver) or the atypical (red cell) soluble arginyl aminopeptidases [12].

References
[1]   Hopsu *et al. Arch. Biochem. Biophys.* **114**: 557-566, 1966.
[2]   Kawata *et al. J. Biochem.* **88**: 1025-1032, 1980.
[3]   Freitas *et al. Int. J. Biochem.* **10**: 81-89, 1979.
[4]   Mäkinen & Mäkinen *Biochem. J.* **175**: 1051-1067, 1978.
[4a] Söderling *Arch. Biochem. Biophys.* **220**: 1-10, 1983.
[5]   Mäkinen & Söderling *Personal communication.*
[6]   Hopsu *et al. Arch. Biochem. Biophys.* **114**: 567-575, 1966.
[7]   Suda *et al. Arch. Biochem. Biophys.* **177**: 196-200, 1976.
[8]   Kawata *et al. J. Biochem.* **88**: 1601-1605, 1980.
[8a] Umezawa & Aoyagi In: *Proteinase Inhibitors* (Katunuma *et al.* eds.), pp. 3-15, Springer-Verlag, New York, 1983.
[9]   Mäkinen *Arch. Biochem. Biophys.* **126**: 803-811, 1968.
[10] Mäkinen *Biochim. Biophys. Acta* **271**: 413-418, 1982.
[11] Parsons *et al. Int. J. Biochem.* **10**: 217-221, 1979.
[12] Söderling & Mäkinen *Arch. Biochem. Biophys.* **220**: 11-21, 1983.
[13] Aoyagi *et al. J. Antibiot.* **31**: 636-638, 1978.
[14] Kirschke *et al. Biochem. J.* **214**: 871-877, 1983.
[15] Söderling & Knuuttila *Life Sci.* **26**: 303-312, 1980.

Bibliography

1964

Behal, F. J., Hamilton, R. D., Dawson, F. B. & Terrell, L. C. A study of the activity of human plasma aminopeptidase components on dipeptides and chromogenic substrates. *Arch. Biochem. Biophys.* **108**: 207-214.

Hopsu, V. K., Kantonen, U.-M. & Glenner, G. G. A peptidase from rat tissues selectively hydrolyzing N-terminal arginine and lysine residues. *Life Sci.* **3**: 1449-1453.

1966

Hopsu, V. K., Mäkinen, K. K. & Glenner, G. G. A peptidase (aminopeptidase B) from cat and guinea pig liver selective for N-terminal arginine and lysine residues. I. Purification and substrate specificity. *Acta Chem. Scand.* **20**: 1225-1230.

Hopsu, V. K., Mäkinen, K. K. & Glenner, G. G. A peptidase (aminopeptidase B) from cat and guinea pig liver selective for N-terminal arginine and lysine residues. II. Modifier characteristics and kinetic studies. *Acta Chem. Scand.* **20**: 1231-1239.

Hopsu, V. K., Mäkinen, K. K. & Glenner, G. G. Characterization of aminopeptidase B: substrate specificity and affector studies. *Arch. Biochem. Biophys.* **114**: 567-575.

Hopsu, V. K., Mäkinen, K. K. & Glenner, G. G. Purification of a mammalian peptidase selective for N-terminal arginine and lysine residues: aminopeptidase B. *Arch. Biochem. Biophys.* **114**: 557-566.

Hopsu-Havu, V. K. & Mäkinen, K. K. Formation of bradykinen from kallidin-10 by aminopeptidase B. *Nature* **212**: 5067-5068.

1967

Mäkinen, K. K. & Hopsu-Havu, V. K. A simplified method for purification of rat liver aminopeptidase B. *Arch. Biochem. Biophys.* **18**: 257-258.

Mäkinen, K. K. & Hopsu-Havu, V. K. The active centre of aminopeptidase B. II. Kinetic studies. *Enzymologia* **32**: 347-363.

Mäkinen, K. K. & Hopsu-Havu, V. K. The presence of enzymes resembling aminopeptidase B in several rat organs. *Ann. Med. Exp. Biol. Fenn.* **45**: 230-234.

1968

Mäkinen, K. K. Effect of certain chemical compounds on rat liver aminopeptidase B acting on L-arginyl-2-naphthylamide with evidence of the occurrence of a similar enzyme in the liver of human fetuses. *Arch. Biochem. Biophys.* **126**: 803-811.

1969

Euranto, E. K., Kankare, J. J. & Mäkinen, K. K. Numerical treatment of kinetic data for enzymatic reactions. Part II. The pH dependence of the maximum rate of hydrolysis of N-L-arginyl-2-naphthylamine catalysed by aminopeptidase B at high substrate concentration. *Suomen Kemistilehti B.* **42**: 246-250.

Mäkinen, K. K. Effect of some alkaline metal salts on the hydrolysis of N-L-arginyl-2-naphthylamine by purified rat liver aminopeptidase B. *Soum. Kemistil. B.* **42**: 434-440.

Mäkinen, K. K., Euranto, E. K. & Kankare, J. J. Numerical treatment of kinetic data for enzymatic reactions. Part I. The dependence of the initial rate of hydrolysis of N-L-arginyl-2-naphthylamine catalysed by aminopeptidase B on the concentration of the substrate. *Suom. Kemistil. B.* **42**: 129-133.

1970

Euranto, E. K., Kankare, J. J. & Mäkinen, K. K. Numerical treatment of kinetic data for enzymic reactions part III. The pH dependence of the kinetic parameters for the hydrolysis of N-L-arginyl-2-naphthylamine catalyzed by aminopeptidase B. *Suom. Kemistil. B.* **43**: 166-170.

Mäkinen, K. K. Effect of tetraphenylboron ions on the rate of the hydrolysis of N-L-arginyl-2-naphthylamine catalyzed by aminopeptidase B. *Suom. Kemistil. B* **43**: 399-401.

Mäkinen, K. K. & Mäkinen, P.-L. Selective effect of domiphen bromide (dodecyldimethyl(2-phenoxyethyl)ammonium bromide) on some hydrolytic enzymes. *Biochim. Biophys. Acta* **206**: 143-151.

Mäkinen, K. K. & Paunio, K. U. Demonstration of aminopeptidase B in human periodontal tissues. *Acta Chem. Scand.* **24**: 1103-1104.

Mäkinen, P.-L., Raekallio, J. & Mäkinen, K. K. On the localization of aminopeptidase B and separation of its two molecular forms by automated recycling chromatography. *Acta Chem. Scand.* **24**: 1101-1102.

Paunio, K. U. & Mäkinen, K. K. Histochemical demonstration of aminopeptidase B in human gingiva. *Suom. Hammaslaakariseuran Toim.* **66**: 265-268.

1971

Mäkinen, K. K. & Mäkinen, P.-L. Effect of sodium chloride on substrate constant and maximal velocity in the enzymic hydrolysis of N-L-aminoacyl-2-naphthylamines and N-L-aminoacyl-p-nitroanilines. *Acta Chem. Scand.* **25**: 969-975.

Mäkinen, K. K. & Mäkinen, P.-L. Evidence on erythrocyte aminopeptidase B. *Int. J. Protein Res.* **3**: 41-47.

1972

Mäkinen, K. K. Evidence for the aggregation of aminopeptidase B during storage and breakdown of the aggregate by substrate and serum albumin. *Biochim. Biophys. Acta* **271**: 413-418.

Mäkinen, K. K. & Paunio, K. U. A histochemical method for the demonstration of aminopeptidase B activity. *J. Histochem. Cytochem.* **20**: 192-194.

Mäkinen, P.-L. & Mäkinen, K. K. Fractionation and properties of aminopeptidase B during purification and storage. *Int. J. Pept. Protein Res.* **4**: 241-255.

1973

Larmas, L. A., Mäkinen, K. K. & Paunio, K. U. A histochemical study of arylaminopeptidases in hydantoin induced hyperplastic, healthy and inflamed human gingiva. *J. Peridont. Res.* **8**: 21-27.

Mäkinen, K. K. & Oksala, E. Evidence on the involvement in inflammation of an enzyme resembling aminopeptidase B. *Clin. Chim. Acta* **49**: 301-309.

1974

Umezawa, H., Hori, S., Sawa, T., Yoshioka, T. & Takeuchi, T. A bleomycin-inactivating enzyme in mouse liver. *J. Antibiot.* **27**: 419-424.

1975

Garner, C. W. & Behal, F. J. Effect of pH on substrate and inhibitor kinetic constants of human liver alanine aminopeptidase. Evidence for two ionizable active center groups. *Biochemistry* **14**: 5084-5088.

Mäkinen, K. K. & Hyyppä, T. A biochemical study of the origin of arginine aminopeptidases in human gingival fluid. *Arch. Oral Biol.* **20**: 509-519.

Mäkinen, K. K., Luostarinen, V., Varrela, J., Rekola, M. & Luoma, S. Arginine aminopeptidase reactions to laser *in vivo* and *in vitro*. *Biochem. Med.* **13**: 192-195.

1976

Mäkinen, K. K. Occurrence and properties of arginine aminopeptidases. In: *Intracellular Protein Catabolism* (Hanson, H. & Bohley, P. eds), pp. 450-459, J. A. Barth, Leipzig.

Mäkinen, K. K. & Virtanen, K. K. Aminopeptidase B in human serum. *Clin. Chim. Acta* **67**: 213-218.

Suda, H., Aoyagi, T., Takeuchi, T. & Umezawa, H. Inhibition of aminopeptidase B and leucine aminopeptidase by bestatin and its stereoisomer. *Arch. Biochem. Biophys.* **177**: 196-200.

Suda, H., Takita, T., Aoyagi, T. & Umezawa, H. The structure of bestatin. *J. Antibiot.* **29**: 100-101.

Umezawa, H., Aoyagi, T., Suda, H., Hamada, M. & Takeuchi, T. Bestatin, an inhibitor of aminopeptidase B, produced by actinomycetes. *J. Antibiot.* **29**: 97-99.

1977

Nishizawa, R., Saino, T., Takita, T., Suda, H., Aoyagi, T. & Umezawa, H. Synthesis and structure-activity relationships of bestatin analogues, inhibitors of aminopeptidase B. *J. Med. Chem.* **20**: 510-515.

Söderling, E., Knuuttila, M. & Mäkinen, K. K. Aminopeptidase B-like enzymes in leukocytes. *FEBS Lett.* **76**: 219-225.

Virtanen, K. K., Mäkinen, K. K. & Oksala, E. Activity of arginine aminopeptidases and phosphatases in inflamed palatal mucosa in denture stomatitis: a histochemical and biochemical study. *J. Dent. Res.* **56**: 674-683.

1978

Aoyagi, T., Tobe, H., Kojima, F., Hamada, M., Takeuchi, T. & Umezawa, H. Amastatin, an inhibitor of aminopeptidase A, produced by actinomycetes. *J. Antibiot. (Tokyo)* **31**: 636-638.

Knuuttila, M., Virtanen, K., Söderling, E. & Mäkinen, K. K. A chloride-activated aminopeptidase in rat inflammatory exudate: properties and evidence of the origin of the enzymes. *Biochem. Biophys. Res. Commun.* **81**: 374-381.

Mäkinen, K. K. & Mäkinen, P.-L. Purification and characterization of two human erythrocyte arylamidases preferentially hydrolysing N-terminal arginine or lysine residues. *Biochem. J.* **175**: 1051-1067.

1979

Freitas, J. O. Jr., Guimaràes, J. A., Borges, D. R. & Prado, J. L. Two arylamidases from human liver and their kinin converting activity. *Int. J. Biochem.* **10**: 81-89.

1980

Kawata, S., Takayama, S., Ninomiya, K. & Makisumi, S. Porcine liver aminopeptidase B. Substrate specificity and inhibition by amino acids. *J. Biochem.* **88**: 1601-1605.

Kawata, S., Takayama, S., Ninomiya, K. & Makisumi, S. Purification and some properties of porcine liver aminopeptidase B. *J. Biochem.* **88**: 1025-1032.

Mäkinen, K. K., Haataja, M., Huusko, P. J., Mäkinen, P.-L., Hämalainen, M., Nieminen, L. & Laine, H. The involvement of the Cl-dependent arginine aminopeptidase in rheumatoid arthritis. *Clin. Chim. Acta* **100**: 71-74.

Söderling, E. A rapid and simple method for purification of the chloride-activated arginine aminopeptidase from biological materials. *Prep. Biochem.* **10**: 191-203.

Söderling, E. & Knuuttila, M. Release of the chloride-dependent arginine aminopeptidase from PMN leukocytes and macrophages during phogocytosis. *Life Sci.* **26**: 303-312.

Yamamoto, K., Suda, H., Ishizuka, M., Takeuchi, T., Aoyagi, T. & Umezawa, H. Isolation of α-aminoacyl arginines in screening of aminopeptidase B inhibitors. *J. Antibiot. (Tokyo)* **33**: 1597-1599.

1981

Söderling, E., Hujanen, E. & Mäkinen, K. K. Variance in the enzymatic properties of the chloride-activated arginine aminopeptidase from erythrocytes of healthy humans. *Biochem. Med.* **26**: 231-238.

1983

Söderling, E. Substrate specificities of Cl⁻-activated arginine aminopeptidases from human and rat origin. *Arch. Biochem. Biophys.* **220**: 1-10.

Söderling, E. & Mäkinen, K. K. Modification of the Cl⁻-activated arginine aminopepidases from rat liver and human erythrocytes: a comparative study. *Arch. Biochem. Biophys.* **220**: 11-21.

Umezawa, H. & Aoyagi, T. Trends in research of low molecular weight protease inhibitors of microbial origin. In: *Proteinase Inhibitors: Medical and Biological Aspects* (Katunuma, N., Umezawa, H. & Holzer, H. eds.), pp. 3-15, Springer-Verlag, New York.

Umezawa, H., Aoyagi, T., Ohuchi, S., Okuyama, A., Suda, H., Takita, T., Hamada, M. & Takeuchi, T. Arphamenines A and B, new inhibitors of aminopeptidase B produced by bacteria. *J. Antibiot.* **36**: 1572-1575.

Prolyl Aminopeptidase

Summary

EC Number: 3.4.11.5

Earlier names: Proline iminopeptidase, proline aminopeptidase, cytosol aminopeptidase V.

Abbreviation: sPAP (s, soluble).

Distribution: Activity characteristic of prolyl aminopeptidase, i.e. hydrolysis of polyproline at pH 8.6 in the presence of 5 mM $MnCl_2$, is detectable in various tissues of the rat, rabbit and hog. The highest levels of activity are present in liver and kidney, with lower levels in spleen, heart, lung, and testis. Prolyl aminopeptidase is apparently a cytosolic enzyme as judged by its presence in the soluble fraction from hog kidney.

Source: Hog kidney [1,2] and beef kidney [6].

Action: Catalyzes the removal of unsubstituted, N-terminal prolyl residues from dipeptide amides, tripeptides, and polypeptides. Dipeptides such as Pro-Gly and Pro-Leu (substrates for prolyl dipeptidase) are very poor substrates [2]. There is some evidence [1] that hydroxyprolyl linkages may be resistant. The highest rates occur when a bulky aliphatic or aromatic residue is present in the penultimate (P_1') position. The approximate relative rates at which peptides are hydrolyzed (at pH 7.8) by the hog kidney enzyme are reported [2] to be as follows: Pro-Leu-NH_2 100, Pro-Val-Gly 56, Pro-Phe-His-Leu 33, Pro-Phe-Gly-Lys 25, Pro-Val-Asp 21, Pro-Gly-Gly 7, Pro-Gly-NH_2 2, and Pro-Pro-Gly 0.1. Hydrolysis rates on Pro-NNap and polyproline were also rated at 0.1, so these do not appear to be useful assay substrates.

In an earlier study conducted with a less pure preparation of the hog kidney enzyme [1], polyproline was reported to a useful, highly-specific substrate, provided Mn^{2+} was available in a pH 8.6 reaction mixture. Indeed, the specific activity of prolyl aminopeptidase was reported in "polyproline units". Additionally, it was reported that sPAP does not remove N-terminal hydroxyprolyl residues; hydrolysis of Hyp-Gly was interpreted as contamination by prolyl dipeptidase. It appears that

cleavage rates are influenced by the nature of the residues located at
the P_1' and P_2' positions. Pro-Leu-Gly-NH$_2$, a hypothalamic peptide
(melanostatin) that inhibits the secretion of melanocyte-stimulating
hormone, was found to be a particularly good substrate [2]. The rate of
proline release from this substrate was estimated to be 68% greater than
the rate observed on Pro-Leu-NH$_2$.

Requirements: pH 7.8 at 40°C in a 25 mM Tris succinate buffer containing 1
mM MnCl$_2$ for activation [2].

Substrate, usual: Pro-Leu-NH$_2$ in conjunction with a procedure for
determining free proline [2]. Pro-NNap is not a useful assay substrate.

Substrate, special: Pro-Leu-Gly-NH$_2$ [2,6] and possibly polyproline [1].

Inhibitors: p-Chloromercuribenzoate and iodoacetamide at 1 mM. Dip-F has
little effect. Cd^{2+} is inhibitory, even in the presence of Mn^{2+}.

Molecular properties: Apparently a Mn^{2+}-metalloprotein, M_r about 300,000,
pI 4.1. Comprised of 5 or 6 subunits of equal size (M_r 56,000) [6]. Only
10-20% inactivated during prolonged storage at 4°C or -20°C. Such
treatment results in the total inactivation of prolyl dipeptidase, an
enzyme with activities that are commonly confused with those of prolyl
aminopeptidase.

Comment: Prolyl aminopeptidase was first isolated from *E. coli* [3], and
was designated EC 3.4.11.5 by the Enzyme Commission. The bacterial
enzyme rapidly hydrolyzed both polyproline and prolyl-NNap, but the hog
kidney enzyme has little action on these substrates. Despite these
specificity differences, the mammalian enzyme does appear to fulfill the
requirements for its systematic classification as a prolyl
aminopeptidase, EC 3.4.11.5. Some endopeptidases may have been improperly
assigned to this group, however. For example, in studies that were
reportedly concerned with prolyl aminopeptidase in rat tissues [4] and
in human serum [5], activities were assayed on
4-phenylazobenzyloxycarbonyl-Pro-Leu-Gly-Pro-D-Arg.

   In an attempt to avoid the confusion that may result from having a
common abbreviation, i.e. PAP, for both prolyl aminopeptidase and proline
aminopeptidase, their different subcellular localizations have been
used. Thus, prolyl aminopeptidase, which is found in the soluble
fraction, is abbreviated sPAP, whereas proline aminopeptidase, which is
membrane-bound, is abbreviated mPAP. As explained in the Introduction,
the "prolyl" designation is used to identify enzymes that catalyze the
cleavage of prolyl linkages.

References
[1] Sarid *et al. J. Biol. Chem.* **237**: 2207-2212, 1962.
[2] Nordwig & Mayer *Hoppe-Seyler's Z. Physiol. Chem.* **354**: 380-383,
     1973.
[3] Sarid *et al. J. Biol. Chem.* **234**: 1740-1746, 1959.
[4] Cutroneo & Fuller *Biochim. Biophys. Acta* **198**: 271-275, 1970.

[5]   Nakano *et al. Clin. Chim. Acta* **81**: 257-260, 1977.
[6]   Khilji & Bailey *Biochim. Biophys. Acta* **527**: 282-288, 1978.

Bibliography

1959

Sarid, S., Berger, A. & Katchalski, E.   Proline iminopeptidase. *J. Biol. Chem.* **234**: 1740-1746.

1962

Sarid, S., Berger, A. & Katchalski, E.   Proline iminopeptidase. II. Purification and comparison with iminodipeptidase (prolinase). *J. Biol. Chem.* **237**: 2207-2212.

1964

Wakabayashi, K.   Enzymatic hydrolysis of L-prolyl-β-naphthylamide and a colorimetric assay method for prolylpeptide hydrolase activity. *J. Biochem.* **55**: 244-253.

1969

Mäkinen, K. K.   The proline iminopeptidases of the human oral cavity. Partial purification and characterization. *Acta Chem. Scand.* **23**: 1409-1437.

1970

Cutroneo, K. R. & Fuller, G. C.   Application of a rapid colorimetric assay to detect alterations in rat proline imino peptidase. *Biochim. Biophys. Acta* **198**: 271-275.

1973

Nordwig, A. & Mayer, H.   The cleavage of prolyl peptides by kidney peptidases. Detection of a new peptidase capable of removing N-terminal proline. *Hoppe-Seyler's Z. Physiol. Chem.* **354**: 380-383.

1977

Nakano, H., Nomoto, S., Ohnishi, S. & Ito, K.   Serum levels of proline imino-peptidase in normal adults and children. *Clin. Chim. Acta* **81**: 257-260.

1978

Khilji, M. A. & Bailey, G. S.   The purification of a bovine kidney enzyme which cleaves melanocyte-stimulating hormone-release inhibiting factor. *Biochim. Biophys. Acta* **527**: 282-288.

Microsomal Alanyl Aminopeptidase

Summary

EC Number: 3.4.11.2

Earlier names: Aminopeptidase M, aminopeptidase N, particle-bound aminopeptidase, micrososomal aminopeptidase, amino-oligopeptidase, pseudo leucine aminopeptidase, membrane aminopeptidase I.

Abbreviation: mAAP.

Distribution: The tissue distribution of mAAP resembles that of LAP, with high levels found in kidney, liver and mucosal cells of the small intestine. However, the subcellular distributions of mAAP and LAP differ greatly, mAAP being strongly membrane-bound. In the kidney and small intestine, the enzyme is localized in the brush border membrane rather than in the endoplasmic reticulum. Cell surface bound mAAP is also found as an ecto-enzyme on such cells as macrophages, leukocytes, and T-lymphocytes. Operationally, mAAP is found in microsomal fractions as a consequence of the vesiculization of brush border (microvillar) membrane and its co-sedimentation with microsomes.

Source: Porcine kidney [1-3] and the intestinal mucosa of the hog [4-6], rabbit [7] and rat [8]. Both hydrophilic [4,5] and amphiphilic [5,6] forms of mAAP have been isolated from small intestinal brush border by means of trypsin treatment or detergent extraction, respectively.

Action: Catalyzes the removal of unsubstituted, N-terminal amino acids from peptides and from amide and arylamide derivatives of amino acids. Although neutral residues are preferred, especially bulky hydrophobic ones, mAAP does appear slowly to cleave acidic residues, and to make a substantial contribution to the release of N-terminal basic residues. In contrast to LAP, mAAP cleaves amino acid amides and p-nitroanilides at similar rates [2]. Alanine is the preferred N-terminal residue, and again in contrast to LAP, substantial rates are exhibited on all neutral amino acids, thus mAAP has been called aminopeptidase N.

As reported [2], the relative rates of hydrolysis (at pH 7.0) of amino acid p-nitroanilides are Ala- 100 (K_m 0.6 mM), Phe- 83 (K_m 3.2 mM), Leu- 71 (K_m 0.24 mM), and Gly- 22 (K_m 1.7 mM). K_m values generally decrease as the hydrophobicity of the amino acid side chain increases. Broad specificity is also exhibited by hog kidney mAAP, at pH 7.0, on 2-naphthylamide derivatives [3]. Ala-NNap is hydrolyzed most rapidly, followed by the NNap derivatives of Phe-, Tyr-, Leu-, Arg-, Thr-, Trp-, Lys-, Ser-, Asp-, His- and Val-. Little or no activity is seen on the NNap derivatives of Pro-, S-benzyl-Cys-, α-Glu-, and γ-Glu-, or on the amide groups of asparagine or glutamine.

Dipeptides such as Met-Lys and Leu-Met are readily hydrolyzed. Of possible physiological significance, dipeptides containing isopeptide bonds also serve as good substrates for mAAP [9]. Such bonds involve the ϵ-amino groups of lysine and/or the ω-carboxyl groups of aspartic and glutamic acid that commonly constitute the covalent crosslinks found in proteins. When compared to α-Met-Lys, ϵ-Met-Lys is a very good substrate [9]. Accordingly, this activity of intestinal mAAP may contribute to the increased nutritional value of casein to which methionine has been covently attached (through isopeptide bonds).

In the hog intestinal brush border membrane, mAAP accounts for virtually all the arylamidase activity, the majority of the aminopeptidase activity, about half of the tripeptidase activity, and a small but significant part of the dipeptidase activity [4]. Thus mAAP most probably plays a vital role in the final stages of intraluminal protein breakdown and absorption.

As described under Comments, mAAP is capable of converting leukotriene D_4 to leukotriene E_4.

Requirements: pH 7.0 at 37°C in 60 mM sodium phosphate buffer. (Tris is a competitive inhibitor of mAAP). The pH optimum can rise to 9.0 as the substrate concentration is increased, but the value of K_m is lowest at pH 7.0-7.5. Reaction rates are sometimes increased (up to 2-fold) by the addition of Co^{2+} at 1-10 mM.

Substrate, usual: -$NPhNO_2$ [2] and -NNap [3] derivatives of alanine or leucine.

Substrate, special: Ala-NMec and Leu-NMec.

Inhibitors: Chelating agents such as 1,10-phenanthroline and 2,2′-dipyridyl are effective at low concentrations. Activity can be restored with Co^{2+}, but concentrations in excess of 10 mM are strongly inhibitory for partially-purified mAAP. The purified enzyme is inhibited by Co^{2+} when preincubated with concentrations as low as 0.1 mM. Aliphatic carboxylic acids and amines exhibit competitive inhibition. The extent of inhibition increases linearly with the length of the aliphatic chain. Hydroxamates of amino acids and aliphatic acids are effective inhibitors [9a]. mAAP is unaffected by Dip-F, Pms-F, and a range of sulfhydryl reagents. Diazonium-1-H-tetrazole, a compound that reacts with the imidazole side chain of histidyl residues, rapidly inactivates mAAP [10]. Nitration of tyrosyl side chains with tetranitromethane also causes inactivation [11].

Inhibition by puromycin is sometimes reported, but is not a consistent characteristic. For example, whereas mAAP from human, beef, and rat kidney is inhibited 75-90% by 1 mM puromycin, the dog and rabbit kidney enzymes are inhibited only 15-30% [12]. By comparison, sAAP (Entry 11.03) is generally strongly inhibited by puromycin at only 0.1 mM.

A new and potentially useful inhibitor is o-(phenylthio)-phenylacetic acid. This compound is a relatively weak anti-inflammatory agent that specifically inhibits mAAP at micromolar concentrations without affecting LAP [13]. A method has been reported [14] for synthesizing this compound. mAAP is unaffected by tertiary butyl threonyl peptides, which are potent inhibitors of LAP. mAAP is also inhibited competitively by "peptide-gap" inhibitors, a new class of inhibitory peptides that contain a thiomethylene linkage substituted for the potentially scissile peptide bond [15]. Such an inhibitor is (S)-2-(S-cysteaminyl)-4-methylpentanoic acid, a non-hydrolyzable analog of the dipeptide Gly-Leu. Dimethylsulfoxide and dimethylformamide, solvents commonly used to carry substrates, are non-competitive inhibitors [16].

Molecular properties: M_r 280,000 for the protease (trypsin)-solubilized enzyme (P-form) isolated from hog kidney [17] and hog intestinal brush border membrane [4]. A metalloglycoprotein containing two atoms of zinc (one per subunit) and about 400 carbohydrate residues that include glucosamine, galactose, mannose, fucose and sialic acid, accounting for about 20% of the molecular mass [17]. Varying amino acid compositions have been reported for the native enzyme [18,19], and there is evidence that tyrosine and histidine side chains participate in catalysis [10,11].

Native mAAP, like glutamyl aminopeptidase ("aminopeptidase A") is a symmetrical dimer (α_2) anchored in the membrane by two hydrophobic peptides [5]. mAAP is dimeric in the hog, rat and man, but is monomeric in the rabbit [20]. It accounts for about 8% of the protein of the hog intestinal brush border. The α-chain of the detergent-solubilized enzyme, the D-form, contains an N-terminal hydrophobic anchor peptide of about 35 amino acids. The β- and γ-chains are the products of a proteolytic cleavage of the α-chain. The M_r values for the α- or β-chains have been placed at 130,000 and 90,000, respectively. The mass of the γ-chain, the C-terminal fragment, is estimated to be about 45,000, uncertainty being the consequence of C-terminal heterogeneity. Proteolytic cleavage of the α-chain gives rise to the $\alpha\beta\gamma$ and $\beta_2\gamma_2$ forms that appear to account for the earlier descriptions of three subunits [4,18]. In contrast to the hog renal and intestinal enzymes, mAAP solubilized from rabbit intestine is monomeric [20]. $A_{280,1\%}$ is 16.3 for the P-form from hog kidney [1] and 15.6 for the P-form from hog intestine [4]. pI approximately 5.

mAAP from human placenta, kidney and renal cell carcinoma are immunologically indistinguishable [21]. However, an antiserum to the human kidney enzyme does not cross-react with mAAP prepared from beef, dog, hog, rabbit or rat kidney [12].

As regards stability, purified mAAP can be lyophilized or stored frozen for many months without loss of activity. It withstands heating to 65°C at pH 7.0, but incubation at 70°C causes a 50% loss of activity in 110 min. At room temperature, rapid losses occur below pH 3.0 and above pH 11.5. Although mAAP is stable in 6 M urea, it is irreversibly denatured in 0.5 M guanidinium chloride. The inactive, Zn-free apoenzyme, like the native enzyme, shows considerable heat-and pH-stabilty. Both forms are stable to a 40 h preincubation at 22°C at pH values ranging from 6.0 to 8.5. At neutral pH their activities are stable for up to 40 h at 37°C. However, differences in stability do become evident at more extreme conditions. The apoenzyme is reactivated by Cu^{2+}, Co^{2+} or Ni^{2+}, in addition to Zn^{2+}, whereas Mg^{2+}, Mn^{2+}, Ca^{2+}, Fe^{2+}, Fe^{3+} and Cd^{2+} are ineffective [17].

Comment: Although mAAP was first isolated from hog kidney microsomes [22] and named aminopeptidase M [23], it was later suggested [20] that it be renamed aminopeptidase N to reflect its preferential action on neutral amino acids. However, as part of our effort to develop a more systematic approach to naming the exopeptidases (consistent with the guidelines given in the Introduction to the present volume) we are proposing that this exopeptidase be named microsomal alanyl aminopeptidase (mAAP). Its membrane association is included in order to distinguish the enzyme from another well-known aminopeptidase that preferentially removes alanyl residues. Because the latter is freely soluble in cells, it has been described (elsewhere in this book) as soluble alanyl aminopeptidase (sAAP; Entry 11.03).

mAAP is inactivated in fractionation procedures that utilize acetone or alcohols [20], and it is probably for this reason that the enzyme was not detected in hog kidney extracts that were commonly used in early attempts to purify leucyl aminopeptidase. Although mAAP and LAP exhibit similarities, they are immunologically distinct.

Because most, if not all of the intestinal and renal exopeptidases are metalloenzymes, it would seem appropriate to avoid many of the earlier published procedures that employ chelating agents in procedures used to isolate brush border membranes as vesicles. Alternative methods have been reported [24].

Bestatin, a strong inhibitor of soluble arginyl aminopeptidase (see Entry 11.04), is also an inhibitor of cell surface bound mAAP, an ecto-enzyme that has been characterized as a bestatin "receptor". Evidence points to the possibility that the immuno-potentiating and antitumor activities of bestatin result from a stimulation of T-lymphocyte proliferation that is probably mediated through the activation of macrophages [25,26].

mAAP activity on the surface of hog polymorphonuclear leukocytes appears to be responsible for the inactivation of tuftsin, a phagocytosis-stimulating tetrapeptide (Thr-Lys-Pro-Arg) that is liberated from leukokinin, the parent molecule, by leukokininase [27]. Some aspects of these studies have been challenged, however [28].

An mAAP-like enzyme on the surface of rat basophilic leukemia (RBL-1) cells is capable of transforming the highly bioactive Slow Reacting Substance of anaphylaxis, SRS-Cys-Gly (leukotriene D_4) into the less active peptidolipid SRS-Cys (leukotriene E_4) [29]. The former is a hydroxylated derivative of arachidonic acid containing the peptide (thiol) substituent linked through a thioether bond to C6 of the lipid moiety. The likelihood of mAAP being the responsible enzyme is greatly enhanced by a study showing that Cys-Gly and S-benzyl-cysteine-NPHNO$_2$ are readily hydrolyzed by mAAP purified from rat kidney brush border membranes [30].

mAAP is very useful for the sequencing of peptides because of its broad specificity, its ability to act on large peptides, its stability, its lack of activity on the amide groups of asparagine and glutamine, and its ability to release derivatized amino acids such as S-carboxymethyl cysteine and monoazotyrosine. Sequencing studies conducted with commercial preparations of mAAP have revealed that most smaller peptides, in amounts up to 0.1 μmole, are completely hydrolyzed within 16 to 24 h when incubated at 40°C with 100 to 200 milliequivalents of enzyme in 0.1 M NH$_4$HCO$_3$, pH 7.5. Prolyl residues are usually removed slowly, and unusual secondary reactions sometimes arise when a prolyl residue is preceded by a bulky, hydrophobic residue, e.g., Leu, Tyr, or Trp. In such cases the X-Pro combination may be liberated as an intact dipeptide. mAAP preparations tend to be contaminated with a carboxypeptidase B-type activity and peptidases such as prolyl and proline dipeptidases. When such preparations are used as sequencing reagents, interfering exopeptidase activity can be eliminated by heating to 65°C, and sometimes also by the judicious use of a chelating agent.

References
[1] Wachsmuth *et al*. *Biochemistry* **5**: 169-174, 1966.
[2] Pfleiderer *Methods Enzymol*. **19**: 514-521, 1970.
[3] Hanson *et al*. *Hoppe-Seyler's Z. Physiol. Chem*. **348**: 680-688, 1967.
[4] Maroux *et al*. *Biochim. Biophys. Acta* **321**: 282-295, 1973.
[5] Benajiba & Maroux *Biochem. J*. **197**: 573-580, 1981.
[6] Sjöström *et al*. *Eur. J. Biochem*. **88**: 503-511, 1978.
[7] Takesue *J. Biochem*. **77**: 103-115, 1975.
[8] Gray & Santiago *J. Biol. Chem*. **252**: 4922-4928, 1977.
[9] Gaertner *et al*. *FEBS Lett*. **133**: 135-138, 1981.
[9a] Wilkes & Prescott *J. Biol. Chem*. **258**: 13517-13521, 1983.
[10] Pfleiderer & Femfert *FEBS Lett*. **4**: 265-268, 1969.
[11] Femfert & Pfleiderer *FEBS Lett*. **4**: 262-264, 1969.
[12] Hiwada *et al*. *Comp. Biochem. Physiol*. **68B**: 485-489, 1981.
[13] Miller & Lacefield *Biochem. Pharmacol*. **28**: 673-675, 1979.
[14] Brannigan *et al*. *J. Med. Chem*. **19**: 798-802, 1976.
[15] Fok & Yankeelov *Biochem. Biophys. Res. Commun*. **74**: 273-278, 1977.

[16] Wachsmuth *et al. Biochemistry* **5**: 175-182, 1966.
[17] Lehky *et al. Biochim. Biophys. Acta* **321**: 274-281, 1973.
[18] Wacker *et al. Helv. Chim. Acta* **54**: 473-485, 1971.
[19] Wachsmuth *Biochem. Z.* **346**: 467-473, 1967.
[20] Feracci & Maroux *Biochim. Biophys. Acta* **599**: 448-463, 1980.
[21] Hiwada *et al. Eur. J. Biochem.* **104**: 155-165, 1980.
[22] Pfleiderer & Celliers *Biochem. Z.* **339**: 186-189, 1963.
[23] Wachsmuth *et al. Biochemistry* **5**: 169-174, 1966.
[24] Louvard *et al. Biochim. Biophys. Acta* **291**: 747-763, 1973.
[25] Müller *et al. Int. J. Immunopharmacol.* **4**: 393-400, 1982.
[26] Leyhausen *et al. Biochem. Pharmacol.* **32**: 1051-1057, 1983.
[27] Nagaoka & Yamashita *Biochim. Biophys. Acta* **675**: 85-93, 1981.
[28] Smith *et al. Biochim. Biophys. Acta* **728**: 222-227, 1983.
[29] Sok *et al. Proc. Natl. Acad. Sci. USA* **77**: 6481-6485, 1980.
[30] Rankin *et al. Biochem. Biophys. Res. Commun.* **96**: 991-996, 1980.

Bibliography

1963

Pfleiderer, G. & Celliers, P. G.   Isolierung einer aminopeptidase aus nierenpartikeln. *Biochem. Z.* **339**: 186-189.

1964

Auricchio, F. & Bruni, C. B.   Bestimmung des Molekulargewichts von Enzymen bei Dextrangelfiltration. *Biochem. Z.* **340**: 321-325.

Pfleider, G., Celliers, P. G., Stanulovic, M., Wachsmuth, E. D., Determann, H. & Braunitzer, G.   Eigenschaften und analytische anwendung der aminopeptidase aus nierenpartikeln. *Biochem. Z.* **340**: 552-564.

1966

Felgenhauer, K. & Glenner, G. G.   The enzymatic hydrolysis of amino acid β-naphthylamides. II. Partial purification and properties of a particle-bound cobalt-activated rat kidney aminopeptidase. *J. Histochem. Cytochem.* **14**: 401-413.

Wachsmuth, E. D.   Verleichende Untersuchungen zur Wirkungsweise von Aminopeptidasen. *Biochem. Z.* **344**: 361-374.

Wachsmuth, E. D., Fritze, I. & Pfleiderer, G.   An aminopeptidase occurring in pig kidney. II. A study on the mechanism of the hydrolysis. *Biochemistry* **5**: 175-182.

Wachsmuth, E. D., Fritze, I. & Pfleiderer, G.   An aminopeptidase ocurring in pig kidney. I. An improved method of preparation. Physical and enzymic properties. *Biochemistry* **5**: 169-174.

1967

Hanson, H., Hütter, H.-J., Mannsfeldt, H.-G., Kretschmer, K. & Sohr, C.
Zur Darstellung und Substratspezifität einer von der Leucinaminopeptidase
unterscheidbaren Aminopeptidase aus Nierenpartikeln. *Hoppe-Seyler's Z.
Physiol. Chem.* **348**: 680-688.

Wachsmuth, E. D. Essentielle Tyrosine der Aminopeptidase aus Partikeln
von Schweinenieren. *Biochem. Z.* **364**: 446-457.

Wachsmuth, E. D. Untersuchungen zur struktur der aminopeptidase aus
partikeln von schweinenieren. *Biochem. Z.* **346**: 467-473.

1969

Femfert, U. & Pfleiderer, G. The tyrosyl residues at the active site of
aminopeptidase M. Modifications by tetranitromethane. *FEBS Lett.* **4**:
262-264.

Pfleiderer, G. & Femfert, U. Histidine residues at the active site of
aminopeptidase M. Modifications by diazonium-1-H-tetrazole (DHT). *FEBS
Lett.* **4**: 265-268.

Plummer, T. H., Jr. Isolation and sequence of peptides at the active
center of bovine carboxypeptidase B. *J. Biol. Chem.* **244**: 5246-5253.

1970

Femfert, U. & Pfleiderer, G. On the mechanism of amide-bond-cleavage
catalyzed by aminopeptidase M. Kinetic studies. *FEBS Lett.* **8**: 65-67.

Pfleiderer, G. Particle-bound aminopeptidase from pig kidney. *Methods
Enzymol.* **19**: 514-521.

1971

DeLange, R. J. & Smith, E. L. Leucine aminopeptidase and other N-terminal
exopeptidases. In: *The Enzymes* (Boyer, P. D. ed.), 3rd edn., vol. 3,
pp. 81-118, (see pp. 102-105), Academic Press, New York.

Femfert, U. On the mechanism of amide bond cleavage catalysed by
aminopeptidase M. Kinetic studies in deuteriumoxide. *FEBS Lett.* **14**:
92-94.

Femfert, U. & Pfleiderer, G. The effect of cations and complexing agents
on the hydrolysis of L-alanine-4-nitroanilide by aminopeptidase M. *FEBS
Lett.* **14**: 89-91.

Wacker, H., Lehky, P., Fischer, E. M. & Stein, E. A. Physical and chemical
characterization of pig kidney particulate aminopeptidase. *Helv. Chim.
Acta* **54**: 473-485.

1972

Femfert, U., Cichocki, P. & Pfleiderer, G. On the mechanism of amide-bond
cleavage catalysed by aminopeptidase M. Enzymatic properties of
nitroaminopeptidase M. *FEBS Lett.* **26**: 39-42.

Kim, Y. S., Birtwhistle, W. & Kim, Y. W. Peptide hydrolases in the brush
border and soluble fractions of small intestinal mucosa of rat and man.

J. Clin. Invest. **51**: 1419-1430.

Kleine, R. & Schubert, J. Eigenschaften eines hochmolekularen arylamidase-phosphatase-komplexes aus rattennierenmikrosomen. *Acta Biol. Med. Ger.* **28**: 269-281.

Thomas, L. & Kinne, R. Studies on the arrangement of aminopeptidase and alkaline phosphatase in the microvilli of isolated brush border of rat kidney. *Biochim. Biophys. Acta* **255**: 114-125.

1973

George, S. G. & Kenny, A. J. Studies on the enzymology of purified preparations of brush border from rabbit kidney. *Biochem. J.* **134**: 43-57.

Lehky, P., Lisowski, J., Wolf, D. P., Wacker, H. & Stein, E. A. Pig kidney particulate aminopeptidase a zinc metalloenzyme. *Biochim. Biophys. Acta* **321**: 274-281.

Louvard, D., Maroux, S., Baratti, J., Desnuelle, P. & Mutaftschiev, S. On the preparation and some properties of closed membrane vesicles from hog duodenal and jejunal brush border. *Biochim. Biophys. Acta* **291**: 747-763.

Maroux, S., Louvard, D. & Baratti, J. The aminopeptidase from hog intestinal brush border. *Biochim. Biophys. Acta* **321**: 282-295.

Tamura, Y., Niinobe, M., Arima, T., Okuda, H. & Fuji, S. Studies on aminopeptidase in rat liver and plasma. *Biochim. Biophys. Acta* **327**: 437-445.

1974

Femfert, U. & Cichocki, P. Hydrophobic areas at the active site of aminopeptidase M. *Hoppe-Seyler's Z. Physiol. Chem.* **355**: 1332-1334.

Scherberich, J. E., Falkenberg, F. W., Mondorf, A. W., Müller, H. & Pfleiderer, G. Biochemical and immunological studies on isolated brush border membranes of human kidney cortex and their membrane surface proteins. *Clin. Chim. Acta* **55**: 179-197.

1975

Louvard, D., Maroux, S. & Desnuelle, P. Topological studies on the hydrolases bound to the intestinal brush border membrane. II. Interactions of free and bound aminopeptidase with a specific antibody. *Biochim. Biophys. Acta* **389**: 389-400.

Louvard, D., Maroux, S., Vannier, Ch. & Desnuelle, P. Topological studies on the hydrolases bound to the intestinal brush border membrane. I. Solubilization by papain and triton X-100. *Biochim. Biophys. Acta* **375**: 236-248.

Takesue, Y. Purification and properties of leucine β-naphthylamidase from rabbit small-intestinal mucosal cells. *J. Biochem.* **77**: 103-115.

1976

Kim, Y. S. & Brophy, E. J. Rat intestinal brush border membrane peptidases. I. Solubilization, purification, and physiochemical properties of two different forms of the enzyme. *J. Biol. Chem.* **251**: 3199-3205.

Kim, Y. S., Brophy, E. J. & Nicholson, J. A. Rat intestinal brush border membrane peptidases. II. Enzymatic properties, immunochemistry, and interactions with lectins of two different forms of the enzyme. *J. Biol. Chem.* **251**: 3206-3212.

Louvard, D., Semeriva, M. & Maroux, S. The brush-border intestinal aminopeptidase, a transmembrane protein as probed by macromolecular photolabelling. *J. Mol. Biol.* **106**: 1023-1035.

Maroux, S. & Louvard, D. On the hydrophobic part of aminopeptidase and maltases which bind the enzyme to the intestinal brush border membrane. *Biochim. Biophys. Acta* **419**: 189-195.

Vannier, C., Louvard, D., Maroux, S. & Desnuelle, P. Structural and topological homology between porcine intestinal and renal brush border aminopeptidase. *Biochim. Biophys. Acta* **455**: 185-199.

Wacker, H., Lehky, P., Vanderhaeghe, F. & Stein, E. A. On the subunit structure of particulate aminopeptidase from pig kidney. *Biochim. Biophys. Acta* **429**: 546-554.

1977

Danielsen, E. M., Sjöström, H., Noren, O. & Dabelsteen, E. Immunoelectrophoretic studies on pig intestinal brush border proteins. *Biochim. Biophys. Acta* **494**: 332-342.

Fok, K.-F. & Yankeelov, J. A., Jr. Peptide-gap inhibitors: I. Competitive inhibition of aminopeptidase M by a hydrolytically resistant dipeptide analogue of glycylleucine. *Biochim. Biophys. Res. Commun.* **74**: 273-278.

Gray, G. M. & Santiago, N. A. Intestinal surface amino-oligopeptidases. I. Isolation of two weight isomers and their subunits from rat brush border. *J. Biol. Chem.* **252**: 4922-4928.

Hiwada, K. & Kokubu, T. Comparison of soluble and membrane-bound neutral arylamidases from renal cell carcinoma. *Clin. Chim. Acta* **80**: 395-401.

Hiwada, K., Terao, M. & Kokubu, T. Placental form of membrane-bound neutral arylamidase found in renal cell carcinoma. *Clin. Chim. Acta* **79**: 569-573.

Hiwada, K., Terao, M., Nishimura, K. & Kokubu, T. Comparison of human membrane-bound neutral arylamidases from small intestine, lung, kidney, liver and placenta. *Clin. Chim. Acta* **76**: 267-275.

Kania, R. K., Santiago, N. A. & Gray, G. M. Intestinal surface amino-oligopeptidases. II. Substrate kinetics and topography of the active site. *J. Biol. Chem.* **252**: 4929-4934.

Kenny, A. J. Proteinases associated with cell membranes. In: *Proteinases in Mammalian Cells and Tissues* (Barrett, A. J. ed.), pp. 393-444 (see pp. 429-430), North-Holland Publishing Co., Amsterdam.

McDonald, J. K. & Schwabe, C. Intracellular exopeptidases. In: *Proteinases in Mammalian Cells and Tissues* (Barrett, A. J. ed.), pp.

311-391 (see pp. 368-369), North-Holland Publishing Co., Amsterdam.

Wachsmuth, E. D. & Staber, F. G. Changes in membrane-bound aminopeptidase on bone marrow-derived macrophages during their maturation in vitro. *Exp. Cell Res.* **109**: 269-276.

Wachsmuth, E. D. & Stoye, J. P. Aminopeptidase on the surface of differentiating macrophages: concentration changes on individual cells in culture. *J. Reticuloendothel. Soc.* **22**: 485-497.

Wachsmuth, E. D. & Stoye, J. P. Aminopeptidase on the surface of differentiating macrophages: induction and characterization of the enzyme. *J. Reticuloendothel. Soc.* **22**: 469-483.

1978

Kenny, A. J. & Booth, A. G. Microvilli: their ultrastructure, enzymology and molecular organization. *Essays Biochem.* **14**: 1-44.

Sjöström, H., Nordén, O., Jeppesen, L., Staun, M., Svensson, B. & Christainsen, L. Purification of different amphiphilic forms of a microvillus aminopeptidase from pig small intestine using immunoadsorbent chromatography. *Eur. J. Biochem.* **88**: 503-511.

1979

Booth, A. G., Hubbard, L. M. L. & Kenny, A. J. Proteins of the kidney microvillar membrane. Immunoelectrophoretic analysis of the membrane hydrolases: identification and resolution of the detergent and proteinase-solubilized forms. *Biochem. J.* **179**: 397-405.

Hiwada, K., Yokoyama, M. & Kokubu, T. Placental type of membrane-bound arylamidase (Shiba isoenzyme) found in human cancerous lung. *Clin. Chim. Acta* **93**: 113-117.

Miller, R. & Lacefield, W. Specific inhibitors of aminopeptidase M - relationship to anti-inflammatory activity. *Biochem. Pharmacol.* **28**: 673-675.

1980

Feracci, H. & Maroux, S. Rabbit intestinal aminopeptidase N. Purification and molecular properties. *Biochim. Biophys. Acta* **599**: 448-463.

Hiwada, K., Ito, T., Yokoyama, M. & Kokubu, T. Isolation and characterization of membrane-bound arylamidases from human placenta and kidney. *Eur. J. Biochem.* **104**: 155-165.

Louvard, D. Apical membrane aminopeptidase appears at site of cell-cell contact in cultured kidney epithelial cells. *Proc. Natl. Acad. Sci. USA* **77**: 4132-4136.

Maze, M. & Gray, G. M. Intestinal brush border aminooligopeptidases: cytosol precursors of the membrane enzyme. *Biochemistry* **19**: 2351-2358.

Nagaoka, I. & Yamashita, T. Leucine aminopeptidase as an ecto-enzyme of polymorphonuclear neutrophils. *Biochim. Biophys. Acta* **598**: 169-172.

Norén, O. & Sjöström, H. The insertion of pig microvillus aminopeptidase into the membrane as probed by [^{125}I]iodonaphthylazide. *Eur. J. Biochem.*

104: 25-31.

Rankin, B. B., McIntyre, T. M. & Curthoys, N. P. Brush border membrane hydrolysis of S-benzyl-cysteine-p-nitroanilide, an activity of aminopeptidase M. *Biochem. Biophys. Res. Commun.* **96**: 991-996.

Sok, D.-E., Pai, J.-K., Atrache, V. & Sih, C. J. Characterization of slow reacting substances (SRSs) of rat basophilic leukemia (RBL-1) cells: effect of cysteine on SRS profile. *Proc. Natl. Acad. Sci. USA* **77**: 6481-6485.

1981

Benajiba, A. & Maroux, S. Subunit structure of pig small-intestinal brush-border aminopeptidase N. *Biochem. J.* **197**: 573-580.

Feracci, H., Benajiba, A., Gorvel, J. P., Doumeng, C. & Maroux, S. Enzymatic and immunological properties of the protease form of aminopeptidases N and A from pig and rabbit intestinal brush border. *Biochim. Biophys. Acta* **658**: 148-157.

Feracci, H., Bernadac, A., Hovsépian, S., Fayet, G. & Maroux, S. Aminopeptidase N is a marker for the apical pole of porcine thyroid epithelial cells in vivo and in culture. *Cell Tissue Res.* **221**: 137-146.

Gaertner, H., Puigserver, A. & Maroux, S. Hydolysis of the isopeptide bond of ε-N-L-methionyl-L-lysine by intestinal aminopeptidase N. *FEBS Lett.* **133**: 135-138.

Hiwada, K., Yokoyama, M. & Kokubu, T. Comparison of membrane-bound arylamidases from human, bovine, hog, dog, rabbit and rat kidneys. *Comp. Biochem. Physiol.* **68B**: 485-489.

Hiwada, K., Yokoyama, M. & Kokubu, T. Isolation and characterization of aminopeptidase N from the human small intestine. *Biomed. Res.* **2**: 517-526.

Nagaoka, I. & Yamashita, T. Inactivation during phagocytosis of leucine aminopeptidase, an ecto-enzyme of polymorphonuclear neutrophils. *Biochim. Biophys. Acta* **678**: 342-351.

Nagaoka, I. & Yamashita, T. Inactivation of phagocytosis-stimulating activity of tuftsin by polymorphonuclear neutrophils. A possible role of leucine aminopeptidase as an ecto-enzyme. *Biochim. Biophys. Acta* **675**: 85-93.

1982

Chan, W. W.-C., Dennis, P., Demmer, W. & Brand, K. Inhibition of leucine aminopeptidase by amino acid hydroxamates. *J. Biol. Chem.* **257**: 7955-7957.

Danielsen, E. M., Norén, O. & Sjöström, H. Biosynthesis of intestinal microvillar proteins. Translational evidence *in vitro* that aminopeptidase N is synthesized as a M_r-115000 polypeptide. *Biochem. J.* **204**: 323-327.

Danielsen. E. M. Biosynthesis of intestinal microvillar proteins. Pulse-chase labelling studies on aminopeptidase N and sucrase-isomaltase. *Biochem. J.* **204**: 639-645.

Erickson, R. H. & Kim, Y. S. Interaction of purified brush border membrane minopeptidase N and dipeptidyl peptidase IV with lectin-Sepharose derivatives. *Biochim. Biophys. Acta* **743**: 37-42.

Gratecos, D., Varesi, L., Knibiehler, M. & Sémériva, M. Photoaffinity labelling of membrane bound porcine aminopeptidase N. *Biochim. Biophys. Acta* **705**: 218-227.

Hovsépian, S., Feracci, H., Maroux, S. & Fayet, G. Kinetic studies of the localization of aminopeptidase N in monolayer and in follicle-associated cultures of porcine thyroid cells. *Cell Tissue Res.* **224**: 601-611.

Kugler, P. Aminopeptidase A is angiotensinase A. II. Biochemical studies on aminopeptidase A and M in rat kidney homogenate. *Histochemistry* **74**: 247-261.

Mizutani, S., Hayakawa, H., Akiyama, H., Sakura, H., Yoshino, M., Oya, M. & Kawashima, Y. Simultaneous determinations of plasma oxytocin and serum placental leucine aminopeptidase (P-LAP) during late pregnancy. *Clin. Biochem.* **15**: 141-145.

Mizutani, S., Okano, K., Hasegawa, E., Sakura, H., Oya, M. & Yamada, M. Human placental leucine aminopeptidase (P-LAP) as a hypotensive agent. *Experientia* **38**: 821-822.

Müller, W. E. G., Schuster, D. K., Zahn, R. K., Maidhof, A., Leyhausen, G., Falke, D., Koren, R. & Umezawa, H. Properties and specificity of binding sites for the immunomodulator bestatin on the surface of mammalian cells. *Int. J. Immunopharmacol.* **4**: 393-400.

Neef, L., Hütter, H. J., Böhme, I., Dotschev, D., Danev, S. T. & Haschen, R. J. A low molecular weight alanine aminopeptidase in urine. Biochemical equivalent of tubular atrophy? *Enzyme* **28**: 348-354.

1983

Chan, W. W.-C. L-Leucinthiol - a potent inhibitor of leucine aminopeptidase. *Biochem. Biophys. Res. Commun.* **116**: 297-302.

Danielsen, E. M., Norén, O. & Sjöström, H. Biosynthesis of intestinal microvillar proteins. Processing of aminopeptidase N by microsomal membranes. *Biochem. J.* **212**: 161-165.

Erickson, R. H. & Kim, Y. S. Interaction of purified brush-border membrane aminopeptidase N and dipeptidyl peptidase IV with lectin-Sepharose derivatives. *Biochim. Biophys. Acta* **743**: 37-42.

Hiwada, K., Yokoyama, M. & Kokubu, T. Placental type of aminopeptidase M found in renal cell carcinoma: frequency, and biochemical and immunological studies. In: *Selected Topics in Clinical Enzymology* (Goldberg, D. M. & Werner, M. eds.), pp. 29-39, Walter de Gruyter, Berlin.

Leyhausen, G., Schuster, D. K., Vaith, P., Zahn, R. K., Umezawa, H., Falke, D. & Müller, W. E. G. Identification and properties of the cell membrane bound leucine aminopeptidase interacting with the potential immunostimulant and chemotherapeutic agent bestatin. *Biochem. Pharmacol.* **32**: 1051-1057.

Shimamura, M., Hazato, T. & Katayama, T. A membrane-bound aminopeptidase isolated from monkey brain and its action on enkephalin. *Biochim. Biophys. Acta* **756**: 223-229.

Smith, G. P., MacGregor, R. R. & Peters, T. J. Subcellular localisation of leucine aminopeptidase in human polymorphonuclear leukocytes. *Biochim. Biophys. Acta* **728**: 222-227.

Wilkes, S. H. & Prescott, J. M. Stereospecificity of amino acid hydroxamate inhibition of aminopeptidases. *J. Biol. Chem.* **258**: 13517-13521.

Yoshimoto, T. & Tsuru, D. Substrate specificity of aminopeptidase M: evidence that the commercial preparation is contaminated by dipeptidyl aminopeptidase IV and prolidase. *J. Biochem.* **94**: 619-622.

1984

Antonov, V. K., Vorotyntseva, T. I., Bessmertnaya, L. Y., Mikhailova, A. G. & Zilberman, M. I. Role of intestinal brush border membrane aminopeptidase N in dipeptide transport. *FEBS Lett.* **171**: 227-232.

Hiwada, K., Tokioka-Terao, M., Nishimura, K. & Kokubu, T. A family with a high serum aminopeptidase (microsomal) activity: properties of the enzyme from serum of the propositus. *Clin. Chem.* **30**: 222-225.

Nagaoka, I. & Yamashita, T. Possible exposure of leucine aminopeptidase on the cell surface of rabbit blood neutrophils by digitonin treatment. *Biochim. Biophys. Acta* **804**: 176-180.

Ward, P. E. Immunoelectrophoretic analysis of vascular, membrane-bound angiotensin I converting enzyme, aminopeptidase M, and dipeptidyl(amino)peptidase IV. *Biochem. Pharmacol.* **33**: 3183-3193.

Glutamyl Aminopeptidase

Summary

EC Number: 3.4.11.7

Earlier names: Aminopeptidase A, angiotensinase A, aspartate aminopeptidase, Ca^{++}-activated glutamate aminopeptidase, glutamyl peptidase, glutamyl-β-naphthylamide hydrolase, acid α-amino peptidase, membrane aminopeptidase II.

Abbreviation: EAP (E, glutamic acid).

Distribution: Found in the serum and various organs of animals. In the rat, the renal cortex exhibits the highest EAP activity. Notable levels of activity also occur in guinea pig pancreatic duct epithelium and islets of Langerhans, porcine duodenum, human serum and red cells, and human parotid gland and kidney. Commonly found in the epithelial brush-border regions of the intestine and the renal proximal tubule. Of possible physiological significance, and in contrast to other brush border aminopeptidases such as microsomal alanyl aminopeptidase (mAAP) and γ-GTP, EAP is also present in the juxtaglomerular apparatus and in vascular segments near the glomerulus where it may act to regulate angiotensin II levels. Like mAAP, it is also plentiful in human placenta.

Source: Small intestine of rabbit [1] and hog [2]; kidney of rat [3], cattle [4] and hog [5,6]; human serum [7] and placenta [7a].

Action: Catalyzes specifically the removal of unsubstituted, N-terminal Glu- and Asp- residues from 2-naphthylamides and peptides, including dipeptides such as α-Glu-Phe and α-Asp-Arg. K_m values for α-L-Glu-NNap vary with Ca^{2+} concentration: 62 μM for 1.25 mM Ca^{2+}, 75 μM for 0.15 mM Ca^{2+}, and 350 μM in the absence of Ca^{2+}. V_{max} values do not change, however [5]. Typically, glutamyl residues are removed more rapidly than are aspartyl residues: for rat kidney EAP, the rate is six times greater [3], and for hog kidney EAP it is four times greater [5,6]. The α-amino and γ-carboxyl groups must be free as shown by the lack of activity on pyroGlu-, γ-Glu- and Gln-NNap. EAP readily removes the N-terminal Asp-residue from α-L-Asp^1-angiotensin II. The stereospecificity exhibited on isomers of angiotensin II is in keeping with the stereospecificity reported for serum angiotensinase activity [8] and the observed lack of activity on α-D-Glu-NNap [3].

Requirements: pH 7.5 at 37°C, 2 mM Ca^{2+}. Characteristically, a 3- to 4-fold activation is attained when $CaCl_2$ is incorporated at 1 to 2 mM into (50 mM) Tris-HCl buffer. Ba^{2+} is slightly less effective [5,6], whereas Sr^{2+} is somewhat more effective [6].

Substrate, usual: α-Glu-NNap [3,5,6].

Substrate, special: α-Glu-NMec.

Inhibitors: Non-competitive inhibitors include chelating agents such as EDTA, EGTA, and 1,10-phenanthroline at 1 mM. Reactivation to about 90% is attainable with Mn^{2+} or Ca^{2+}, the former being more effective. Mg^{2+} shows only a slight effect [5,6]. The activity of EAP is almost completely inhibited by Zn^{2+}, Ni^{2+}, Cu^{2+}, Hg^{2+}, and Cd^{2+} at 1 mM. $IAcNH_2$, p-chloromercuribenzoate, Dip-F and phosphoramidon show no inhibition, but thiol compounds do. Exposure of EAP to 10 mM 2-mercaptoethanol or 10 mM dithiothreitol results in 71% and 97% inhibition, respectively. Amastatin [(2S,3R)-3-amino-2-hydroxy-5-methylhexanoyl]-L-Val-L-Val-L-Asp, a peptide derived from a strain of *Streptomyces* [9], is a potent and relatively specific competitive inhibitor of EAP (K_i 2.5×10^{-7} M). Inhibition by amastatin is not reversed by Ca^{2+} [6]. LAP is the only other protease thus far reported to be inhibited by amastatin. Soluble arginyl aminopeptidase ("aminopeptidase B") is unaffected.

Molecular properties: M_r 350,000-400,000 for the (amphiphilic) detergent-solubilized (D-form) enzyme (D-EAP); M_r 250,000-270,000 for the (hydrophilic) protease-solubilized P-form (P-EAP). The D-form can be converted to the P-form by tryptic cleavage. Asp (or Asn) is N-terminal on the P-form, and Leu- on the D-form. Both forms of the enzyme are symmetrical dimers, with the D-form possessing a hydrophobic anchor of about 42 residues (at the N-terminus of each subunit), by which the enzyme is held at the membrane surface. Each subunit contains one Ca^{2+} atom, but no detectable Zn^{2+} or Mn^{2+}. Hog kidney D-EAP is comprised of 4 glycopeptides with M_r values estimated to be 155,000, 110,000, 90,000 and 45,000 [5], but the smaller chains are almost certainly fragments of the 155,000 parent chain.

In contrast to EAP from hog kidney [5] and small intestine [2], the beef kidney enzyme is a much smaller molecule (M_r 65,000). It consists of two subunits (M_r 33,000), each comprised of two polypeptide chains [4]. EAP from the rabbit intestine differs from that of other species thus far studied [1]. Typically, it contains one atom of calcium per monomer, but it exists only in a monomeric form (M_r 170,000 for P-EAP), as does mAAP in this species. $A_{280,1\%}$ is 11.0 (hog kidney P-form) and 12.1 (rabbit intestinal P-form).

Comment: Glutamyl aminopeptidase is a microvillar enzyme that can be solubilized with detergents or proteinases, the former method always yielding the higher-M_r forms [10]. Like several other brush border hydrolases, EAP is anchored to the membrane via an N-terminal hydrophobic sequence. It has been proposed that the subunits are probably synthesized by the same route and independently inserted into the membrane matrix. Of the total protein in the intestinal brush border, EAP accounts for about 3.5% in the rabbit [1] and about 4% in the hog [2]. EAP and mAAP are closely-related, but immunologically-distinct, membrane-bound aminopeptidases. Although they tend to co-purify, they can be effectively separated on immobilized concanavalin A [2] and by amastatin-affinity chromatography [6]. α-L-Glu-NNap is a relatively specific substrate for the assay of EAP [2]. It has been reported that EAP is devoid of activity on aminopeptidase substrates such as Leu-NNap, Ala-NNap, Phe-NNap, Leu-NH$_2$, Leu-Gly, and Gly-Gly [3]. On the other hand, hog intestinal EAP exhibits a low level of activity on N-terminal neutral residues that is not thought to be attributable to contamination by mAAP. Circulating EAP is most probably the serum aminopeptidase responsible for the rapid destruction of angiotensin II [8,11]. The increased serum angiotensinase activity seen during pregnancy is probably attributable to EAP arising from the placenta [7a,12].

The Enzyme Nomenclature list [13] indicates that EC 3.4.11.7 preferentially hydrolyzes L-α-aspartyl peptides, and accordingly the name "aspartate aminopeptidase" is applied. This specificity does not appear to be consistent with the results of numerous studies (including the reference [3] cited in the EN list) all of which show that L-α-glutamyl derivatives are the preferred substrates. We therefore consider it more appropriate to refer to EC 3.4.11.7 as "glutamyl aminopeptidase".

References
[1]   Gorvel *et al. Biochim. Biophys. Acta* **615**: 271-274, 1980.
[2]   Benajiba & Maroux *Eur. J. Biochem.* **107**: 381-388, 1980.
[3]   Glenner *et al. Nature* **194**: 867 only, 1962.
[4]   Chulkova & Orekhovich *Biochemistry USSR* **43**: 964-969, 1978.
[5]   Danielsen *et al. Biochem. J.* **189**: 591-603, 1980.
[6]   Tobe *et al. Biochim. Biophys. Acta* **613**: 459-468, 1980.
[7]   Nagatsu *et al. Biochim. Biophys. Acta* **198**: 255-270, 1970.
[7a] Mizutani *et al. Biochim. Biophys. Acta* **678**: 168-170, 1981.
[8]   Regoli *et al. Biochem. Pharmacol.* **12**: 637-646, 1963.
[9]   Aoyagi *et al. J. Antibiot. (Tokyo)* **31**: 636-638, 1978.
[10] George & Kenny *Biochem. J.* **134**: 43-57, 1973.
[11] Khairallah *et al. Science* **140**: 672-674, 1963.
[12] Sakura *et al. Biochem. Int.* **6**: 609-515, 1983.
[13] Nomenclature Committee of the International Union of Biochemistry.
      In: *Enzyme Nomenclature, 1978*, p.302, Academic Press, 1979.

Bibliography

1961

Glenner, G. G. & Folk, J. E. Glutamyl peptidases in rat and guinea pig kidney slices. *Nature* **192**: 338-340.

1962

Glenner, G. G. & McMillan, P. J. A mammalian peptidase specific for the hydrolysis of N-terminal α-L-glutamyl and aspartyl residues. *Nature* **194**: 867 only.

1963

Khairallah, P. A., Bumpus, F. M., Page, I. H. & Smeby, R. R. Angiotensinase with a high degree of specificity in plasma and red cells. *Science* **140**: 672-674.

Regoli, D., Riniker, B. & Brunner, H. The enzymatic degradation of various angiotensin II derivatives by serum, plasma or kidney homogenate. *Biochem. Pharmacol.* **12**: 637-646.

1965

Hess, R. Arylamidase activity related to angiotensinase. *Biochim. Biophys. Acta* **99**: 316-324.

Nagatsu, I., Gillespie, L., Folk, J. E. & Glenner, G. G. Serum aminopeptidases, "angiotensinase", and hypertension. I. Degradation of angiotensin II by human serum. *Biochem. Pharmacol.* **14**: 721-728.

Nagatsu, I., Gillespie, L., George, J. M., Folk, J. E. & Glenner, G. G. Serum aminopeptidases, "angiotensinase," and hypertensin - II. Amino acid β-naphthylamide hydrolysis by normal and hypertensive serum. *Biochem. Pharmacol.* **14**: 853-861.

1966

Nagatsu, I. & Hara, J. Changes in the activities of serum aminopeptidase A and leucine aminopeptidases by the injection of parathormone in rats. *Endocrinol. Jpn.* **13**: 216-222.

1967

Khairallah, P. A. & Page, I. H. Plasma angiotensinases. *Biochem. Med.* **1**: 1-8.

Nagatsu, I. & Hara, J. Relationship between parathyroid function and serum aminopeptidase A activity. *Nature* **213**: 206 only.

1969

Nagatsu, I. & Mehl, J. W. Separation of amino acid naphthylamidases in human plasma on Sephadex G.200 and DEAE-cellulose. *Biochem. Pharmacol.* **18**: 2453-2467.

Yman, L. & Kulling, B. An aminopeptidase from human serum specific for hydrolysis of α-L-dicarboxylic amino acid residues. I. Purification from retroplacental serum by preparative polyacrylamide-gel electrophoresis. *Acta Pharm. Suec.* **6**: 561-568.

Yman, L. & Kulling, B. An aminopeptidase from human serum specific for hydrolysis of α-L-dicarboxylic amino acid residues. II. Some properties and its relationship to the serum angiotensinase activity. *Acta Pharm. Suec.* **6**: 569-578.

1970

Nagatsu, I., Nagatsu, T., Yamamoto, T., Glenner, G. G. & Mehl, J. W. Purification of amidopeptidase A in human serum and degradation of angiotensin II by the purified enzyme. *Biochim. Biophys. Acta* **198**: 255-270.

1973

George, S. G. & Kenny, A. J. Studies on the enzymology of purified preparations of brush border from rabbit kidney. *Biochem. J.* **134**: 43-57.

1976

Andria, G., Marzi, A. & Auricchio, S. α-Glutamyl-β-naphthylamide hydrolase of rabbit small intestine. Localization in the brush border and separation from other brush border peptidases. *Biochim. Biophys. Acta* **419**: 42-50.

1977

Danielsen, E. M., Sjöström, H., Norén, O. & Dabelsteen, E. Immunoelectrophoretic studies on pig intestinal brush border proteins. *Biochim. Biophys. Acta* **494**: 332-342.

McDonald, J. K. & Schwabe, C. Intracellular exopeptidases. In: *Proteinases in Mammalian Cells and Tissues* (Barrett, A. J. ed.), pp. 311-391 (see p. 370), North-Holland Publishing Co., Amsterdam.

1978

Aoyagi, T., Tobe, H., Kojima, F., Hamada, M., Takeuchi, T. & Umezawa, H. Amastatin, an inhibitor of aminopeptidase A, produced by actinomycetes. *J. Antibiot. (Tokyo)* **31**: 636-638.

Chulkova, T. M. & Orekhovich, V. N. Isolation and properties of aminopeptidase A from bovine kidneys. *Biochemistry U.S.S.R.* **43**: 964-969.

Kenny, A. J. & Booth, A. G. Microvilli: their ultrastructure, enzymology and molecular organization. *Essays Biochem.* **14**: 1-44.

Konttinen, A., Murros, J., Ojala, K., Salaspuro, M., Somer, H. & Räsänen, J. A new cause of increased serum aspartate aminotransferase activity. *Clin. Chim. Acta* **84**: 145-147.

1979

Booth, A. G., Hubbard, L. M. L. & Kenny, A. J. Proteins of the kidney microvillar membrane. Immunoelectrophoretic analysis of the membrane hydrolases: identification and resolution of the detergent- and proteinase-solubilized forms. *Biochem. J.* **179**: 397-405.

Chulkova, T. M. & Orekhovich, V. N. Molecular structure of aminopeptidase A from bovine kidneys. *Biochemistry U.S.S.R.* **44**: 1211-1213.

1980

Benajiba, A. & Maroux, S. Purification and characterization of an aminopeptidase A from hog intestinal brush-border membrane. *Eur. J. Biochem.* **107**: 381-388.

Danielsen, E. M., Norén, O., Sjöström, H., Ingram, J. & Kenny, A. J. Proteins of the kidney microvillar membrane. Aspartate aminopeptidase: purification by immunoadsorbent chromatography and properties of the detergent- and proteinase-solubilized forms. *Biochem. J.* **189**: 591-603.

Del Vecchio, P. J., Ryan, J. W., Chung, A. & Ryan, U. S. Capillaries of the adrenal cortex possess aminopeptidase A and angiotensin-converting-enzyme activities. *Biochem. J.* **186**: 605-608.

Gorvel, J. P., Benajiba, A. & Maroux, S. Purification and characterization of the rabbit intestinal brush border aminopeptidase A. *Biochim. Biophys. Acta* **615**: 271-274.

Lojda, Z. & Gossrau, R. Study on aminopeptidase A. *Histochemistry* **67**: 267-290.

Tobe, H., Kojima, F., Aoyagi, T. & Umezawa, H. Purification by affinity chromatography using amastatin and properties of aminopeptidase A from pig kidney. *Biochim. Biophys. Acta* **613**: 459-468.

1981

Feracci, H., Benajiba, A., Gorvel, J. P., Doumeng, C. & Maroux, S. Enzymatic and immunological properties of the protease form of aminopeptidases N and A from pig and rabbit intestinal brush border. *Biochim. Biophys. Acta* **658**: 148-157.

Kugler, P. Localization of aminopeptidase A (angiotensinase A) in the rat and mouse kidney. *Histochemistry* **72**: 269-278.

Mizutani, S., Okano, K., Hasegawa, E., Sakura, H. & Yamada, M. Aminopeptidase A in human placenta. *Biochim. Biophys. Acta* **678**: 168-170.

1982

Kugler, P. Aminopeptidase A is angiotensinase A. I. Quantitative histochemical studies in the kidney glomerulus. *Histochemistry* **74**: 229-245.

Kugler, P. Aminopeptidase A is angiotensinase A. II. Biochemical studies on aminopeptidase A and M in rat kidney homogenate. *Histochemistry* **74**: 247-261.

Kugler, P. Ultracytochemistry of aminopeptidase A (angiotensinase A) in the kidney glomerulus and juxtaglomerular apparatus. *Histochemistry* **74**: 199-212.

1983

Danielsen, E. M., Sjöström, H. & Norén, O. Biosynthesis of intestinal microvillar proteins. Pulse-chase labelling studies on maltase-glucoamylase, aminopeptidase A and dipeptidyl peptidase IV. *Biochem. J.* **210**: 389-393.

Kugler, P. Angiotensinase A in the renomedullary interstitial cells. *Histochemistry* **77**: 105-115.

Sakura, H., Kobayashi, H., Mizutani, S., Sakura, N., Hashimoto, T. & Kawashima, Y. Kinetic properties of placental aminopeptidase A: N-terminal degradation of angiotensin II. *Biochem. Int.* **6**: 609-615.

1984

Lalu, K., Lampelo, S., Nummelin-Kortelainen, M. & Vanha-Perttula, T. Purification and partial characterization of aminopeptidase A from the serum of pregnant and non-pregnant women. *Biochim. Biophys. Acta* **789**: 324-333.

Microsomal Proline Aminopeptidase

Summary

EC Number: 3.4.11.9

Earlier names: Aminopeptidase P and X-prolyl aminopeptidase.

Abbreviation: mPAP (m, microsomal).

Distribution: Only studied in microsomal fractions derived from hog kidney. An enzyme with similar specificity has been isolated from *E. coli.*

Source: Hog kidney cortex [1].

Action: Catalyzes the removal of any unsubstituted, N-terminal amino acid, including proline, that is adjacent to a penultimate prolyl residue. If the penultimate residue is hydroxyproline, as in Gly-Hyp-Ala, hydrolysis rates are greatly reduced. As reported for the hog kidney enzyme [1], the relative rates of release of the N-terminal amino acid from a range of small peptides, at pH 7.8, are as follows: Gly-Pro-Hyp 100, Gly-Pro-Ala 100, Gly-Pro-Gly-Gly 25, Gly-Pro 22, Gly-Pro-Met-Gly-Pro-Ala 16, Ala-Pro-Gly 6, polyproline 3, and Gly-Hyp-D,L-Ala 2. The enzyme has no action on Pro-Gly-Phe. A lack of activity on Leu-Gly-Pro-Ala illustrates the exopeptidase character of mPAP.

Requirements: pH 7.8 at 40°C, with full activation requiring preincubation for 60 min at 40°C in a buffer consisting of 24 mM sodium barbiturate (veronal)-25 mM sodium acetate-108 mM NaCl, adjusted to pH 7.8 with 1 M HCl and supplemented with $MnCl_2$ (2 mM) and bovine serum albumin (2%). Reactions are initiated by combining equal volumes of the activated enzyme and a 1% solution of peptide in the activation buffer [1].

Substrate, usual: Gly-Pro-Hyp in conjunction with an assay procedure for free glycine [2]. A substrate concentration of 17 mM reportedly gives linear reaction kinetics up to 30% hydrolysis [1].

Substrate, special: Phe(NO_2)-Pro-Pro-NH-CH_2-CH_2-NH-Abz, an intramolecularly quenched fluorogenic substrate in which the Phe(NO_2) group quenches the fluorescence of the Abz (o-aminobenzoyl) group. Upon the release of p-nitrophenylalanine, a proportional increase in fluorescence occurs

[3].

Inhibitors: None reported, but chelating agents would be expected to inhibit this Mn^{2+}-dependent exopeptidase.

Molecular properties: A Mn^{2+}-dependent enzyme that is separable into two forms on the basis of molecular size. The unusually high M_r values (1 and 2 million) suggest that the enzyme is most probably associated with membrane fragments of two different sizes. Indeed, the purified enzyme contains lipid components characteristic of phosphoinositide, sphingomyelin, lecithin, cephalin and cholesterol. These lipids, to which the enzyme is apparently bound, are not removed by the butanol solubilization step used in the purification procedure.

Comment: An enzyme having the specificity of proline aminopeptidase was first purified from *E. coli* and named aminopeptidase P [4,5]. It was subsequently designated EC 3.4.11.9 by the Enzyme Commission. It would appear that the substrate specificity of the mammalian enzyme described in this entry also meets the requirements for its systematic assignment to EC 3.4.11.9.

The 5-fold activation of mPAP, a lipoprotein complex, that is obtained by the addition of albumin or a microsomal membrane fraction to the preincubation mixture, is most probably a consequence of the ability of these additives to substitute for particles that are normally present *in vivo* and generally maintain the conformational integrity of the enzyme.

Although Gly-Pro is hydrolyzed at a substantial rate by mPAP, it is not an appropriate assay substrate because proline dipeptidase (prolidase) also splits this dipeptide.

The breakdown of polypeptides by exopeptidases such as leucyl aminopeptidase and tissue carboxypeptidase A typically stops one residue away from a prolyl residue. The action of proline aminopeptidase followed by prolyl aminopeptidase provides a possible mechanism by which the aminopeptidase action could be permitted to continue.

References

[1]   Dehm & Nordwig *Eur. J. Biochem.* **17**: 364-371, 1970.
[2]   Alexander *et al. J. Biol. Chem.* **160**: 51-59, 1945.
[3]   Fleminger *et al. Eur. J. Biochem.* **125**: 609-615, 1982.
[4]   Yaron & Mlynar *Biochem. Biophys. Res. Commun.* **32**: 658-663, 1968.
[5]   Yaron & Berger *Methods Enzymol.* **19**: 521-534, 1970.

Bibliography

1968

Nordwig, A. & Dehm, P. Breakdown of protein: cleavage of peptides of the X-Pro-Y type by kidney peptidases. *Biochim. Biophys. Acta* **160**: 293-295.

Yaron, A. & Mlynar, D. Aminopeptidase-P. *Biochem. Biophys. Res. Commun.* **32**: 658-663.

1970

Dehm, P. & Nordwig, A. Influence of serum albumin on the activity of a microsomal aminopeptidase. *FEBS Lett.* **9**: 225-228.

Dehm, P. & Nordwig, A. The cleavage of prolyl peptides by kidney peptidases. Partial purification of a "X-prolyl-aminopeptidase" from swine kidney microsomes. *Eur. J. Biochem.* **17**: 364-371.

Yaron, A. & Berger, A. Aminopeptidase-P. *Methods Enzymol.* **19**: 521-534.

1977

Kenny, A. J. Proteinases associated with cell membranes. In: *Proteinases in Mammalian Cells and Tissues* (Barrett, A. J. ed.), pp. 393-444 (see pp. 424-425), North-Holland Publishing Co., Amsterdam.

1978

Kenny, A. J. & Booth, A. G. Microvilli: their ultrastructure, enzymology and molecular organization. *Essays Biochem.* **14**: 1-44.

1982

Fleminger, G., Carmel, A., Goldenberg, D. & Yaron, A. Fluorogenic substrates for bacterial aminopeptidase P and its analogs detected in human serum and calf lung. *Eur. J. Biochem.* **125**: 609-615.

1983

Fleminger, G. & Yaron, A. Sequential hydrolysis of proline-containing peptides with immobilized aminopeptidases. *Biochim. Biophys. Acta* **743**: 437-446.

Entry 11.09

Cystinyl Aminopeptidase

Summary

EC Number: 3.4.11.3

Earlier names: Oxytocinase, cystine aminopeptidase, cystyl aminopeptidase.

Abbreviation: CAP.

Distribution: Found only in pregnancy plasma and placenta from women and other primate mammals. Fetal serum and amniotic fluid are essentially free of CAP activity. Only traces occur in the blood of non-pregnant women, and none in the blood of men. Although CAP activity has been identified in the lysosomal-mitochondrial, microsomal and supernatant fractions from human placenta, serum CAP appearing during pregnancy has the electrophoretic mobility of the lysosomal enzyme [1]. The difference in electrophoretic patterns of CAP from the various subcellular compartments suggest the existence of multiple molecular forms. Immunofluorescence histochemical studies of CAP in human placenta show the enzyme to be located in the syncytio-trophoblastic cells, with detectable levels in the outer margins of the chorion and to a lesser extent in the amnion [2].

Source: Retroplacental serum collected at normal deliveries [2-4] and from term placentas [5].

Action: Cleaves the peptide bond between an N-terminal cystine and the adjacent residue. In oxytocin, the presumed physiological substrate, the adjacent residue is tyrosine, and the terminal half-cystine is disulfide bonded to a half-cystine located at position six. Cleavage of the cystinyl bond in oxytocin results in an opening of the disulfide ring structure with a concomitant destruction of the biological activity of this neurohypophyseal hormone.

   CAP purified to apparent homogeneity from retroplacental serum exhibits a broad aminopeptidase specificity. It is reported [6] to hydrolyze a wide range of aminoacyl-NNap substrates at pH 7.4, the relative rates being: Leu-NNap 100, Arg-NNap 102, Lys-NNap 78, Ala-NNap 58, Tyr-NNap 56, Cys(-NNap)$_2$ 35, Trp-NNap 13, His-NNap 12, Phe-NNap 6, and Gly-NNap 4. No action occurs on Asp- and Glu-NNap. K_m values range from 0.029 mM for Trp-NNap to 0.59 mM for Leu-NNap, at pH 7.4 and 37°C [6].

Requirements: Activity ranges from pH 4.0-8.7, with a maximum at pH 7.3 at 37°C. Assays with Cys(-NNap)$_2$ as substrate are done at pH 6.0, because it is insoluble at higher pH.

Substrate, usual: Cys(-NNap)$_2$ [7], Cys(Bzl)-NPhNO$_2$ [7,8] or Leu-NNap [6], the latter requiring the presence of methionine, as explained under Comment.

Substrate, special: Cys(Bzl)-NMec at pH 7.0 and 37°C; K$_m$ 5.4 μM [9]. Oxytocin for test of ultimate specificity.

Inhibitors: Chelating agents such as 8-hydroxyquinoline (1 mM) and 1,10-phenanthroline (1 mM) are strong inhibitors, whereas EDTA is less effective. Strong inhibition is also obtained with 1 mM Zn^{2+}, Cu^{2+} or Cd^{2+}. Zn^{2+} behaves as a competitive inhibitor (K$_i$ 83 μM). CAP is insensitive to bestatin and hydrophobic amino acids [6]. Serum CAP is unaffected by ovomucoid, soybean trypsin inhibitor, iodoacetate, fluoride and cystine.

Molecular properties: A Zn^{2+}-containing sialoglycoprotein (M_r 280,000) with a pI of about 3.7, a value that rises to 4.5 upon desialylation [10,11]. The carbohydrate moieties, which comprise 44% of the molecule, are attached to serine and threonine through O-glycosidic linkages. Galactose and galactosamine comprise about 67% of the carbohydrate, and fucose, glucosamine and sialic acid account for the remainder [10]. The native enzyme consists of two identical subunits (M_r 140,000), each containing two tightly bound atoms of zinc [6].

Consistent with the view that serum CAP is of placental origin, enzymes derived from both sources show a common molecular mass and charge. Immunohistochemical studies provide additional evidence for a common identity [2].

Two forms of CAP, designated CAP-1 and CAP-2, can be separated electrophoretically. Between pH 8.2 and 9.6 these forms exist in a pH-dependent equilibrium; however, only CAP-2 occurs in the circulation [12], and it seems to be formed *via* an intramolecular bridge between the carboxyl group of a neuraminic acid residue and a positively-charged amino acid side chain. Only the CAP-1 form exists after treatment of the enzyme with neuraminidase [11].

CAP exhibits some loss of activity upon freezing and freeze-drying, and is completely inactivated by exposure to conditions of pH above 11.0 or below 3.5 [11].

Comment: The placenta is a rich source of enzymes, but CAP is one of the few that is relatively specific to this tissue. During pregnancy, the placenta releases CAP into the maternal circulation, where levels remain low for the first trimester, and then rise progressively, possibly exponentially, during the second and third trimesters to a maximum at about term, whereupon they decline following parturition [13]. It is generally believed that these elevated blood levels of CAP serve to prevent a premature onset of uterine contractions by destruction of oxytocin. In abnormal pregnancies the rise of CAP activity is often inadequate and erratic.

A growing body of evidence indicates that measurements of serum CAP activity in pregnant women over the course of gestation can serve as a means of assessing feto-placental function and development, and that such measurements could perhaps replace more laborious methods requiring the determination of total urinary estrogens or pregnanediol [14,15].

It has been reported that CAP can be assayed with a high degree of specificity by use of Leu-NNap in the presence of 20 mM methionine, and that this assay has greater sensitivity than that with $Cys(-NNap)_2$ [16]. Leucyl arylamidase activity normally present in the plasma of non-pregnant women is generally believed to be due to soluble alanyl aminopeptidase from liver and intestine, and this is inhibited by methionine and other hydrophobic amino acids. Bestatin should also prove useful in this regard because it is a potent inhibitor of several of the aminopeptidases that are capable of hydrolyzing Leu-NNap, but does not affect CAP.

References
[1]   Oya *et al. Physiol. Chem. Phys.* **8**: 327-335, 1976.
[2]   Small & Watkins *Cell Tiss. Res.* **162**: 531-539, 1975.
[3]   Tuppy & Wintersberger *Monatsh. Chem.* **91**: 1001-1010, 1960.
[4]   Sjöholm & Yman *Acta Pharm. Suec.* **3**: 377-388, 1966.
[5]   Cornell *Biochem. Int.* **1**: 10-18, 1980.
[6]   Sakura *et al. Biochem. Int.* **2**: 173-179, 1981.
[7]   Chapman *et al. Clin. Chim. Acta* **51**: 335-339, 1974.
[8]   Tovey *et al. Clin. Chem.* **19**: 756-761, 1973.
[9]   Suzuki *et al. Clin. Chim. Acta* **115**: 223-228, 1981.
[10] Yman *Acta Pharm. Suec.* **7**: 29-36, 1970.
[11] Yman & Sjöholm *Acta Pharm. Suec.* **4**: 13-22, 1967.
[12] Sjöholm & Yman *Acta Pharm. Suec.* **3**: 389-396, 1966.
[13] Majkic-Singh *et al. Clin. Biochem.* **15**: 152-153, 1982.
[14] Josephides & Turkington *J. Obstet. Gynaecol. Br. Commonw.* **74**: 258-261, 1967.
[15] Chapman *et al. J. Obstet. Gynaecol. Br. Commonw.* **78**: 435-443, 1971.
[16] Mizutani *et al. Clin. Biochem.* **9**: 16-18, 1976.

Bibliography

1930

Fekete, K.   Studies on the physiology of pregnancy. *Endokrinologie* **7**: 364-369.

1932

Bergmann, M., Zervas, L. & Schleich, H. Proteolytic enzymes. II. The nature of the proline linkages in gelatin. *Chem. Ber.* **65**: 1747-1750.

Bergmann, M., Zervas, L., Schleich, H. & Leinert, F. Proteolytic enzymes, behaviour of proline peptides. *Z. Physiol. Chem.* **212**: 72-84.

Fekete, K. Does an active hypophysis-posterior lobe hormone occur in the blood during pregnancy? *Endokrinologie* **10**: 16-23.

1956

Hawker, R. W. Inactivation of antidiuretic hormone and oxytocin during pregnancy. *Q. J. Exp. Physiol.* **41**: 301-308.

Werle, E. & Semm, K. Oxytocinase in pregnancy blood. *Arch. Gynaekol.* **187**: 449-457.

1957

Tuppy, H. & Nesvadba, H. Über die Aminopeptidaseaktivität des Schwangerenserums und ihre Beziehung zu dessen Vermögen, Oxytocin zu inaktivieren. *Monatsh. Chem.* **88**: 977-988.

1959

Berankova, Z., Rychlik, I. & Sorm, F. Inhibition of oxytocin inactivation by some peptides. *Experientia* **15**: 298-299.

Hooper, K. C. & Jessup, D. C. The distribution of enzymes destroying oxytocin and vasopressin in human placentae. *J. Physiol.* **146**: 539-549.

Müller-Hartburg, W., Nesvadba, H. & Tuppy, H. [The use of a chemical method for the determination of oxytocinase in serum of pregnant women.] *Arch. Gynäkol.* **191**: 442-456.

Tuppy, H. & Nesvadba, H. Cystine di-β-naphthylamide. *Austrian Patents* **204**: 195 only.

1960

Berankova, Z., Rychlik, I. & Sorm, F. Enzymic inactivation of oxytocin. I. Selective inhibitors of oxytocin inactivation. *Coll. Czech. Chem. Commun.* **25**: 2575-2580.

Titus, M. A., Reynolds, D. R., Glendening, M. B. & Page, E. W. Plasma aminopeptidase activity (oxytocinase) in pregnancy and labor. *Am. J. Obstet. Gynecol.* **80**: 1124-1128.

Tuppy, H. & Wintersberger, E. Purification and properties of serum oxytocinase. *Monatsh. Chem.* **91**: 1001-1010.

1961

Berankova, Z., Rychlik, I. & Sorm, F. Enzymic inactivation of oxytocin. II. Fission of some peptide fragments of the oxytocin structure and their derivatives by pregnancy serum and liver cell sap. *Coll. Czech. Chem. Commun.* **26**: 1708-1715.

Mendez-Bauer, C. J., Carballo, M. A., Cabot, H. M., De Paiva, C. E. N. & Gonzalez-Panizza, V. H. Studies on plasma oxytocinase. In: *Oxytocin* (Caldeyro-Barcia, R. & Heller, H. eds.), pp. 325-335, Pergamon Press, New York.

Page, E. W., Titus, M. A., Mohun, G. & Glendening, M. B. The origin and distribution of oxytocinase. *Am. J. Obstet. Gynecol.* **82**: 1090-1095.

Semm, K. The significance of oxytocinase in pregnancy and labour. In: *Oxytocin* (Caldeyro-Barcia, R. & Heller, H. eds.), pp. 336-343, Pergamon Press, New York.

Tuppy, H. Biochemical studies of oxytocinase. In: *Oxytocin* (Caldeyro-Barcia, R. & Heller, H. eds.), pp. 315-324, Pergamon Press, New York.

1962

Riad, A. M. Studies on pregnancy serum cystine aminopeptidase activity "Oxytocinase". *J. Obstet. Gynaecol. Br. Commonw.* **69**: 409-416.

Semm, K. & Waidl, E. [Histochemical studies on the formation of serum-oxytocinase in the human trophoblast.] *Z. Geburtshilfe Gynaekol.* **158**: 165-171.

Tuppy, H., Wiesbauer, U. & Wintersberger, E. p-Nitroanilides of amino acids as substrates for aminopeptidases and other proteolytic enzymes. *Hoppe-Seyler's Z. Physiol. Chem.* **329**: 278-288.

1963

Tuppy, H., Wiesbauer, U. & Wintersberger, E. The action of neuraminidase on serum oxytocinase. *Monatsh. Chem.* **94**: 321-328.

1965

Melander, S. Plasma oxytocinase activity. A methodological and clinical study with special reference to pregnancy. *Acta Endocrinol. Suppl.* **96**: 1-94.

Sjöholm, I., Yman, L. & Sandberg, F. Purification of oxytocinase (cystine aminopeptidase) by ethanol fractionation. *Acta Pharm. Suec.* **2**: 261-266.

1966

Babuna, C. & Yenen, E. Further studies on serum oxytocinase in pathologic pregnancy. *Am. J. Obstet. Gynecol.* **94**: 868-875.

James, N. T. Histochemical demonstration of oxytocinase in the human placenta. *Nature* **210**: 1276-1277.

Rydén, G. Cystine aminopeptidase and oxytocinase activity in pregnancy. A comparative study in human and rat tissue. *Acta Obstet. Gynecol. Scand.* **45**, Suppl. 3: 4-83.

Sjöholm, I. & Yman, L. Electrophoretic studies on oxytocinase (cystine aminopeptidase). *Acta Pharm. Suec.* **3**: 389-396.

Sjöholm, I. & Yman, L. Preparation of highly purified oxytocinase (cystine aminopeptidase) from retroplacental serum. *Acta Pharm. Suec.* **3**: 377-388.

Wintersberger, E., Müller-Hartburg, W. & Tuppy, H. [A simple method for

the chemical determination of oxytocinase activity in pregnancy sera.]
Clin. Chim. Acta **14**: 786-792.

1967

Barth, T., Pliška, V., Rychlík, I. & Šorm, F. Enzymic inactivation of oxytocin. V. Purification and some properties of enzymes from human retroplacental serum. *Coll. Czech. Chem. Commun.* **32**: 2327-2336.

Barth, T., Pliska, V. & Rychlik, I. Chymotryptic and tryptic cleavage of oxytocin and vasopressin. *Coll. Czech. Chem. Commun.* **32**: 1058-1063.

Josephides, E. C. H. & Turkington, V. E. Serum cystine aminopeptidase "oxytocinase" as an index of placental function. *J. Obstet. Gynaecol. Br. Commonw.* **74**: 258-261.

Sjöholm, I. Biochemical studies on oxytocin and oxytocinase. *Acta Pharm. Suec.* **4**: 81-96.

Sjöholm, I. & Yman, L. Degradation of oxytocin, lysine vasopressin, angiotensin II, and angiotensin II amide by oxytocinase (cystine aminopeptidase). *Acta Pharm. Suec.* **4**: 65-76.

Yman, L. & Sjöholm, I. Some physico-chemical properties of oxytocinase (cystine aminopeptidase). *Acta Pharm. Suec.* **4**: 13-22.

1969

Sjöholm, I. Oxytocinase and its possible significance in the degradation of oxytocin during pregnancy. *FEBS Lett.* **4**: 135-139.

Tovey, J. E. Serum oxytocinase. *Clin. Biochem.* **2**: 289-310.

Yman, L. & Kulling, B. Convenient method for purification of an aminopeptidase fraction, including oxytocinase (cystine aminopeptidase) from retroplacental serum. *Acta Pharm. Suec.* **6**: 313-319.

1970

Kulling, B. & Yman, L. Purification and some properties of leucine aminopeptidases from retroplacental serum. *Acta Pharm. Suec.* **7**: 65-74.

Yman, L. Composition of oxytocinase (cystine aminopeptidase). Amino acid and sugar analysis and some studies on the carbohydrate-polypeptide linkage and orientation of sialic acid. *Acta Pharm. Suec.* **7**: 29-36.

Yman, L. Studies on human serum aminopeptidase. Some properties of oxytocinase, human serum aminopeptidase A and leucine aminopeptidase and their purification from retroplacental serum. *Acta Pharm. Suec.* **7**: 75-86.

1971

Barth, T., Rychlik, I. & Mannsfeldt, H. G. Human pregnancy oxytocinase isolation and substrate characteristics. *Coll. Czech. Chem. Commun.* **36**: 2540-2546.

Chapman, L., Silk, E., Skupny, A., Tooth, E. A. & Barnes, A. Spectrofluorimetric assay of serum cystine aminopeptidase in normal and diabetic pregnancy compared with total oestrogen excretion. *J. Obstet.*

Gynaecol. Br. Commonw. **78**: 435-443.

Skramovsky, V., Večerek, B., Bendová, M., Trnka, V. Fluorometric determination of cystine aminopeptidase-oxytocinase in blood serum. *Clin. Chim. Acta* **35**: 353-357.

Small, C. W. & Watkins, W. B. An improved method for the determination of human pregnancy serum oxytocinase activity. *Enzymologia* **41**: 121-128.

Small, C. W. & Watkins, W. B. Immunohistochemical localisation of some placental enzymes. *N.Z. Med. J.* **74**: 338 only.

Van Oudheusden, A. P. M. Kinetic determination of serum oxytocinase in last trimester of pregnancy. *Clin. Chim. Acta* **32**: 140-141.

1972

Chapman, L., Rege, V. P., Jowett, T. P., Silk, E. & Tooth, E. A. An automated method for the assay of serum cystine aminopeptidase in pregnancy. *Clin. Chim. Acta* **38**: 339-345.

Peeters, J. A. B. M. Automated determination of serum oxytocinase activity. *Clin. Chem.* **18**: 563-564.

Rydén, G. Cystine aminopeptidase activity in pregnancy. II. Its clinical application as an index of placental function. *Obstet. Gynecol. Scand.* **51**: 329-334.

Tovey, J. E., Sykes, J. R., Robinson, D. A., Hurry, D. J., Cooper, K. & Beynon, C. L. Comparison of total urinary estrogens and serum cystine aminopeptidase estimations in pregnancy. *Clin. Biochem.* **5**: 104-108.

Watkins, W. B. & Small, C. W. An automated method for the measurement of serum aminopeptidase activity with particular reference to cystine aminopeptidase (oxytocinase). *Biochem. Med.* **6**: 82-89.

Watkins, W. B. & Small, C. W. Immunologic inactivation of human pregnancy serum oxytocinase activity. *Am. J. Obstet. Gynecol.* **113**: 973-978.

1973

Chapman, L., Burrows-Peakin, R., Jowett, T. P., Rege, V. P. & Silk, E. The normal serum cystine aminopeptidase (CAP) range in pregnancy. *Clin. Chim. Acta* **47**: 89-92.

Egan, S. M. & Livingston, A. Breakdown of [^{3}H]oxytocin by rat tissues in vitro. *J. Endocrinol.* **59**: 305-312.

Kleiner, H. & Brouet-Yager, M. Separation of L-cystinyl-di-β-naphthylamide hydrolase (oxytocinase) isoenzymes of human amniotic fluid by acrylamide gel electrophoresis. *Clin. Chim. Acta* **45**: 109-112.

Tovey, J. E., Dawson, P. J. G. & Fellowes, K. P. Evaluation of S-benzyl-L-cystine-4'-nitroanilide as a substrate for serum cystine aminopeptidase. *Clin. Chem.* **19**: 756-761.

Usategui-Gomez, M., Tarbutton, P. & Yeager, F. A colorimetric assay for oxytocinase in pregnancy serum. *Clin. Chim. Acta* **47**: 409-415.

Watson, D. & Gibbard, S. Automated determination of serum cystine aminopeptidase. *Clin. Biochem.* **6**: 60-61.

1974

Chapman, L., Burrows-Peakin, R., Rege, V. P. & Silk, E. Comparison of L-cystine-di-β-naphthylamide and S-benzyl-L-cystine-4-p-nitroanilide as substrates for the assay of serum cystine aminopeptidase in pregnancy. *Clin. Chim. Acta* **51**: 335-339.

Oya, M., Yoshino, M. & Asano, M. Human placental aminopeptidase isozymes. *Experientia* **30**: 985-986.

Small, C. W. & Watkins, W. B. S-benzyl-L-cysteine-p-nitroanilide. A new substrate for the determination of oxytocinase with improved specificity. *Biochem. Med.* **9**: 103-112.

Van Buul, T. & Van Oudheusden, A. P. M. A comparative study of recent assays for the determination of cystine aminopeptidase (oxytocinase). *Clin. Chim. Acta* **54**: 263-268.

1975

Oya, M., Yoshino, M. & Mizutani, S. Molecular heterogeneity of human placental aminopeptidase isozymes. *Experientia* **31**: 1019-1020.

Small, C. W. & Watkins, W. B. Oxytocinase - immunohistochemical demonstration in the immature and term human placenta. *Cell Tiss. Res.* **162**: 531-539.

1976

Durham, B. H. Spectrophotometric end-point method for assay of serum cystyl-aminopeptidase in pregnancy. *Clin. Chem.* **22**: 79-82.

Mizutani, S., Yoshino, M. & Oya, M. A comparison of oxytocinase and L-methionine-insensitive leucine aminopeptidase during normal pregnancy. *Clin. Biochem.* **9**: 228 only.

Mizutani, S., Yoshino, M. & Oya, M. Placental and non-placental leucine aminopeptidases during normal pregnancy. *Clin. Biochem.* **9**: 16-18.

Oya, M., Wakabayashi, T., Yoshino, M. & Mizutani, S. Subcellular distribution and electrophoretic behaviour of aminopeptidase in human placenta. *Physiol. Chem. Phys.* **8**: 327-335.

1977

Blum, M. & Sirota, P. Serum cystine aminopeptidase and leucine aminopeptidase activity in women with benign and malignant uterine and ovarian tumors. *Israel J. Med. Sci.* **13**: 875-880.

Morton, D. B. The occurrence and function of proteolytic enzymes in the reproductive tract of mammals. In: *Proteinases in Mammalian Cells and Tissues* (Barrett, A. J. ed.), pp. 445-500 (see pp. 488-491), North-Holland Publishing Co., Amsterdam.

Uete, T., Morikawa, M., Shimizu, S., Shimano, N. & Konishi, A. Simple refined fluorometric method for measuring cystyl-amino peptidase activity. *Clin. Biochem.* **10**: 193-196.

1978

Hiwada, K., Saeki-Yamaguchi, C., Inaoka, Y. & Kokubu, T.   Cystine aminopeptidases from pregnancy serum and placenta. *Biochem. Med.* **20**: 296-304.

Tiderström, G. & Heinegård, D.   Determination of cystyl-aminopeptidase. Isoenzymes in seminal plasma and serum of different origin. *Clin. Chim. Acta* **88**: 293-304.

1979

Durham, B. H. & Rewell, R. E.   Serum cystyl aminopeptidase activity in the 36th week of pregnancy. *J. Clin. Pathol.* **32**: 318-320.

Lampelo, S. & Vanha-Perttula, T.   Fractionation and characterization of cystine aminopeptidase (oxytocinase) and arylamidase of the human placenta. *J. Reprod. Fertil.* **56**: 285-296.

Minato, S.   A new colorimetric method for the determination of serum enzyme γ-glutamyl transpeptidase, cystine aminopeptidase, and leucine aminopeptidase. *Clin. Chim. Acta* **92**: 249-255.

1980

Cornell, J. S.   Oxytocinase (L-cystine aminopeptidase) from term human placenta. *Biochem. Int.* **1**: 10-18.

Unger, T. & Struck, H.   Hydrolysis of some cystine aminopeptidase and aminopeptidase substrates by human placental enzymes after isoelectric focusing. *J. Clin. Chem. Clin. Biochem.*. **18**: 631-635.

1981

Sakura, H., Lin, T. Y., Doi, M., Mizutani, S. & Kawashima, Y.   Purification and properties of oxytocinase, a metalloenzyme. *Biochem. Int.* **2**: 173-179.

Suzuki, M., Ueno, T., Takahashi, T., Kanaoka, Y., Okuyama, T., Furuya, H. & Sekine, T.   A new fluorometric ultramicro determination of serum cystine aminopeptidase using an aminocoumarine derivative. *Clin. Chim. Acta* **115**: 223-228.

Uete, T., Motokura, H., Kitano, Y., Fukutani, C., Uenishi, N. & Ando, N.   Comparison of substrates for measuring cystyl-amino peptidase activity in serum of pregnancy and hepatic disease and in various tissues. *J. Clin. Chem. Clin. Biochem.* **19**: 145-154.

1982

Majkić-Singh, N., Vuković, A., Spasić, S., Ruzic, A., Stojanov, M. & Berkeš, I.   Oxytocinase (CAP) activity in serum during normal pregnancy. *Clin. Biochem.* **15**: 152-153.

1984

Gopalaswamy, G., Krishna, K. S. & Kanagasabapathy, A. S.   Staining of cystyl aminopeptidase ("oxytocinase") isoenzymes on polyacrylamide gels. *Clin. Chem.* **30**: 1115-1116.

Lysosomal Arginyl Aminopeptidase

Summary

EC Number: This enzyme is cathepsin H, EC 3.4.22.16, an "endoaminopeptidase" [7]. It is included here in recognition of its strong aminopeptidase activity.

Earlier names: Cathepsin III, leucyl naphthylamidase, amino acid naphthylamidase, lysosomal aminopeptidase, arylamidase, cathepsin B_3, cathepsin B_1A, endoaminopeptidase, benzoylarginine-2-naphthylamide (BANA) hydrolase, benzoylarginine-2-naphthylamide amidohydrolase.

Abbreviation: *l*RAP (*l*, lysosomal; R, arginine).

Distribution: Appears to be widely distributed in mammalian tissues [1], and is most plentiful in lysosome-rich organs such as beef spleen [2] and beef and hog kidney [1,3]. Has a lysosomal distribution in rat liver and kidney [4].

Source: Rat skin [5], rat liver lysosomes [6,7], rabbit lung [8] and human liver [9,10].

Action: N-Terminal amino acids are released from synthetic substrates as well as from peptides of various lengths [7,9,11,12]. Most preparations hydolyze Arg-NNap at a rate that is roughly twice those obtained on Leu-NNap and Bz-DL-Arg-NNap. Rat liver *l*RAP hydrolyzes naphthylamides at pH 6.0 at the following relative rates: Arg-NNap 100, Thr-NNap 87, Met-NNap 73, Ser-NNap 69, His-NNap 66, Lys-NNap 64, Ala-NNap 60, Leu-NNap 52, and Glu-NNap 25 [7].

    The relative rates at which rabbit lung *l*RAP hydrolyzes amino acid naphthylamides are as follows: Arg-NNap 100, Ser-NNap 98, His-NNap 90, Lys-NNap 88, Met-NNap 82, Ala-NNap 55, Phe-NNap 46, Tyr-NNap 35, Glu-NNap 29, Leu-NNap 26, Val-NNap 20, Pro-NNap 0, and Ile-NNap 0 [12].

    Endopeptidase substrates such as Bz-Arg-NNap and Bz-Arg-OEt are hydrolyzed at lower rates than their unblocked counterparts. Bz-Arg-$NPhNO_2$ and Bz-Arg-NH_2 are generally poor substrates [7,9,11,12]. Z-Arg-Arg-NNap, an excellent assay substrate for cathepsin B [13], shows only a trace of activity with a rate that is 0.1% of that of cathepsin B [9].

    Peptides such as Phe-Phe and Leu-Gly-Gly are hydrolyzed at appreciable rates [7], but Z-Gly-Phe, Z-Glu-Tyr and Bz-Gly-Arg (commonly used for the assay of lysosomal carboxypeptidases A and B) are not attacked [7].

The endopeptidase activity of /RAP, although pronounced on small synthetic substrates, is generally weak on proteins and highly variable between species. Whereas neither the rat nor the human enzyme has appreciable action on hemoglobin or histones [7], or collagen in solution [14], the rabbit lung enzyme shows marked activity on collagen at pH 3.5 [12]. On the other hand, rat liver /RAP, unlike the human enzyme, exhibits significant activity on azocasein at pH 5.5 (7,10]. In contrast to other cysteine proteases such as cathepsins B and L, /RAP does not inactivate aldolase [10].

As described under Comments, /RAP is capable of converting leukotriene D_4 to leukotriene E_4.

Requirements: pH 6.8 at 37°C. Full activation requires preincubation for 5 min in buffer containing 1 mM EDTA and 3 mM cysteine [9] or 5 mM dithiothreitol [15]. A low pH (5.5) is required on protein (azocasein) substrates [10].

Substrate, usual: Arg-NNap [9]. Bz-Arg-NNap does not discriminate from cathepsin B. (See Comment for suggested use of inhibitors to exclude interferring activities.)

Substrate, special: Arg-NMec [16]. Although its kinetic properties are similar to the naphthylamide, it provides a 75-fold greater sensitivity.

Inhibitors: Sulfhydryl-blocking agents such as 4-chloromercuribenzoate, IAcOH, and MalNEt are powerful inhibitors of /RAP [7,9,11,12]. Leu-CH$_2$Cl (1 μM) is a potent active-site directed inhibitor of both exopeptidase and endopeptidase activities [7], whereas only moderate inhibition is produced by (10 μM) Tos-Lys-CH$_2$Cl and Tos-Phe-CH$_2$Cl [7,9], general thiol-blocking agents. Z-Phe-Ala-CHN$_2$(10 μM), a potent inhibitor of cathepsin B, produces only slow inhibition of /RAP [9]. Only the rat liver enzyme shows some sensitivity to Dip-F [7].

Leupeptin (1 μM), which completely inhibits cathepsins B and L, has little or no effect on /RAP [9,14]. Other microbial inhibitors such as (14 μM) pepstatin [9], (100 μM) puromycin [16], and (20 μM) bestatin [15] are also ineffective. E-64 [L-3-carboxy-2,3-*trans*-epoxypropionyl-leucylamido(4-guanidino)butane] at 10 μM completely inhibits /RAP whereas other aminopeptidases active on Arg-NNap are unaffected by E-64 [18].

α_2-Macroglobulin has been reported to inhibit both the aminopeptidase and endopeptidase activities of /RAP [7,8]. In contrast to its relatively low reactivity with synthetic inhibitors, /RAP is powerfully inhibited by tight-binding protein inhibitors including egg-white cystatin [25], cystatins A and B of human liver [26], and similar inhibitors from other tissues [1,17] as well as α-cysteine proteinase inhibitor [27].

Molecular properties: A glycoprotein having an affinity for concanavalin A (unlike cathepsin B, Entry 16.02). M_r values range from 26,000 to 28,000 for /RAP from various organs and species. The 220-residue sequence of rat liver /RAP shows striking homology with the 252-residue sequence of rat liver cathepsin B [19], but the sequence is more closely related to those of the plant cysteine proteinases, papain and actinidin, than that of cathepsin B.

One major isoelectric form (pI 7.1) occurs in rat liver [7], two (pI 6.0 and 6.4) in human liver [9], and several (ranging from pI 5.8-6.5) in rabbit lung [12]. $A_{280.1\%} - 12.2$ [9].

All forms are irreversibly inactivated above pH 7.0, and show maximal stability in storage at about pH 5.0, in the presence of a reversible sulfhydryl-blocking agent [10]. Rat liver /RAP shows the greater resistance to heat.

Comment: Lysosomal arginyl aminopeptidase is most probably identical to the thiol-dependent cathepsin III originally detected hydrolyzing Leu-NH$_2$ [20], and to the leucyl naphthylamidase activity detected cytochemically in lysosomes and attributed to cathepsin B [21]. However, the subsequent separation of these activities on DEAE cellulose showed that cathepsin B does not possess leucyl naphthylamidase activity [22].

The assay of /RAP in impure preparations can be made more selective by use of inhibitors that suppress the arginine-releasing activity of other aminopeptidases. For example (10 μM) bestatin inhibits soluble arginyl aminopeptidase (sRAP; Entry 11.04) and leucyl aminopeptidase (LAP; Entry 11.01), whereas (100 μM) puromycin inhibits soluble alanyl aminopeptidase (sAAP; Entry 11.03). At these concentrations, the inhibitors have little or no effect on /RAP [15,16].

On the other hand, because (10 μM) E-64 inhibits the cysteine proteinases without affecting aminopeptidases such as soluble arginyl aminopeptidase, it should be possible to determine /RAP, in impure preparations, as the E-64-sensitive fraction of the activity assayed on Arg-NNap or Arg-NMec at pH 6.8.

Rat liver /RAP is capable, by virtue of its exopeptidase activity, of catalyzing the conversion of leukotriene D$_4$, a slow reacting substance of anaphylaxis, to leukotriene E$_4$ [23]. The former peptidolipid is a hydroxylated derivative of arachidonic acid containing a Cys-Gly substituent linked through a thioether bond to C6 of the lipid moiety. /RAP catalyzes the stoichiometric release of glycine. Such a reaction could occur during the inflammatory process as a consequence of leukocyte degranulation.

A cysteine protease with a substrate specificity resembling that of /RAP has been purified from rabbit skeletal muscle and named hydrolase H [24]. It shows endopeptidase activity on protamine and Bz-Arg-NNap substrates, exopeptidase activity on Leu-NNap, Leu-Gly-Gly, and Phe-Gly-Ile-Ala, and a sensitivity to leupeptin that is low compared to cathepsin B. However, unlike /RAP, its activity is maximal at pH 7.5 - 8.0, it is stable at alkaline pH, and it is a large (340,000 M_r) protein comprised of three different subunits. The naming of this enzyme, as an exopeptidase at least, will have to await the characterization of its aminopeptidase activity. However, in a recent comparison of its activities on a range of oligopeptides and proteins, hydrolase H displayed a broad specificity as an aminopeptidase, and an extremely limited specificity as an endopeptidase [28].

References

[1] Lenney *et al.* *Eur. J. Biochem.* **101**: 153-161, 1979.
[2] Fruton *et al.* *J. Biol. Chem.* **138**: 249-262, 1941.
[3] Fruton *et al.* *J. Biol. Chem.* **141**: 763-774, 1941.
[4] Mahadevan & Tappel *J. Biol. Chem.* **242**: 2369-2374, 1967.
[5] Järvinen & Hopsu-Havu *Acta Chem. Scand. B* **29**: 772-780, 1975.
[6] Davidson & Poole *Biochim. Biophys. Acta* **397**: 437-442, 1975.
[7] Kirschke *et al.* *Acta Biol. Med. Ger.* **36**: 185-199, 1977.
[8] Singh & Kalnitsky *J. Biol. Chem.* **253**: 4319-4326, 1978.
[9] Schwartz & Barrett *Biochem. J.* **191**: 487-497, 1980.
[10] Barrett & Kirschke *Methods Enzymol.* **80**: 535-561, 1982.
[11] Järvinen *Acta Chem. Scand. B* **30**: 53-60, 1976.
[12] Singh & Kalnitsky *J. Biol. Chem.* **255**: 369-374, 1980.
[13] McDonald & Ellis *Life Sci.* **17**: 1269-1276, 1975.
[14] Kirschke *et al.* *Ciba Found. Symp.* **75**: 15-35 1980.
[15] Kirschke *et al.* *Biochem. J.* **214**: 871-877, 1983.
[16] Barrett *et al.* *Biochem. J.* **187**: 909-912, 1980.
[17] Järvinen *Acta Chem. Scand. B* **30**: 933-940, 1976.
[18] Barrett *et al.* *Biochem. J.* **201**: 189-198, 1982.
[19] Takio *et al.* *Proc. Natl Acad. Sci. USA* **80**: 3666-3670, 1983.
[20] Fruton & Bergmann *J. Biol. Chem.* **130**: 19-27, 1939.
[21] Sylvén *Histochemie* **15**: 150-159, 1968.
[22] McDonald *et al.* *Nature* **225**: 1048-1049, 1970.
[23] Yokota *et al.* *J. Biochem.* **94**: 1173-1178, 1983.
[24] Okitani *et al.* *Eur. J. Biochem.* **115**: 269-274, 1981.
[25] Anastasi *et al.* *Biochem. J.* **211**: 129-138, 1983.
[26] Green *et al.* *Biochem. J.* **218**: 939-946, 1984.
[27] Pagano *FEBS Lett.* **166**: 62-66,1984.
[28] Nishimura *et al.* *Eur. J. Biochem.* **137**: 23-27, 1983.

Bibliography

1949

Krimsky, I. & Racker, E. Protection of glycolysis in mouse brain homogenates by amides and esters of amino acids. *J. Biol. Chem.* **179**: 903-914.

1965

Smith. E. E., Kaufman, J. T. & Rutenburg, A. M. The partial purification of an amino acid naphthylamidase from human liver. *J. Biol. Chem.* **240**:

1718-1721.

1966

Nagatsu, I., Mason, S. & Glenner, G. G. Tissue fixation by diazonium salts of a particle-bound rat kidney aminopeptidase. *J. Histochem. Cytochem.* **14**: 663-668.

1967

Mahadevan, S. & Tappel, A. L. Arylamidases of rat liver and kidney. *J. Biol. Chem.* **242**: 2369-2374.

1968

Mäkinen, P.-L. & Raekallio, J. Purification and properties of an arylaminopeptidase of rat wound tissue, acting chiefly on the 2-naphthylamides of L-methionine and L-valine. *Acta Chem. Scand.* **22**: 3111-3119.

Sylvén, B. Studies on the histochemical "leucine aminopeptidase" reaction. VI. The selective demonstration of cathepsin B activity by means of the naphthylamide reaction. *Histochemie* **15**: 150-159.

Sylvén, B. & Snellman, O. Studies on the histochemical "leucine aminopeptidase" reaction. V. Cathepsin B as a potential effector of LNA hydrolysis. *Histochemie.* **12**: 240-243.

1969

Mäkinen, P.-L. & Raekallio, K. Characterization of four arylaminopeptidases of rat skin. *Enzymologia* **36**: 93-110.

1970

Kaulen, D., Henning, R. & Stoffel, W. Comparison of some enzymes of the lysosomal and the plasma membrane of the rat liver cell. *Hoppe-Seyler's Z. Physiol. Chem.* **351**: 1555-1563.

McDonald, J. K., Zeitman, B. B. & Ellis, S. Leucine naphthylamide: an inappropriate substrate for the histochemical detection of cathepsins B and B'. *Nature* **225**: 1048-1049, and **226**: 90 only.

1971

Sandström, B. Some observations on a histochemically demonstrable leucyl-naphthylamidase activity (cathepsin B?) in chicken liver. *Histochemie* **26**: 35-39.

1972

de Lumen, B. O. & Tappel, A. L. α-N-Benzoylarginine-β-naphthylamide amidohydrolase of rat liver lysosomes. *J. Biol. Chem.* **247**: 3552-3557.

Kirschke, H., Langner, J., Wiederanders, B., Ansorge, S. & Bohley, P. Intrazellulärer Proteinabbau. IV. Isolierung und Charakterisierung von peptidasen aus Rattenleberlysosomen. *Acta Biol. Med. Ger.* **28**: 305-322.

1974

Lundgren, E. & Roos, G. Plasma membrane-bound and lysosomal isozymes of amino acid naphthylamidase. *Biochim. Biophys. Acta* **358**: 208-218.

1975

Albrechtsen, R. & Jensen, H. Histochemical demonstration of an LNA-splitting enzyme in the cerebellum of the rat. *Acta Pathol. Microbiol. Scand.* **83**: 503-510.

Davidson, E. & Poole, B. Fractionation of the rat liver enzymes that hydrolyze benzoyl-arginine-2-naphthylamide. *Biochim. Biophys. Acta* **397**: 437-442.

1976

Kirschke, H., Langner, J., Wiederanders, B., Ansorge, S., Bohley, P. & Broghammer, U. Intrazellulärer Proteinabbau. VII. Kathepsine L und H: zwei neue Proteinasen aus Rattenleberlysosomen. *Acta Biol. Med. Ger.* **35**: 285-299.

Kirschke, H., Langner, J., Wiederanders, B., Ansorge, S., Bohley, P. & Hanson, H. Cathepsin L and proteinases with cathepsin B1-like activity from rat liver lysosomes. In: *Intracellular Protein Catabolism* (Hanson, H. & Bohley, P. eds), pp. 210-217, J. A. Barth, Leipzig.

1977

Kirschke, H. Cathepsin H: an endoaminopeptidase. *Acta Biol. Med. Ger.* **36**: 1547-1548.

Kirschke, H., Langner, J., Wiederanders, B., Ansorge, S., Bohley, P. & Hanson, H. Cathepsin H: an endoaminopeptidase from rat liver lysosomes. *Acta Biol. Med. Ger.* **36**: 185-199.

Kirschke, H., Langner, J., Wiederanders, B., Ansorge, S., Bohley, P. & Hanson, H. Cathepsin L and cathepsin B₃ from rat liver lysosomes. In: *Intracellular Protein Catabolism II* (Turk, V. & Marks, N. eds), pp. 299-303, Plenum Press, New York.

1978

Singh, H. & Kalnitsky, G. Separation of a new α-N-benzoylarginine-β-naphthylamide hydrolase from cathepsin B1. Purification, characterization, and properties of both enzymes from rabbit lung. *J. Biol. Chem.* **253**: 4319-4326.

Singh, H., Kuo, T. & Kalnitsky, G. Collagenolytic activity of lung BANA hydrolase and cathepsin B1. In: *Protein Turnover and Lysosome Function* (Segal, H. L. & Doyle, D. J. eds), pp. 315-331, Academic Press, New York.

1980

Barrett, A. J. Fluorimetric assays for cathepsin B and cathepsin H with methylcoumarylamide substrates. *Biochem. J.* **187**: 909-912.

Kirschke, H., Langner, J., Riemann, S., Wiederanders, B., Ansorge, S. & Bohley, P. Lysosomal cysteine proteinases. In: *Protein Degradation in Health and Disease.* Ciba Found. Symp. **75**: 15-35.

Schwartz, W. N. & Barrett, A. J. Human cathepsin H. *Biochem. J.* **191**: 487-497.

Singh, H. & Kalnitsky, G. α-N-Benzoylarginine-β-naphthylamide hydrolase, an aminoendopeptidase from rabbit lung. *J. Biol. Chem.* **255**: 369-374.

1981

Barrett, A. J. & Kirschke, H. Cathepsin B, cathepsin H and cathepsin L. *Methods Enzymol.* **80**: 535-561.

Okitani, A., Nishimura, T. & Kato, H. Characterization of hydrolase H, a new muscle protease possessing aminoendopeptidase activity. *Eur. J. Biochem.* **115**: 269-274.

1982

Barrett, A. J., Kembhavi, A. A., Brown, M. A., Kirschke, H., Knight, C. G., Tamai, M. & Hanada, K. L-*Trans*-epoxysuccinyl-leucylamido(4-guanidino)butane (E-64) and its analogues as inhibitors of cysteine proteinases including cathepsins B, H and L. *Biochem. J.* **201**: 189-198.

Kominami, E. & Katunuma, N. Immunological studies on cathepsins B and H from rat liver. *J. Biochem.* **91**: 67-71.

Stauber, W. T. & Ong, S. H. Fluorescence demonstration of a cathepsin H-like protease in cardiac, skeletal and vascular smooth muscles. *Histochem. J.* **14**: 585-591.

1983

Kalnitsky, G., Chatterjee, R., Singh, H., Lones, M. & Paszkowski, A. Bifunctional activities and possible modes of regulation of some lysosomal cysteinyl proteases. In: *Proteinase Inhibitors. Medical and Biological Aspects* (Katunuma, N., Umezawa, H. & Holzer, H. eds), pp.263-274, Japan Scientific Societies Press, Tokyo.

Nishimura, T., Okitani, A., Katakai, R. & Kato, H. Mode of action towards oligopeptides and proteins of hydrolase H, a high molecular weight aminoendopeptidase from rabbit skeletal muscle. *Eur. J. Biochem.* **137**: 23-27.

Perez, H. D., Ohtani, O., Banda, D., Ong, R. & Fukuyama, K. Generation of biologically active, complement- (C5) derived peptides by cathepsin H. *J. Immunol.* **131**: 397-402.

Takio, K., Towatari, T., Katunuma, N., Teller, D. C. & Titani, K. Homology of amino acid sequence of rat liver cathepsins B and H with that of papain. *Proc. Natl Acad. Sci. USA* **80**: 3666-3670.

Yokota, K., Shono, F., Yamamoto, S., Kominami, E. & Katunuma, N. Transformation of leukotriene D$_4$ catalyzed by lysosomal cathepsin H of rat liver. *J. Biochem.* **94**: 1173-1178.

Methionyl Aminopeptidase

Summary

EC Number: 3.4.11.-

Earlier names: Methionine aminopeptidase.

Abbreviation: MAP.

Distribution: MAP activity is present in most tissues of the rat, but is especially plentiful in kidney. Detergent-containing (pH 6.0) extracts of most rat tissues show rates of hydrolysis of Met-NNap (at pH 7.0) that exceed those seen on Lys-, Arg-, Leu-, and Ala-NNap. Subcellular fractionation of liver, the second richest source, shows MAP activity to be primarily membrane-bound in both mitochondrial and microsomal fractions. Activity is not detectable in nuclear or lysosomal fractions [1].

Methionyl naphthylamidase activity is also the most prominant aminoacyl arylamidase activity in rat plasma [1,2]. However, human and monkey plasmas show rates on Ala-NNap that are 76% and 52% higher, respectively, than rates seen on Met-NNap [2].

Source: Rat liver [1].

Action: The greatest rate occurs on Met-NNap (100), followed by Leu-NNap (57), Ala-NNap (15), Arg-NNap (7) and Lys-NNap (4).

Kinin-converting activity occurs on Met-Lys-bradykinin. Methionine is released much more rapidly than is the second residue, lysine.

Requirements: pH 7.0 sodium phosphate buffer at 37°C. (Not activated by Cl⁻.)

Substrate, usual: Met-NNap (K_m 61 μM).

Substrate, special: Met-Lys-bradykinin.

Inhibitors: Inhibition (73%) is obtained with p-hydroxymercuribenzoate (1 αM), 61% with 1,10-phenanthroline (3 mM) and 50% with puromycin (1.8 mM).

Molecular properties: The M_r 74,000 enzyme is estimated to contain about 5 sulfhydryl groups per molecule. Purified preparations are notably unstable: about 25% of the activity is lost after 7 days storage at -20°C. Stabilization is not provided by mercaptoethanol (2 mM) or glycerol (2.2 M).

Comment: In addition to its potential role in kinin conversion, MAP is of special interest because of its possible contribution to the post-initiation cleavage of the terminal methionine from nascent peptides. In eukaryotic cells the initiator methionine is generally not formylated [3]. However, within rabbit reticulocytes, where both formylmethionine and methionine occur at the N-termini of nascent hemoglobin chains, both MAP and acetylaminoacyl peptidase (Entry 18.06) activities are prominent [4]. MAP that is loosely associated with brain ribosomes is similarly believed to account for the removal of methionine from the N-termini of nascent proteins [5].

References
[1]   Freitas *et al. Int. J. Biochem.* **13**: 991-997, 1981.
[2]   McDonald *et al. Biochem. Biophys. Res. Commun.* **16**: 135-140, 1964.
[3]   Chatterjee *et al. Proc. Natl. Acad. Sci. USA* **69**: 1375-1379, 1972.
[4]   Yoshida & Lin *J. Biol. Chem.* **247**: 952-957, 1972.
[5]   Kerwar *et al. Arch. Biochem. Biophys.* **143**: 336-337, 1971.

Bibliography

1971

Kerwar, S. S., Weissbach, H. & Glenner, G. G. An aminopeptidase activity associated with brain ribosomes. *Arch. Biochem. Biophys.* **143**: 336-337.

1972

Yoshida, A. & Lin, M. NH_2-Terminal formylmethionine- and NH_2-terminal methionine-cleaving enzymes in rabbits. *J. Biol. Chem.* **247**: 952-957.

1981

Freitas, J. O., Jr., Termignoni, C., Borges, D. R., Sampaio, C. A. M., Prado, J. L. & Guimaraes, J. A. Methionine aminopeptidase associated with liver mitochondria and microsomes. *Int. J. Biochem.* **13**: 991-997.

Section 13
DIPEPTIDYL PEPTIDASES

Tripeptide Aminopeptidase

Summary

EC Number: 3.4.11.4

Earlier names: Imidoendipeptidase, aminoexotripeptidase, lymphopeptidase, aminotripeptidase, tripeptidase.

Abbreviation: TAP.

Distribution: Widely distributed in animal tissues. Activity is especially high in absorptive cells of the kidney and intestinal mucosa. Enterocytes (intestinal epithelial cells) from rat and guinea pig jejunum are rich in peptidases active against tripeptides. Although this activity is prominent in both the brush border and cytosolic compartments of these cells [1], the breakdown of tripeptides within the latter compartment (of rabbit enterocytes) is almost entirely attributable to TAP [2]. That of the membrane fraction is attributable to microsomal alanyl aminopeptidase (mAAP).

Source: Hog [3], beef [4] and rat [5] kidneys; beef [6], monkey [7] and rat [8] brain; hog [9] and rabbit [2] intestinal mucosa; horse erythrocytes [10]; and calf thymus [11]. Horse kidney and calf thymus contain only 5-10% of the levels found in hog kidney.

Action: Catalyzes the release of the N-terminal amino acid from a wide range of tripeptides [3]. Its action, which is restricted to tripeptides, requires the presence of a free N-terminal α-amino group and a free C-terminal α-carboxyl group.

Relative rates of hydrolysis of tripeptides by hog kidney TAP are as follows: Pro-Gly-Gly 100, Ala-Gly-Gly 57, Leu-Gly-Gly 53, Gly-Gly-Gly 33, Phe-Gly-Gly 18, Tyr-Gly-Gly 11, and Trp-Gly-Gly 0.7 [3]. Hydrolysis of many other tripeptides shows that TAP has a very broad specificity, however.

Relative rates for rabbit intestinal TAP are as follows: Pro-Gly-Gly 100, Ala-Gly-Gly 95, Ala-Ala-Ala 77, Leu-Gly-Gly 49, and Gly-Gly-Gly 23 [2]. Rates for rat brain TAP are: Pro-Gly-Gly 100, Ala-Gly-Gly 314, Leu-Gly-Gly 286, Phe-Gly-Gly 220, Gly-Gly-Gly 188, Gly-Gly-Leu 105, Gly-Gly-Phe 74, Tyr-Gly-Gly 57, and Arg-Gly-Gly 46. Ala-Ala-Ala-OMe is hydrolyzed, but not Ala-Ala-Ala-Ala, Pro-Gly-Gly-NH_2, Gly-Gly-Phe-NH_2, Pro-Leu-Gly-NH_2 (melanostatin), or γ-Glu-Cys-Gly (glutathione) [8].

Residues at the central or C-terminal positions of tripeptides can be of any class. Highest rates are obtained on tripeptides with N-terminal neutral residues. Rates are generally slower when there is a basic amino acid residue at the N-terminus. Only tripeptides possessing an N-terminal glutamic acid or aspartic acid are completely resistant to hydrolysis [3] or show very low rates [2]. Those lacking a peptide hydrogen at the sensitive bond, e.g. Gly-Pro-Gly [2,3] and Ala-Sar-Gly [12], also are resistant to hydrolysis. Further, D-residues are tolerated only in the C-terminal position [2,12]. N-terminal asparagine and glutamine are released, but not terminal pyroglutamic acid [3].

Requirements: pH 7.8 at 37°C. Activity is sometimes significantly enhanced by 1 mM Mn^{2+} [4,6,8].

Substrate, usual: Gly-Gly-Gly. Rates of hydrolysis may be measured colorimetrically following the addition of ninhydrin [3] or L-amino acid oxidase and o-dianisidine [7], fluorometrically using o-phthalaldehyde [8], or spectrophotometrically by following the decrease in peptide absorption at 230 nm [2] or the decrease in absorption of a tripeptide-Cu^{2+} complex at 550 nm [4,13]. The reaction mixture may also be sampled periodically for amino acid analysis [2].

Substrate, special: Pro-Gly-Gly, which offers greater specificity because of its resistance to attack by most aminopeptidases.

Inhibitors: Sulfhydryl-blocking agents such as 4-chloromercuribenzoate and metal-binding agents such as 1,10-phenanthroline are strongly inhibitory [3,8]. Generally, MalNEt, IAcOH and EDTA are less effective. Dip-F and Pms-F have little or no effect.

Bestatin (K_i 0.1-0.5 μM) is a potent inhibitor of TAP [7,8]. Chymostatin and antipain are active in the 0.3-0.6 mM range. Bacitracin, puromycin, and leupeptin have no effect. Captopril, a mercaptoprolyl derivative known to inhibit peptidyl dipeptidase (angiotensin-converting enzyme) is also a potent inhibitor of TAP (IC_{50} 1.0 μM) [8]. The chemotactic factor Met-Leu-Tyr is the most inhibitory tripeptide thus far reported, whereas the N-formylated chemotactic factor, fMet-Leu-Phe is inactive [8].

TAP is also inhibited by a group of compounds possessing antihistamine, local anesthetic and anticholinesterase activity at concentrations as low as 0.1 mM. They include pyribenzamine, benadryl, procaine, cocain, and eserine [14]. All approximate the chain length and charge characteristics of the tripeptide substrate.

Cd^{2+} has long been recognized as a particularly inhibitory metal ion [11,12].

Molecular properties: M_r values range from 49,900 to 71,000. Three forms of hog kidney TAP have been described that differ in their chromatographic behaviour on DEAE-cellulose [15]. They consist of a stable monomeric form (M_r 71,000), a stable dimeric form (M_r 142,000) comprised of identical subunits, and an unstable dimeric form that readily dissociates to the monomeric subunits. The first two forms have identical amino acid compositions.

The stable M_r 71,000 monomer consists of a single polypeptide chain that is rich in tryptophan. About 11 thiol groups are present, but no disulfide bonds. The enzyme also contains zinc, but at the non-stoichiometric level of 1 atom per M_r 142,000 dimer. This suggests a possible role in stabilization rather than at the active site. $A_{280.1\%}$ - 17.5; pI - 4.94.

Kinetic constants for the hydrolysis of triglycine by the stable dimer are reported to be: K_m - 20 mM and k_{cat} - 122,000 min^{-1} [3].

Rabbit intestinal TAP is a glycoprotein containing a very small amount of sugar and a N-terminal proline [2]. Neither amino sugars nor sialic acids are present. The M_r 50,000 enzyme contains one atom of zinc. $A_{280.1\%}$ - 11.5.

Comment: Although studies with peptide substrates of various lengths have shown that the action of TAP is limited to tripeptides, the C-terminal peptide bond is not absolutely essential. The hydrolysis of glycyl-δ-aminovaleric acid [16] and the slight susceptibility of Gly-Gly-β-Ala [10] and glycyl-p-aminobenzoic acid [16] serve to indicate that the -NH₂ and -COOH groups must be separated by the correct distance.

More than 50 years ago it was reported that extracts of hog intestinal mucosa hydrolyzed glycyl-p-aminobenzoic acid but not the ortho derivative. This activity was then attributed to a dipeptidase [17], but it has since been shown to result from the action of TAP [16].

The intestinal absorption of amino acids from di- and tri-peptides is frequently more rapid than the rates observed for the consituent free amino acids [18]. It has thus been postulated that there is an uptake of these peptides across the brush-border membrane followed by intracellular hydrolysis within the cytoplasm of these cells [19]. If this is the case, the high levels of cytosolic TAP in these cells would most probably play an important role in the complete breakdown of absorbed tripeptides.

References
[1] Wells *et al. Biochim. Biophys. Acta* **569**: 82-88, 1979.
[2] Doumeng & Maroux *Biochem. J.* **177**: 801-808, 1979.
[3] Chenoweth *et al. J. Biol. Chem.* **248**: 1672-1683, 1973.
[4] Khilji *et al. Mol. Cell. Biochem.* **23**: 45-52, 1979.
[5] Kirschke *et al. Hoppe-Seyler's Z. Physiol. Chem.* **350**: 1449-1459, 1969.
[6] Sobel & Brecher *Can. J. Biochem.* **49**: 676-685, 1971.
[7] Hayashi & Oshima *J. Biochem.* **87**: 1403-1411, 1980.
[8] Sachs & Marks *Biochim. Biophys. Acta* **706**: 229-238, 1982.
[9] Smith & Bergmann *J. Biol. Chem.* **153**: 627-651, 1944.
[10] Adams *et al. J. Biol. Chem.* **199**: 845-856, 1952.

[11] Ellis & Fruton *J. Biol. Chem.* **191**: 153-159, 1951.
[12] Fruton *et al.* *J. Biol. Chem.* **173**: 457-469, 1948.
[13] Eccleston *Biochim. Biophys. Acta* **132**: 186-187, 1967.
[14] Ziff & Smith *Proc. Soc. Exp. Biol. Med.* **80**: 761-764, 1952.
[15] Chenoweth *et al.* *J. Biol. Chem.* **248**: 1684-1686, 1973.
[16] Davis & Smith *J. Biol. Chem.* **214**: 209-213, 1955.
[17] Balls & Köhler *Chem. Ber.* **64B**: 34-45, 1931.
[18] Mathews *Physiol. Rev.* **55**: 537-608, 1975.
[19] Rubino *et al.* *J. Biol. Chem.* **246**: 3542-3548, 1971.

Bibliography

1931

Balls, A. K. & Köhler, F. The specificity of animal proteases. XX.
Mechanism of enzymic dipeptide cleavage. *Chem. Ber.* **64B**: 34-45.

1944

Smith, E. L. & Bergmann, M. The peptidases of intestinal mucosa. *J. Biol. Chem.* **153**: 627-651.

1948

Fruton, J. S., Smith, V. A. & Driscoll, P. E. On the proteolytic enzymes
of animal tissues. VII. A peptidase of calf thymus. *J. Biol. Chem.* **173**:
457-469.

1949

Smith, E. L. Proteolytic enzymes. *Annu. Rev. Biochem.* **18**: 35-58 (see p.
38).

1951

Ellis, D. & Fruton, J. S. On the proteolytic enzymes of animal tissues.
IX. Calf thymus tripeptidase. *J. Biol. Chem.* **191**: 153-159.
Smith, E. L. The specificity of certain peptidases. *Adv. Enzymol.* **12**:
191-257 (see pp. 206-210).
Stern, K., Birmingham, M. K., Cullen, A. & Richer, R. Peptidase activity
in leucocytes, erythrocytes and plasma of young adult and senile
subjects. *J. Clin. Invest.* **30**: 84-89.

1952

Adams, E., Davis, N. C. & Smith, E. L. Peptidases of erythrocytes. III.
Tripeptidase. *J. Biol. Chem.* **199**: 845-856.
Adams, E., McFadden, M. & Smith, E. L. Peptidases of erythrocytes. I.
Distribution in man and other species. *J. Biol. Chem.* **198**: 663-670.
Ziff, M. & Smith, A. A. Inhibition of aminotripeptidase. *Proc. Soc. Exp. Biol. Med.* **80**: 761-764.

1955

Davis, N. C. & Smith, E. L. Action of tripeptidase on dipeptides of appropriate structure. *J. Biol. Chem.* **214**: 209-213.

Smith, E. L. Aminopeptidases. A. Aminotripeptidase (tripeptidase). *Methods Enzymol.* **2**: 83-87.

1957

Tsuboi, K. K., Penefsky, Z. J. & Hudson, P. B. Enzymes of the human erythrocyte. III. Tripeptidase, purification and specific properties. *Arch. Biochem. Biophys.* **68**: 54-68.

1966

Hanson, H. Hydrolasen: Peptidasen (Exopeptidasen). In: *Hoppe-Seyler/Thierfelder-Handbuch der physiologisch und pathologisch chemischen Analyse* (Lang, K., Lehnartz, E., Hoffmann-Ostenhof, O. & Siebert, G. eds.), 10th edn., vol. 6, part C, pp. 1-229 (see pp. 172-182), Springer-Verlag, Berlin.

1967

Eccleston, J. F. The determination of aminotripeptidase activity by a rapid spectrophotometric method. *Biochim. Biophys. Acta* **132**: 186-187.

1969

Kirschke, H., Lasch, J. & Hanson, H. Untersuchungen über Vorkommen und Kinetik von Tripeptidasen in der Rattenniere. *Hoppe-Seyler's Z. Physiol. Chem.* **350**: 1449-1459.

1970

Peters, T. J. The subcellular localization of di- and tri-peptide hydrolase activity in guinea-pig small intestine. *Biochem. J.* **120**: 195-203.

1971

Sobel, R. E. & Brecher, A. S. Ox brain aminopeptidases. Further purification of three bovine enzymes. *Can. J. Biochem.* **49**: 676-685.

1973

Chenoweth, D., Brown, D. M., Valenzuela, M. A. & Smith, E. L. Aminotripeptidase of swine kidney. II. Amino acid composition and molecular weight. *J. Biol. Chem.* **248**: 1684-1686.

Chenoweth, D., Mitchel, R. E. J. & Smith, E. L. Aminotripeptidase of swine kidney. I. Isolation and characterization of three different forms: utility of the enzyme in sequence work. *J. Biol. Chem.* **248**: 1672-1683.

Palmer, F. B. St. C. & Russell, D. W. Extraction of 2,4,6-trinitrophenyl derivatives of amino acids and peptides from aqueous acid. A simple method for determining aminotripeptidase activity. *Anal. Biochem.* **55**: 222-235.

1977

McDonald, J. K. & Schwabe, C. Intracellular exopeptidases. In: *Proteinases in Mammalian Cells and Tissues* (Barrett, A. J. ed), pp. 311-391 (see pp. 360-361), North-Holland Publishing Co., Amsterdam.

1979

Doumeng, C. & Maroux, S. Aminotripeptidase, a cytosol enzyme from rabbit intestinal mucosa. *Biochem. J.* **177**: 801-808.

Khilji, M. A., Akrawi, A. F. & Bailey, G. S. Purification and partial characterisation of a bovine kidney aminotripeptidase (capable of cleaving prolyl-glycylglycine). *Mol. Cell. Biochem.* **23**: 45-52.

Wells, G. P., Nicholson, J. A. & Peters, T. J. Subcellular localisations of di- and tripeptidases in guinea pig and rat enterocytes. *Biochim. Biophys. Acta* **569**: 82-88.

1980

Hayashi, M. & Oshima, K. Isolation and characterization of aminotripeptidase from monkey brain. *J. Biochem.* **87**: 1403-1411.

1982

Sachs, L. & Marks, N. A highly specific aminotripeptidase of rat brain cytosol. Substrate specificity and effects of inhibitors. *Biochim. Biophys. Acta* **706**: 229-238.

Section 12
TRIPEPTIDASES

Dipeptidyl Peptidase I

Summary

EC Number: 3.4.14.1

Earlier names: Cathepsin C, dipeptidyl aminopeptidase I (DAP I), dipeptidyl arylamidase I, dipeptidyl transferase, glucagon-degrading enzyme.

Abbreviation: DPP I.

Distribution: Typically lysosomal and widely distributed in the tissues. In man, most abundant in spleen and then kidney; in the rat, the order is liver followed by spleen [6].

Source: Beef pituitary [11] and spleen [12], rat liver [1], and other lysosome-rich tissues [6].

Action: Removes N-terminal dipeptides sequentially from polypeptides having unsubstituted N-termini. Activity ceases when either Arg- or Lys- arises in an N-terminal position, or a -Pro- residue appears on either side of the penultimate peptide bond. Peptides of virtually any chain length are subject to attack. Five dipeptides are removed from β-corticotropin (a 39-residue peptide), 8 from glucagon (a 29-residue peptide, 9 from secretin (a 27-residue peptide), 10 from the A chain of oxidized beef insulin (a 21-residue peptide), 13 from the B chain of oxidized beef insulin (a 30-residue peptide), 2 from angiotensin II (an octapeptide) and 2 from the tetrapeptide amide of gastrin [6].

Among the fluorogenic naphthylamide substrates, the highest rate is seen on the 2-naphthylamide of Gly-Arg- 100, followed by Ala-Arg- 72, Pro-Arg- 55, Gly-Trp- 50, Ser-Met- 33, Ala-Ala- 15, Glu-His- 11, Ser-Tyr- 9.0, Phe-Arg- 8.0, Gly-Phe- 5.6, His-Ser- 5.3, and Leu-Ala- 3.4. Substrates used for the assay of DPP II (Lys-Ala-NNap), DPP III (Arg-Arg-NNap), and DPP IV (Gly-Pro-NNap) are not attacked, nor are N-acylated substrates, e.g. Z-Gly-Arg-NNap [1].

The polymerase (dipeptidyl transferase) activity of DPP I is maximal at pH 7.7. The polymerization of hydrophobic dipeptide amides results in the formation of relatively small, insoluble polymers, e.g. (Gly-Phe)$_3$NNap [5], whereas hydrophilic dipeptide naphthylamides give rise to relatively large soluble polymers, e.g. (Gly-Arg)$_n$NNap [1].

Requirements: pH 5.0-6.0 at 37°C, a thiol compound and halide ions (Cl⁻ or Br⁻). The halide requirement is virtually absolute, and both needs are usually met with 10 mM mercaptoethylamine HCl [11,13].

Substrate, usual: Gly-Phe-NNap.

Substrate, special: Gly-Arg-NNap, for which k_{cat} is about 16 times that of the usual assay substrate [1].

Inhibitors: Sulfhydryl reagents such as iodoacetate and 4-chloromercuriphenyl sulfonate, and the amide derivatives of hydrophobic amino acids [10]. DPP I is slowly, irreversibly inactivated by 0.5 mM E-64 [14] and by Gly-Phe-CHN₂ [14a]. Reversibly inhibited by cystatins (protein inhibitors of cysteine proteinases) at the same inhibitory site as the proteinases [19], and by the homologous plasma α-cysteine proteinase inhibitor [20].

Molecular properties: M_r about 200,000; 1.8S. pI 6.0 (beef) or 5.0 (rat). The protein seems to consist of eight subunits arranged in two sets of four. Each subunit probably contains one thiol group, but two types may be present: 6 with N-terminal Asx and 2 with N-terminal Leu. Hog DPP I has been reported to resemble papain and related cysteine proteinases in amino acid composition [21].

Comment: Useful as a protein sequencing reagent [2-4]. Pro-Arg-NNapOMe has been employed for subcellular localization by light and electron microscopy [6,7] and fluorescence enzyme histochemistry [15,16]. Activity attributed to this enzyme has been detected in human skeletal muscle [8], and has been shown to be significantly elevated in patients with muscular dystrophies and polymyositis [9].

"Cathepsin C" was the name given to an enzyme discovered in hog kidney [17] and characterized as a SH-dependent enzyme with a very restricted specificity. "Dipeptidyl aminopeptidase I" was the name give to an enzyme discovered in the beef pituitary gland [11] and characterized as a Cl⁻- and SH-dependent exopeptidase with a remarkably broad specificity. Both enzymes were subsequently shown to have a common identity [1]. The enzyme has been named dipeptidyl peptidase I by the Enzyme Nomenclature Committee [18].

References
   [1]   McDonald, *et al. J. Biol. Chem.* **244**: 2693-2709, 1969.
   [2]   Callahan *et al. Methods Enzymol.* **25**: 282-298, 1972.
   [3]   Caprioli & Seifert *Biochem. Biophys. Res. Commun.* **64**: 295-303, 1975.
   [4]   Schwabe *et al. Recent Prog. Horm. Res.* **34**: 123-211, 1978.
   [5]   Nilsson & Fruton *Biochemistry* **3**: 1220-1224, 1964.
   [6]   McDonald *et al.* In: *Tissue Proteinases* (Barrett & Dingle, eds), pp. 69-107, North-Holland Publishing Co., Amsterdam, 1971.
   [7]   Smith & Van Frank In: *Lysosomes in Biology and Pathology* (Dingle & Dean, eds), vol. 4, pp. 193-249, North-Holland Publishing Co.,

Amsterdam, 1975.

[8] Bury & Pennington *Biochem. J.* **145**: 413-416, 1975.
[9] Kar & Pearson *Clin. Chim. Acta* **82**: 185-192, 1978.
[10] Metrione & MacGeorge *Biochemistry* **14**: 5249-5252, 1975.
[11] McDonald *et al. J. Biol. Chem.* **241**: 1494-1501, 1966.
[12] Metrione *et al. Biochemistry* **5**: 1597-1604, 1966.
[13] McDonald *et al. Biochem. Biophys. Res. Commun.* **24**: 771-775, 1966.
[14] Barrett *et al. Biochem. J.* **201**: 189-198, 1982.
[14a]Green & Shaw *J. Biol. Chem.* **256**: 1923-1928, 1981.
[15] Stauber & Ong *J. Histochem. Cytochem.* **30**: 162-164, 1982.
[16] McDonald & Schwabe *Ann. N. Y. Acad. Sci.* **380**: 178-186, 1982.
[17] Gutmann & Fruton *J. Biol. Chem.* **174**: 851-858, 1948.
[18] Enzyme Nomenclature Committee. *Eur. J. Biochem.* **116**: 423-435, 1981.
[19] Nicklin & Barrett *Biochem. J.* **223**: 245-253, 1984.
[20] Gounaris *et al. Biochem. J.* **221**: 445-452, 1984.
[21] Lynn & Labow *Can. J. Biochem. Cell Physiol.* **62**: 1301-1308, 1985.

Bibliography

1948

Gutmann, H. R. & Fruton, J. S. On the proteolytic enzymes of animal tissues VII. An intracellular enzyme related to chymotrypsin. *J. Biol. Chem.* **174**: 851-858.

1952

Jones, M. E., Hearn, W. R., Fried, M. & Fruton, J. S. Transamidation reactions catalyzed by cathepsin C. *J. Biol. Chem.* **195**: 645-656.
Tallan, H. H., Jones, M. E. & Fruton, J. S. On the proteolytic enzymes of animal tissues. X. Beef spleen cathepsin C. *J. Biol. Chem.* **194**: 793-805.

1953

Fruton, J. S., Hearn, W. R., Ingram, V. M., Wiggans, D. S. & Winitz, M. Synthesis of polymeric peptides in proteinase-catalyzed transamidation reactions. *J. Biol. Chem.* **204**: 891-902.

1954

Wiggins, D. S., Winitz, M. & Fruton, J. S. Action of cathepsin C on dipeptide esters. *Yale J. Biol. Med.* **27**: 11-19.

1956

Fruton, J. S. & Mycek, M. J. Studies of beef spleen cathepsin C. *Arch. Biochem. Biophys.* **65**: 11-20.

Izumiya, N. & Fruton, J. S. Specificity of cathepsin C. *J. Biol. Chem.*
218: 59-76.

1957

Finkenstaedt, J. T. Intracellular distribution of proteolytic enzymes in
rat liver tissue. *Proc. Soc. Exp. Biol. Med.* **95**: 302-304.

1959

de la Haba, G., Cammerata, P. S. & Timasheff, S. N. The partial
purification and some physical properties of cathepsin C from beef
spleen. *J. Biol. Chem.* **234**: 316-319.

1960

Greenbaum, L. M. Cathepsins and kinin-forming and -destroying enzymes.
In: *The Enzymes* (Boyer, P. D. ed.), 3rd edn., vol. 3, pp. 475-483 (see
pp. 479-480), Academic Press, New York.

1961

Planta, R. J. & Gruber, M. Specificity of cathepsin C. *Biochim. Biophys.*
Acta **53**: 443-444.

1962

Fruton, J. S. & Knappenberger, M. H. Polymerization reactions catalyzed
by intracellular proteinases. III. Action of cathepsin C on a
tetrapeptide amide. *Biochemistry* **1**: 674-676.
Würz, H., Tanaka, A. & Fruton, J. S. Polymerization of dipeptide amides
by cathepsin C. *Biochemistry* **1**: 19-29.

1963

Planta, R. J. & Gruber, M. A simple estimation of cathepsin C using a new
chromogenic substrate. *Anal. Biochem.* **5**: 360-367.
Purcell, G. M. & Barnhart, M. I. Prothrombin activation with cathepsin
C. *Biochim. Biophys. Acta* **78**: 800-802.

1964

Bouma, J. M. W. & Gruber, M. The distribution of cathepsins B and C in
rat tissue. *Biochim. Biophys. Acta* **89**: 545-547.
Nilsson, K. K. & Fruton, J. S. Polymerization reactions catalyzed by
intracellular proteinases. IV. Factors influencing the polymerization of
dipeptide amides by cathepsin C. *Biochemistry* **3**: 1220-1224.
Planta, R. J. & Gruber, M. Chromatographic purification of the thiol
enzyme cathepsin C. *Biochim. Biophys. Acta* **89**: 503-510.
Planta, R. J., Gorter, J. & Gruber, M. The catalytic properties of
cathepsin C. *Biochim. Biophys. Acta* **89**: 511-519.

1965

McDonald, J. K., Reilly, T. J. & Ellis, S. A chloride-activated dipeptidyl-β-naphthylamidase of the pituitary gland. *Life Sci.* 4: 1665-1668.

Vanha-Perttula, T., Hopsu, V. K. & Glenner, G. G. A dipeptide naphthylamidase from hog kidney. *Histochemie* 5: 448-449.

Vanha-Perttula, T., Hopsu, V. K., Sonninen, V. & Glenner, G. G. Cathepsin C activity as related to some histochemical substrate. *Histochemie* 5: 170-181.

1966

Bouma, J. M. E. & Gruber, M. Intracellular distribution of cathepsin B and cathepsin C in rat liver. *Biochem. Biophys. Acta* 113: 350-358.

McDonald, J. K., Ellis, S. & Reilly, T. J. Properties of dipeptidyl arylamidase I of the pituitary. Chloride and sulfhydryl activation of seryltyrosyl-β-naphthylamide hydrolysis. *J. Biol. Chem.* 241: 1494-1501.

McDonald, J. K., Reilly, T. J., Zeitman, B. B. & Ellis, S. Cathepsin C: a chloride-requiring enzyme. *Biochem. Biophys. Res. Commun.* 24: 771-775.

Metrione, R. M., Neves, A. G. & Fruton, J. S. Purification and properties of dipeptidyl transferase (cathepsin C). *Biochemistry* 5: 1597-1604.

1968

Heinrich, C. P. & Fruton, J. S. The action of dipeptidyl transferase as a polymerase. *Biochemistry* 7: 3556-3565.

Hopsu-Havu, V. K. & Rintola, P. A sulphydryl-dependent and chloride-activated peptidase (cathepsin C) that hydrolyses alanyl-alanine naphthylamide. *Histochem. J.* 1: 1-17.

Voynik, I. M. & Fruton, J. S. The specificity of dipeptidyl transferase. *Biochemistry* 7: 40-44.

1969

McDonald, J. K., Callahan, P. X., Zeitman, B. B. & Ellis, S. Inactivation and degradation of glucagon by dipeptidyl aminopeptidase I (cathepsin C) of rat liver. Including a comparative study of secretin degradation. *J. Biol. Chem.* 244: 6199-6208.

McDonald, J. K., Zeitman, B. B., Reilly, T. J. & Ellis, S. New observations on the substrate specificity of cathepsin C (dipeptidyl aminopeptidase I). Including the degradation of β-corticotropin and other peptide hormones. *J. Biol. Chem.* 244: 2693-2709.

1970

Callahan, P. X., Shepard, J. A., Reilly, T. J., McDonald, J. K. & Ellis, S. Separation and identification of dipeptides by paper and column chromatography. *Anal. Biochem.* 38: 330-356.

Gorter, J. & Gruber, M. Cathepsin C: an allosteric enzyme. *Biochim. Biophys. Acta* 198: 546-555.

Metrione, R. M., Okuda, Y. & Fairclough, G. F. Subunit structure of dipeptidyl transferase. *Biochemistry* **9**: 2427-2432.

1971

McDonald, J. K., Callahan, P. X., Ellis, S. & Smith, R. E. Polypeptide degradation by dipeptidyl aminopeptidase I (cathepsin C) and related peptidases. In: *Tissue Proteinases* (Barrett, A. J. & Dingle, J. T., eds), pp. 69-107, North-Holland Publishing Co., Amsterdam.

Valyulis, R.-A. A. & Stepanov, V. M. Cathepsin C used for determining the amino acid sequence in peptides. *Biochemistry USSR* **36**: 731 only.

1972

Calam, D. H. & Thomas, H. J. Water-insoluble enzymes for peptide sequencing: dipeptidyl aminopeptidase I (cathepsin C), an enzyme with subunit structure. *Biochim. Biophys. Acta* **276**: 328-332.

Callahan, P. X., McDonald, J. K. & Ellis, S. Sequencing of peptides with dipeptidyl aminopeptidase I. *Methods Enzymol.* **25**: 282-298.

Huang, F. L. & Tappel, A. L. Properties of cathepsin C from rat liver. *Biochim. Biophys. Acta* **268**: 527-538.

Lindley, H. The specificity of dipeptidyl aminopeptidase I (cathepsin C) and its use in peptide sequence studies. *Biochem. J.* **126**: 683-688.

McDonald, J. K., Callahan, P. X. & Ellis, S. Preparation and specificity of dipeptidyl aminopeptidase I. *Methods Enzymol.* **25**: 272-281.

McDonald, J. K., Zeitman, B. B. & Ellis, S. Detection of a lysosomal carboxypeptidase and a lysosomal dipeptidase in highly-purified dipeptidyl aminopeptidase I (cathepsin C) and the elimination of their activitites from preparations used to sequence peptides. *Biochem. Biophys. Res. Commun.* **46**: 62-70.

Ovchinnikov, Y. A. & Kiryushkin, A. A. Determination of the amino acid sequence of peptides using dipeptidyl aminopeptidases. *FEBS. Lett.* **21**: 300-302.

1973

Caprioli, R. M., Seifert, W. E., Jr. & Sutherland, D. E. Polypeptide sequencing: use of dipeptidylaminopeptidase I and gas chromatography/mass spectrometry. *Biochem. Biophys. Res. Commun.* **55**: 67-75.

Corran, P. H. & Waley, S. G. The amino acid sequence of rabbit muscle triose phosphate isomerase. *FEBS Lett.* **30**: 97-99.

Paukovits, W. R. A simple ultra-micro method for the separation and identification of dipeptides in mixtures obtained during polypeptide sequence determination with dipeptidylaminopeptidase I. *J. Chromatogr.* **85**: 154-158.

Vanha-Perttula, T. & Kalliomäki, J. L. Comparison of dipeptide arylamidase I and II. Amino acid arylamidase and acid phosphatase activities in normal and pathological human sera. *Clin. Chim. Acta* **44**: 249-258.

1974

Lindley, H. & Davis, P. C. Gas chromatography of some dipeptide derivatives. *J. Chromatogr.* **100**: 117-121.

McDonald, J. K., Zeitman, B. B., Callahan, P. X. & Ellis, S. Angiotensinase activity of dipeptidyl aminopeptidase I (cathepsin C) of rat liver. *J. Biol. Chem.* **249**: 234-240.

Schauer, P., Hren-Vencelj, H. & Likar, M. On the activity of cathepsin C in human embryonic kidney cell cultures infected with *herpesvirus hominis*. *Experientia* **30**: 232-233.

1975

Bury, A. F. & Pennington, R. J. Hydrolysis of dipeptide 2-naphthylamides by human muscle enzymes. *Biochem. J.* **145**: 413-416.

Caprioli, R. M. & Seifert, W. E. Hydrolysis of polypeptides and proteins utilizing a mixture of dipeptidyl aminopeptidases with analysis by GC/MS. *Biochem. Biophys. Res. Commun.* **64**: 295-303.

Krutzsch, H. C. & Pisano, J. J. The use of dipeptidyl aminopeptidase (DAP) and gas chromatography-mass spectrometry (GC-MS) in polypeptide sequencing. In: *Peptides: Chemistry, Structure and Biology; Proceedings of the 4th American Peptide Symposium* (Walter, R. & Meienhofer, J. eds.), Ann Arbor Science Publishers, Inc., Ann Arbor, Michigan.

Metrione, R. M. & MacGeorge, N. L. The mechanism of action of dipeptidyl aminopeptidase. Inhibition by amino acid derivatives and amines; activation by aromatic compounds. *Biochemistry* **14**: 5249-5252.

Smirnov, Y. V. & Potapenko, N. A. Use of cathepsin C for determination of the structure of long peptides. *Bioorg. Khim.* **1**: 1528-1530.

Smith, R. E. & Van Frank, R. M. The use of amino acid derivatives of 4-methoxy-β-naphthylamine for assay and subcellular localization of tissue proteinases. In: *Lysosomes in Biology and Pathology* (Dingle, J. T. & Dean, R. T. eds), pp. 193-249, North-Holland Publishing Co., Amsterdam.

1976

Parsons, M. E. & Pennington, R. J. T. Separation of rat muscle aminopeptidases. *Biochem. J.* **155**: 375-381.

Young, M. A. & Desiderio, D. M. Detection of asparagine and glutamine in peptides sequenced by dipeptidyl aminopeptidase I via gas chromatography-mass spectrometry. *Anal. Biochem.* **70**: 110-123.

1977

Krutzsch, H. C. & Pisano, J. J. Analysis of dipeptides by gas chromatography-mass spectrometry and application to sequencing with dipeptidyl aminopeptidases. *Methods Enzymol.* **47**: 391-404.

McDonald, J. K. & Schwabe, C. Intracellular exopeptidases. In: *Proteinases in Mammalian Cells and Tissues* (Barrett, A. J. ed.), pp. 311-391 (see pp. 314-322), North-Holland Publishing Co., Amsterdam.

1978

Kar, N. C. & Pearson, C. M. Dipeptidyl peptidases in human muscle disease. *Clin. Chim. Acta.* **82**: 185-192.

Metrione, R. M. Chromatography of dipeptidyl aminopeptidase I on inhibitor-sepharose columns. *Biochim. Biophys. Acta* **526**: 531-536.

Schwabe, C., Steinetz, B., Weiss, G., Segaloff, A., McDonald, J. K., O'Byrne, E., Hochman, J., Carriere, B. & Goldsmith, L. Relaxin. *Recent Progress in Hormone Research* **34**: 123-211.

Seifert, W. E., Jr. & Caprioli, R. M. Hydrolysis of proteins using dipeptidyl aminopeptidases: analysis of the N-terminal portion of spinach plastocyanin. *Biochemistry* **17**: 436-441.

1979

Paszkowski, A., Singh, H. & Kalnitsky, G. Studies on lung dipeptidylpeptidase I. In: *Proceedings of the 11th International Congress of Biochemistry*, p. 231 only, Toronto, Canada.

Peralta, E., Yang, H.-Y. T. & Costa, E. Identification of tissue [Met^5]-enkephalin by a combination of dipeptidyl-aminopeptidase I (DAP I) digestion and selected ion monitoring. *Fed. Proc. Fed. Am. Soc. Exp. Biol.* **38**: 363 only.

1980

Gelman, B. B., Papa, L., Davis, M. H. & Gruenstein, E. Decreased lysosomal dipeptidyl aminopeptidase I activity in cultured human skin fibroblasts in Duchenne's muscular dystrophy. *J. Clin. Invest.* **65**: 1398-1406.

Obled, C., Arnal, M. & Valin, C. Variations through the day of hepatic and muscular cathepsin A (carboxypeptidase A; EC 3.4.12.2), C (dipeptidyl peptidase; EC 3.4.14.1) and D (endopeptidase D; EC 3.4.23.5) activities and free amino acids of blood in rats: influence of feeding schedule. *Br. J. Nutr.* **44**: 61-69.

1981

Green, G. D. J. & Shaw, E. Peptidyl diazomethyl ketones are specific inactivators of thiol proteinases. *J. Biol. Chem.* **256**: 1923-1928.

1982

Barrett, A. J., Kembhavi, A. A., Brown, M. A., Kirschke, H., Knight, C. G., Tamai, M. & Hanada, K. L-*trans*-epoxysuccinyl-leucylamido(4-guanidino)butane (E-64) and its analogues as inhibitors of cysteine proteinases including cathepsins B, H and L. *Biochem. J.* **201**: 189-198.

McDonald, J. K. & Schwabe, C. Relaxin-induced elevations of cathepsin B and dipeptidyl peptidase I in the mouse pubic symphysis, with localization by fluorescence enzyme histochemistry. Ann. N. Y. Acad. Sci. **380**: 178-186.

Stauber, W. T. & Ong, S.-H. Fluorescence demonstration of dipeptidyl peptidase I (cathepsin C) in skeletal, cardiac, and vascular smooth muscles. *J. Histochem. Cytochem.* **30**: 162-164.

1983

Jadot, M., Wattiaux-De Coninck, S. & Wattiaux, R.   Latence de la cathepsine
  C du foie de rat et dipeptidylnaphthylamides. *Arch. Int. Physiol.
  Biochem.* **91B**: 60-61.
Paszkowski, A., Singh, H. & Kalnitsky, G.   Properties of dipeptidyl
  peptidase I from rabbit (*Oryctolagus cuniculus*) lungs. *Acta Biochim.
  Pol.* **30**: 363-380.

1984

Jadot, M., Colmant, C., Wattiaux-De Coninck, S. & Wattiaux, R.
  Intralysosomal hydrolysis of glycyl-L-phenylalanine 2-naphthylamide.
  Biochem. J. **219**: 965-970.
Nicklin, M. J. H. & Barrett, A. J.   Inhibition of cysteine proteinases and
  dipeptidyl peptidase I by egg-white cystatin. *Biochem. J.* **223**: 245-253.

1985

Lynn, K. R. & Labow, R. S.   A comparison of four sulfhydryl cathepsins (B,
  C, H and L) from porcine spleen. *Can. J. Biochem. Cell Biol.* **62**: 1301-
  1308.

Dipeptidyl Peptidase II

Summary

EC Number: 3.4.14.2

Earlier names: Dipeptidyl arylamidase II, carboxytripeptidase, dipeptidyl
aminopeptidase II (DAP II), dipeptidyl aminopeptidase A (of human brain),
and "dipeptidyl peptidase V".

Abbreviation: DPP II.

Distribution: Shown to have a lysosomal localization and to be widely
distributed amongst tissues [1-5]. Although species variations occur,
the highest levels of activity are generally found in the thyroid,
adenohypophysis, spleen and kidney of man, rat, guinea pig and cattle.
The ovary of the pregnant hog is also a plentiful source [6]. Levels are
relatively low in the liver. In the rat and guinea pig, DPP II activity
is especially abundant in the epididymis, where it is concentrated in
the acrosomes of spermatozoa [5,7]. In the brain of the rat, DPP II has
a unique neuronal localization [24]. DPP II is also plentiful in beef
dental pulp [8] (a fibroblast-rich connective tissue) and in rat
peritoneal and alveolar macrophages [9,10] and mast cells [11].

Source: Beef anterior pituitary glands [12] and dental pulp [8], rat skin
[2] and kidney [13], and pregnant hog ovaries [14].

Action: Removes N-terminal dipeptides from oligopeptides, especially
tripeptides, and from 2-naphthylamides, amides, and methyl esters of
dipeptides, provided the N-termini are unsubstituted [1,12]. Virtually
any residue may reside at the terminal position, but Ala- and Pro- are
the preferred penultimate residues.

    Rates of hydrolysis of naphthylamides by beef pituitary DPP II are
highest on the NNap derivatives of Lys-Ala- 100 and Lys-Pro- 100,
followed by Phe-Pro- 75, Arg-Pro- 62, Ala-Pro- 58, Arg-Ala- 30, Leu-Ala-
10, Ala-Ala- 7, and Gly-Pro- 5. No action occurs on α-N-Z, ϵ-N-Z-Lys-
Ala-NNap or Boc-Arg-Pro-NNap, Ala-NNap, or Pro-NNap [8,12]. A similar
order of specificity is exhibited by DPP II derived from beef dental
pulp [8].

    DPP II purified from the tissues of other species shows an altered

order of specificity on naphthylamides. Relative rates for the enzyme
from rat skin are: Leu-Ala- 117, Lys-Ala- 100, Ala-Ala- 39, Gly-Pro- 29
[2]. Relative rates for DPP II purified from hog ovaries are: Phe-Pro-
700, Lys-Pro- 300, Arg-Pro- 180, Lys-Ala- 100, Arg-Ala- 73, Gly-Pro- 20,
Ile-Ala- 13, and Ala-Ala- 7 [6,14].

Relative rates of hydrolysis of p-nitroanilides by rat kidney DPP II
are Ala-Ala- 100, Gly-Pro- 94, Lys-Pro- 66, and Glu-Pro- 60 [13].

Relative rates of tripeptide hydrolysis by beef pituitary DPP II are
as follows: Ala-Ala-Ala 100, Met-Met-Ala 62, Met-Met-Met 33, Ser-Met-Glu
28, Ser-Met-Gln 23, Met-Gly-Met 9, Gly-Gly-Met 3, and Phe-Phe-Phe 1. No
action occurs on D-Ala-D-Ala-D-Ala, Gly-Gly-Gly, Val-Val-Val, Ala-Gly-
Gly, or Gly-Phe-Ala [12]. Some relative rates for the beef dental pulp
enzyme are: Lys-Ala-Ala 125, Lys-Ala-Pro 125, Ala-Ala-Ala 100, Lys-Pro-
Ala 44, Gly-Pro-Ala 36, and Ala-Pro-Ala 31 [8]. Relative rates for rat
kidney DPP II: Ala-Ala-Ala 100, Gly-Pro-Ala 54, and Gly-Ala-Ala 27. Only
a trace of activity occurs on tetrapeptides such as Ala-Ala-Ala-Ala and
Gly-Pro-Leu-Gly. Gly-Gly-Gly, Ala-Gly-Gly, and Gly-Pro-Hyp are not
attacked [13].

Relative rates of hydrolysis of dipeptide esters by beef pituitary
DPP II are as follows: Lys-Ala-OMe 100, Ala-Ala-OMe 56, and Ser-Met-OMe
41. Activity is not detectable on Ala-Ala-Ala-OMe, Gly-Gly-OMe, Lys-Lys-
OMe, Ser-Tyr-OMe, Met-OMe or Ala-OMe [12].

Requirements: pH 5.5 for naphthylamide derivatives and pH 4.5 for
tripeptides [12]. No activators are required, but the buffer (cation)
concentration should be minimal. Higher-M_r cations such as Tris should
be avoided (see Inhibitors) [1].

Substrate, usual: Lys-Ala-NNap [1].

Substrate, special: Phe-Pro-NNap for DPP II derived from hog tissues [14]
and Lys-Pro-NNap for DPP II from all other sources [4,8,13]. The
proline-containing substrates provide maximum sensitivity and increased
resistence to attack by aminopeptidases, including other dipeptidyl
peptidases, provided the pH is maintained at 5.0-5.5.

Inhibitors: Dip-F, Pms-F, and p-nitrophenyl-p′-guanidinobenzoate; heavy
metals, e.g. Hg^{2+}; and Lys-Ala-CH_2Cl [8,12,14]. Competitive inhibition
is exhibited by a wide range of cations: the greater the atomic or
molecular mass of the cation the more pronounced the inhibition.
Accordingly, Na^+ (K_i 1.8 mM) shows minimal inhibition, whereas $Tris^+$ (K_i
0.32 mM), $puromycin^{2+}$ (K_i 20 μM) and Hyamine 10-X (a cationic detergent)
are increasingly inhibitory [1,4,8].

Molecular properties: M_r 130,000; pI 4.8-5.0 for DPP II from beef pituitary
[12], rat kidney [13] and human brain [22]. The kidney enzyme is
comprised of M_r 65,000 subunits. Hog ovarian DPP II (M_r 108,000) is also
comprised of two subunits of equal size, but they are smaller M_r 54,000).
The hog enzyme contains about 2% carbohydrate (hexose), primarily
mannose, and has a pI of 4.9 [6]. Extracts of rat peritoneal macrophages

contain ten isozymes, whereas alveolar macrophages contain only four [15].

Comment: Because DPP II has an exclusively lysosomal distribution, is relatively stable, and does not need sulfhydryl activation (a requirement that is somewhat incompatible with the use of reactive diazonium salts), it has been adopted as a useful lysosomal marker in cytochemical studies. Thus, Lys-Ala-NNapOMe has been employed as a specific substrate for both light and electron microscopy [3,16]. Lys-Pro-NNapOMe, designed as a special cytochemical substrate [8], has been used for the fluorescence enzyme histochemical localization of DPP II [17].

The activity of the enzyme has been demonstrated in extracts of human skeletal muscle [18], and is significantly elevated in patients with muscular dystrophies and polymyosites [19]. Elevated blood levels have been observed in a variety of human diseases, which include thromboembolism, myocardial infarction, diabetes and alcoholism [4].

In the rat testis, DPP II is present in both the interstitial and seminiferous tubular tissues, whereas DPP I is present in only the interstitial tissue [20].

A DPP II-like lysosomal enzyme, with a somewhat different substrate specificity and a lower M_r (50,000-80,000) has been purified from bovine spleen and shown to possess a pronounced carboxypeptidase activity at pH 5.0 that is virtually restricted to tripeptides [21].

DPP II of human brain has been separated into two distinct activities, dipeptidyl aminopeptidases A and B. The former is a M_r 130,000 enzyme optimally active at pH 5.5 on Lys-Ala-NNap, and is believed to constitute the human form of DPP II. The latter is a M_r 160,000 enzyme optimally active at pH 4.8. It too hydrolyzes both Lys-Ala- and Ala-Pro-arylamides, but distinct differences in properties suggest that it may represent a new dipeptidyl peptidase [22].

During a survey of dipeptidyl peptidases present in the ovary of the pregnant hog, it was observed that Phe-Pro-NNap was hydrolyzed many times faster than Lys-Ala-NNap at pH 5.5. This unusual order of specificity was interpreted as evidence for the possible existence of a previous unrecognized dipeptidyl peptidase, tentatively termed "DPP V" [14]. However, subsequent attempts to isolate a distinct "DPP V" were unsuccessful; a single enzyme with DPP II-like properties was purified to apparent homogeneity. The purified ovarian enzyme retained the same marked preference for Phe-Pro-NNap over Lys-Ala-NNap. This preference is also manifested by extracts from a wide range of other hog tissues. Thus, it appears that the substrate specificity characteristics of hog DPP II are in contrast to the more typical specificities exhibited by extracts of tissues from a wide range of other species [6].

DPP II activity fluctuates in the ovary of the sow during the estrous cycle. Levels are markedly elevated ($p < 0.001$) during the luteal and proestrous phases compared to those seen in the postovulatory phase and in the ovaries of prepubertal (noncycling) gilts. Highest mean specific activities of DPP II occur during the proestrous phase when follicular collagen is undergoing rapid degradation [23].

References

[1] McDonald *et al*. *J. Biol. Chem.* **243**: 2028-2037, 1968.

[2] Hopsu-Havu *et al*. *Arch. Klin. Exp. Dermatol.* **236**: 282-296, 1970.

[3] McDonald *et al*. In: *Tissue Proteinases* (Barrett, A. J. & Dingle J. T. eds.), pp. 69-107, North-Holland Publishing Co., Amsterdam, 1971.

[4] Vanha-Perttula & Kalliomäki *Clin. Chim. Acta* **44**: 249-258, 1973.

[5] Gossrau & Lojda *Histochemistry* **70**: 53-76, 1980.

[6] Eisenhauer & McDonald *Unpublished observations*.

[7] McDonald & Owers *Unpublished observations*.

[8] McDonald & Schwabe *Biochim. Biophys. Acta* **616**: 68-81, 1980.

[9] Sannes *et al*. *Lab. Invest.* **37**: 243-253, 1977.

[10] Sannes *J. Histochem. Cytochem.* **31**: 684-690, 1983.

[11] Sannes *et al*. *J. Histochem. Cytochem.* **27**: 1496-1498, 1979.

[12] McDonald *et al*. *J. Biol. Chem.* **243**: 4143-4150, 1968.

[13] Fukasawa *et al*. *Biochim. Biophys. Acta* **745**: 6-11, 1983.

[14] Eisenhauer & McDonald *Fed. Proc. Fed. Am. Soc. Exp. Biol.* **41**: 507 only,1982.

[15] Allen *et al*. *J. Histochem. Cytochem.* **28**: 947-952, 1980.

[16] Smith & Van Frank In: *Lysosomes in Biology and Pathology* (Dingle, J.T. & Dean, R.T. eds.), vol. 4, pp. 193-249, North-Holland Publishing Co., Amsterdam, 1975.

[17] Stauber & Ong *J. Histochem. Cytochem.* **29**: 672-677, 1981.

[18] Bury & Pennington *Biochem. J.* **145**: 413-416, 1975.

[19] Kar & Pearson *Clin. Chim. Acta* **82**: 185-192, 1978.

[20] Vanha-Perttula *J. Reprod. Fertil.* **32**: 55-63, 1973.

[21] Stein *et al*. *Hoppe-Seyler's Z. Physiol. Chem.* **349**: 472-484, 1968.

[22] Kato *et al*. *J. Neurochem.* **34**: 602-608, 1980.

[23] Eisenhauer *et al*. *Endocrinology* **112S**: 295 only, 1983.

[24] Gorenstein *et al*. *J. Neurosci.* **1**: 1096-1102, 1981.

Bibliography

1968

McDonald, J. K., Leibach, F. H., Grindeland, R. E. & Ellis, S. Purification of dipeptidyl aminopeptidase II (dipeptidyl arylamidase II) of the anterior pituitary gland; peptidase and dipeptide esterase activities. *J. Biol. Chem.* **243**: 4143-4150.

McDonald, J. K., Reilly, T. J., Zeitman, B. B. & Ellis, S. Dipeptidyl arylamidase II of the pituitary; properties of lysyl-alanyl-β-naphthylamide hydrolysis: inhibition by cations, distribution in tissues, and subcellular localization. *J. Biol. Chem.* **243**: 2028-2037.

Stein, V. U., Weber, U. & Buddecke, E.   Über eine neue Carboxypeptidase (EC 3.4.2.?) aus der Milz. *Hoppe-Seyler's Z. Physiol. Chem.* **349**: 472-484.

1969

Hopsu-Havu, V. K. & Jansén, C. T.   Peptidases in the skin. II. demonstration and partial separation of several specific dipeptide naphthylamidases in the rat and human skin. *Arch. Klin. Exp. Dermatol.* **235**: 53-62.

1970

Hopsu-Havu, V. K., Jansén, C. T. & Jarvinen, M.   Partial purification and characterization of an acid dipeptide naphthylamidase (carboxytripeptidase) of the rat skin. *Arch. Klin. Exp. Dermatol.* **236**: 282-296.

1971

McDonald, J. K., Callahan, P. X., Ellis, S. & Smith, R. E.   Polypeptide degradation by dipeptidyl aminopeptidase I (cathepsin C) and related peptidases. In: *Tissue Proteinases* (Barrett, A. J. & Dingle, J. T., eds), pp. 69-107 (see pp. 70-86), North-Holland Publishing Co., Amsterdam.

1973

Vanha-Perttula, T.   Aminopeptidases of rat testis: III. activity of dipeptidyl aminopeptidases I and II in normal and experimental conditions. *J. Reprod. Fert.* **32**: 55-63.

Vanha-Perttula, T. & Kalliomäki, J. L.   Comparison of dipeptide arylamidase I and II. Amino acid arylamidase and acid phosphatase activities in normal and pathological human sera. *Clin. Chim. Acta* **44**: 249-258.

1975

Bury, A. F. & Pennington, R. J.   Hydrolysis of dipeptide 2-naphthylamides by human muscle enzymes. *Biochem. J.* **145**: 413-416.

Smith, R. E. & Van Frank, R. M.   The use of amino acid derivatives of 4-methoxy-β-naphthylamine for assay and subcellular localization of tissue proteinases. In: *Lysosomes in Biology and Pathology* (Dingle, J. T. & Dean, R. T., eds), pp. 193-249, North-Holland Publishing Co., Amsterdam.

1976

Parsons, M. E. & Pennington, R. J. T.   Separation of rat muscle aminopeptidases. *Biochem. J.* **155**: 375-381.

1977

McDonald, J. K. & Schwabe, C. Intracellular exopeptidases. In: *Proteinases in Mammalian Cells and Tissues* (Barrett, A. J., ed), pp. 311-391, (see pp. 322-327), North-Holland Publishing Co., Amsterdam.

Sannes, P. L., McDonald, J. K. & Spicer, S. S. Dipeptidyl aminopeptidase II in rat peritoneal wash cells; cytochemical localization and biochemical characterization. *Lab. Invest.* **37**: 243-253.

1978

Kar, N. C. & Pearson, C. M. Dipeptidyl peptidases in human muscle disease. *Clin. Chim. Acta* **82**: 185-192.

1979

Dolbeare, F. & Vanderlaan, M. A fluorescent assay of proteinases in cultured mammalian cells. *J. Histochem. Cytochem.* **27**: 1493-1495.

Sannes, P. L., McDonald, J. K., Allen, R. C. & Spicer, S. S. Cytochemical localization and biochemical characterization of dipeptidyl aminopeptidase II in macrophages and mast cells. *J. Histochem. Cytochem.* **27**: 1496-1498.

Smith, R. E. & Dean, P. N. A study of acid phosphatase and dipeptidyl aminopeptidase II in monodispersed anterior pituitary cells using flow cytometry and electron microscopy. *J. Histochem. Cytochem.* **27**: 1499-1504.

1980

Allen, R. C., Sannes, P. L., Spicer, S. S. & Hong, C. C. Comparisons of alveolar and peritoneal macrophages: soluble protein, esterase, dipeptidyl aminopeptidase II, and proteinase inhibitor. *J. Histochem. Cytochem.* **28**: 947-952.

Gossrau, R. & Lojda, Z. Study on dipeptidylpeptidase II (DPP II). *Histochemistry* **70**: 53-76.

Kato, T., Hama, T. & Nagatsu, T. Separation of two dipeptidyl aminopeptidases in the human brain. *J. Neurochem.* **34**: 602-608.

McDonald, J. K. & Schwabe, C. Dipeptidyl peptidase II of bovine dental pulp. Initial demonstration and characterization as a fibroblastic, lysosomal peptidase of the serine class active on collagen-related peptides. *Biochim. Biophys. Acta* **616**: 68-81.

1981

Gorenstein, C., Tran, V. T. & Snyder, S. H. Brain peptidase with a unique neuronal localization: the histochemical distribution of diptidyl-aminopeptidase II. *J. Neurosci.* **1**: 1096-110.

Stauber, W. T. & Ong, S.-H. Fluorescence demonstration of dipeptidyl peptidase II in skeletal, cardiac, and vascular smooth muscles. *J. Histochem. Cytochem.* **29**: 672-677.

Swanson, A. A., Davis, R. M., Albers-Jackson, B. & McDonald, J. K. Lens exopeptidases. *Exp. Eye Res.* **32**: 163-173.

1983

Eisenhauer, D., Chilton, B. S. & McDonald, J. K. Characterization of a
new ovarian dipeptidyl peptidase and its increased levels of activity
during proestrous in the hog. *Endocrinology* **112**: (Suppl) 295 only.
Fukasawa, K., Fukasawa, K. M., Hiraoka, B. Y. & Harada, M. Purification
and properties of dipeptidyl peptidase II from rat kidney. *Biochim.
Biophys. Acta* **745**: 6-11.
Sannes, P. L. Subcellular localization of dipeptidyl peptidases II and IV
in rat and rabbit alveolar macrophages. *J. Histochem. Cytochem.* **31**:
684-690.

Dipeptidyl Peptidase III

Summary

EC Number: (3.4.14.4)

Earlier names: Dipeptidyl aminopeptidase III (DAP III), dipeptidyl arylamidase III, red cell angiotensinase, neutral angiotensinase, enkephalinase B.

Abbreviation: DPP III.

Distribution: Present in the cytosol in most cells. A comparison of tissues in the rat showed the highest concentration in the pancreas, followed by ventral prostate, thymus, submaxillary gland, and spleen [1,6]. Red cells, including those from rabbit and man, contain high levels of activity [2]; significant levels have also been detected in beef eye lens [3], and in human and rat skin [4]. The regional distribution of DPP III throughout the rat brain has been reported. In some regions it is typically cytosolic, and in others, membrane associated [6].

Source: Originally purified from beef pituitary glands, where DPP III was first detected [5]. An apparently identical enzyme has been purified from rat skin [4] and brain [6], and rabbit red cells [2].

Action: The activity of pituitary DPP III on naphthylamides is virtually limited to Arg-Arg-NNap (K_m 8.3 μM) and is optimal at pH 9.0 [5]. Z-Arg-Arg-NNap and Arg-NNap are not attacked, and nor are the corresponding derivatives of Lys-Lys-, Ala-Ala-, Leu-Ala, or the assay substrates for DPP I, II and IV (Gly-Arg-, Lys-Ala-, and Gly-Pro-). DPP III from rat skin hydrolyzes Arg-Arg-NNap (K_m 0.1 mM) optimally at pH 8.0, and exhibits a significant (44%) level of activity on Leu-Ala-NNap at this pH [4].

Removes N-terminal dipeptides sequentially from polypeptides having unsubstituted N-termini. Optimal activity occurs between pH 6.8 and 7.5 [5]. The purified enzyme appears to have a relatively broad specificity on oligopeptides containing four or more residues, i.e. Ala$_4$, Ala$_6$, Lys$_4$, Phe$_4$, Val-Leu-Ser-Glu-Gly [5]. Glu$_4$, Gly$_4$ and all tripeptides tested are resistant to attack, as are some large polypeptides, e.g. corticotropin, glucagon, and the S-peptide of ribonuclease.

The angiotensins [1,2,6] and enkephalins [6] are especially good substrates. The first two N-terminal dipeptides, Asn-Arg and Val-Tyr, are removed in succession (at pH 7.5) from (Asn^1,Val^5)angiotensin II by beef pituitary DPP III [1] and rabbit red cell DPP III [2]. Human (Asp^1,Ile^5)angiotensin II is also degraded, but, characteristic of an aminopeptidase, no action occurs on desamino-(Asp^1,Val^5)angiotensin II [2], nor on acetylated peptides [5]. Bonds involving proline appear to be resistant, e.g. Val-His-Pro-Phe (the C-terminal fragment of angiotensin II) and Arg-Pro-NNap. Tyr-Gly is released from Leu-enkephalin (K$_m$ 5.7 μM), and Asp-Arg and Val-Tyr from human angiotensin II (K$_m$ 0.35 μM) by DAP III from rat brain [6].

Requirements: pH 6.8-7.5 at 37°C for most peptide substrates, and pH 9.0 for the naphthylamide substrate [5]. Activators are not generally required, but the purified enzyme may benefit from added Cl$^-$ (0.1 M) and mercaptoethanol (2 mM) [4]. Puromycin (0.1 mM) has been used to selectively inhibit arginyl aminopeptidase in crude extracts [6].

Substrate, usual: Arg-Arg-NNap. Marked substrate inhibition occurs above 40 μM for beef DPP III [5] and above 200 μM for the rat skin enzyme [4].

Substrate, special: A special assay has been described for DPP III in human red cells using Arg-Arg-NNap [7].

Inhibitors: Dip-F is a potent inhibitor [2], but other agents such as 4-chloromercuribenzenesulfonate, Hg^{2+} and EDTA are also inhibitory [5]. The effect of EDTA appears to be direct, since it is reversed by dialysis as well as by addition of Zn^{2+}. Rat DPP III is unaffected by EDTA [4]. Aromatic dipeptides (Tyr-Tyr and Tyr-Phe) are potent inhibitors [6].

Molecular properties: M_r is about 80,000, and pI about 5.5 for beef pituitary DPP III [5]. Enzymes from rat brain [6] and human eye lens [8] show pI values closer to 4.5.

Comment: This is probably a serine peptidase, despite its sensitivity to sulfhydryl reagents. Unpublished studies by McDonald & Ellis (see [2]) have demonstrated that dipeptidyl peptidase III purified from rabbit red cells has a pronounced angiotensinase activity that could account for the observed angiotensinase activity of hemolyzed red cells [9] and for the "endopeptidase" activity that cleaves angiotensin II with the combined specificity of trypsin and chymotrypsin [10].

The abundance of the enzyme in red cells gives rise to a certain ambiguity regarding the origin of activity detected in extracts of tissues containing entrained blood.

Activity has been detected in extracts of rat [11] and human skeletal muscle, but, unlike DPP I and DPP II, levels are not significantly elevated in the muscles of humans with a variety of muscle-wasting diseases [12].

Enkephalinase B [13] and dipeptidyl peptidases A and B [14] of brain, which cleave the N-terminal dipeptide from Leu-enkephalin, are most probably DPP III or closely related species variants [6].

References

[1] McDonald *et al.* In: *Tissue Proteinases* (Barrett, A.J. & Dingle, J.T. eds.), pp. 69-107, North-Holland Publishing Co., Amsterdam, 1971.

[2] McDonald & Schwabe In: *Proteinases in Mammalian Cells and Tissues* (Barrett, A. J. ed.), pp.311-391, North-Holland Publishing Co., Amsterdam, 1977.

[3] Swanson *et al. Biochem. Biophys. Res. Commun.* **84**: 1151-1159, 1978.

[4] Hopsu-Havu *et al. Arch. Klin. Exp. Dermatol.* **236**: 267-281, 1970.

[5] Ellis & Nuenke *J. Biol. Chem.* **242**: 4623-4629, 1967.

[6] Lee & Snyder *J. Biol. Chem.* **257**: 12043-12050, 1982.

[7] Jones & Kapralou *Anal. Biochem.* **119**: 418-423, 1982.

[8] Swanson *et al. Curr. Eye Res.* **3**: 287-291, 1984.

[9] Braun-Menéndez *et al. Renal Hypertension*, pp. 166-177, Thomas, Springfield, Ill., 1946.

[10] Kokubu *et al. Biochim. Biophys. Acta* **191**: 668-676, 1969.

[11] Parsons & Pennington *Biochem. J.* **155**: 375-381, 1976.

[12] Kar & Pearson *Clin. Chim. Acta* **82**: 185-192, 1978.

[13] Gorenstein & Snyder *Proc. R. Soc. London Ser. B* **210**: 123-132, 1980.

[14] Hazato *et al. J. Biochem.* **95**: 1265-1271, 1984.

Bibliography

1967

Ellis, S. & Nuenke, J. M. Dipeptidyl arylamidase III of the pituitary. Purification and characterization. *J. Biol. Chem.* **242**: 4623-4629.

1969

Hopsu-Havu, V. K. & Jansén, C. T. Peptidases in the skin. II. Demonstration and partial separation of several specific dipeptide naphthylamidases in the rat and human skin. *Arch. Klin. Exp. Dermatol.* **235**: 53-62.

Kokubu, T., Akutsu, H., Fujimoto, S., Ueda, E., Hiwada, K. & Yamamura, Y. Purification and properties of endopeptidase from rabbit red cells and its process of degradation of angiotensin. *Biochim. Biophys. Acta* **191**: 668-676.

1970

Hopsu-Havu, V. K., Jansén, C. T. & Järvinen, M. Partial purification and characterization of an alkaline dipeptide naphthylamidase (Arg-Arg-NAase) of the rat skin. *Arch. Klin. Exp. Dermatol.* **236**: 267-281.

1971

McDonald, J. K., Callahan, P. X., Ellis, S. & Smith, R. E. Polypeptide degradation by dipeptidyl aminopeptidase I (cathepsin C) and related peptidases. In: *Tissue Proteinases* (Barrett, A. J. & Dingle, J. T. eds), pp. 69-107 (see pp. 92-96), North-Holland Publishing Co., Amsterdam.

1976

Parsons, M. E. & Pennington, R. J. T. Separation of rat muscle aminopeptidases. *Biochem. J.* **155**: 375-381.

1977

McDonald, J. K. & Schwabe, C. Intracellular exopeptidases. In: *Proteinases in Mammalian Cells and Tissues* (Barrett, A. J. ed), pp. 311-391 (see pp. 361-364), North-Holland Publishing Co., Amsterdam.

1978

Kar, N. C. & Pearson, C. M. Dipeptidyl peptidases in human muscle disease. *Clin. Chim. Acta* **82**: 185-192.

Swanson, A. A., Albers-Jackson, B. & McDonald, J. K. Mammalian lens dipeptidyl aminopeptidase III. *Biochem. Biophys. Res. Commun.* **84**: 1151-1159.

1979

Gorenstein, C. & Snyder, S. H. Two distinct enkephalinases: solubilization, partial purification and separation from angiotensin converting enzyme. *Life Sci.* **25**: 2065-2070.

1980

Gorenstein, C. & Snyder, S. H. Enkephalinases. *Proc. R. Soc. Lond.* **B 210**: 123-132.

1982

Hazato, T., Inagaki-Shimamura, M., Katayama, T. & Yamamoto, T. Separation and characterization of a dipeptidyl aminopeptidase that degrades enkephalins from monkey brain. *Biochem. Biophys. Res. Commun.* **105**: 470-475.

Jones, T. H. D. & Kapralou, A. A rapid assay for dipeptidyl aminopeptidase III in human erythrocytes. *Anal. Biochem.* **119**: 418-423.

Lee, C.-M. & Snyder, S. H. Dipeptidyl-aminopeptidase III of rat brain. *J. Biol. Chem.* **257**: 12043-12050.

1984

Hazato, T., Shimamura, M., Ichimura, A. & Katayama, T. Purification and characterization of two distinct dipeptidyl aminopeptidases in soluble fraction from monkey brain and their action on enkephalins. *J. Biochem.* **95**: 1265-1271.

Nishikiori, T., Kawahara, F., Naganawa, H., Muraoka, Y., Aoyagi, T. & Umezawa, H. Production of acetyl-L-leucyl-L-argininal, inhibitor of dipeptidyl aminopeptidase III by bacteria. *J. Antibiot.* **37**: 680-681.

Swanson, A. A., Davis, R. M. & McDonald, J. K. Dipeptidyl peptidase III of human cataractous lenses. Partial purification. *Curr. Eye Res.* **3**: 287-291.

Dipeptidyl Peptidase IV

Summary

EC Number: 3.4.14.5

Earlier names: Dipeptidyl aminopeptidase IV (DAP IV), postproline dipeptidyl aminopeptidase IV, X-Pro dipeptidyl aminopeptidase, Gly-Pro naphthylamidase.

Abbreviation: DPP IV.

Distribution: Most vertebrate tissues contain the enzyme, but the activities vary widely [1,2]. In the kidney, where the enzyme is exceptionally concentrated, it is located primarily in the cortex and is reportedly abundant in both brush-border and microvillus-membrane fractions. The enzyme is also plentiful in brush-border membrane preparations from the small intestine.

In the liver, DPP IV is located primarily in cell membranes around bile canaliculi. It is also concentrated in the cell membranes and secretory granules or serous acinar cells of the parotid and submaxillary glands [3].

In the myocardium, striated muscle, aorta and lung, its activity is located primarily in the endothelial cells of the venous portion of the capillary bed and small venules [4]. In the human submaxillary gland, about half of the enzyme is present in the microsomal fraction, and in the bovine parotid gland, too, the bulk of the activity is associated with this membranous fraction.

In lymphatic tissues and in blood and bone marrow, DPP IV occurs only in lymphocytes, and predominantly in Tμ lymphocytes [5]. In certain DPP IV-rich human cell lines it occurs primarily as a plasma membrane ectoenzyme [6]. DPP IV can be readily localized in tissue macrophages [7], but not in monocytes, B lymphocytes, or other mononuclear cells [5].

In addition, the enzyme has been reported in human saliva, serum and cerebrospinal fluid, and in bovine dental pulp.

Source: Hog kidney [8,9], liver [10], pancreas [11] and small intestine [12], lamb kidney [13], rat liver [14] and intestine [21], human placenta

[15] and kidney [22].

Action: Removes N-terminal dipeptides sequentially from polypeptides having unsubstituted N-termini. The penultimate residue must be Pro, Hyp or Ala, with the greatest rates occurring on Pro, provided the third residue is neither Pro nor Hyp. The N-terminal position may be occupied by any one of several residues, including Pro, Phe, Leu and Arg [2]. Examples include Gly-Pro-Ala, Gly-Pro-Leu, Gly-Pro-Glu, Gly-Pro-Gly-Gly, Gly-Pro-Leu-Gly-Pro, and Leu-Pro-Gly-Gly. Gly-Pro-Pro and Gly-Pro-Hyp are resistant to attack [2]. Specificity studies with human placental DPP IV [15] show that Arg-Pro and Lys-Pro are released from substance P, Tyr-Pro and Phe-Pro from casomorphin-5, Thr-Pro from kentsin (Thr-Pro-Arg-Lys), His-Pro from His-Pro-Phe and Pro-Pro from Pro-Pro-Gly at pH 8.0. Resistant peptides include bradykinin (Arg-Pro-Pro-Gly...) and tuftsin (Thr-Lys-Pro-Arg).

Rat liver DPP IV hydrolyzes naphthylamides at the following relative rates: Gly-Pro-NNap 100, Phe-Pro-NNap 167, Arg-Pro-NNap 126, Gly-Ala-NNap 2.5, and Arg-Ala-NNap 5.8. Z-Gly-Pro-NNap and Pro-NNap are inactive [2]. K_m values for Gly-Pro-NNap (pH 7.8, 37°C) are 0.15 mM for the rat liver enzyme and 0.3 mM for the hog kidney enzyme.

The rates of hydrolysis of a wide range of dipeptide 4-(phenylazo)-phenylamides have also been reported [16].

Requirements: pH 7.8-8.0 at 37°C for the renal and intestinal enzymes, and pH 8.5-9.0 for the enzyme purified from human submaxillary glands. The enzyme is relatively stable, but is reported to lose activity below pH 5.0 when purified.

Substrate, usual: Gly-Pro-NNap.

Substrate, special: Gly-Pro-NMec [17], Gly-Pro-NNapOMe (see Comments).

Inhibitors: Dip-F is a good inhibitor, whereas Pms-F is relatively ineffective. Partial inhibition has been reported for E600 (diethyl-p-nitrophenyl phosphate). Metal ions such at Pb^{2+}, Hg^{2+} and Zn^{2+} are strongly inhibitory, and anions with high charge density, e.g. phosphate and citrate, are also inhibitory. Dipeptides such as Gly-Pro and Gly-Leu are competitive inhibitors. Diprotin A, a microbial peptide, selectively and competitively inhibits DPP IV (K_i 3.5 μM) [18]. EDTA, MalNEt and bestatin are ineffective.

Molecular properties: A glycoprotein containing about 18% carbohydrate and having a dimeric structure. DPP IV from a great many sources displays a native M_r between 220,000 and 280,000, and contains two identical subunits (M_r 110,000-137,000), each containing an active site. It is commonly found on the outer cell-surface, and, as shown for the hog kidney enzyme [19], possesses a 4000 M_r hydrophobic anchor. Human fibroblast DPP IV is unique, even compared to the enzyme from human kidney and placenta, in that its native M_r is about 400,000 and it is comprised of different subunits (M_r 125,000 and 135,000) that are disulfide linked [20]. As with DPP IV from other sources, it is present on the cell surface, and is Dip-F sensitive and stable in 8 M urea.

Comment: Caution is required when assaying impure preparations for dipeptidyl peptidase activity at alkaline pH on non-acylated peptide substrates having Gly or Ala at their N-termini. As a consequence of misrecognition, the non-protonated (uncharged) forms of these substrates may be hydrolyzed by acylaminoacyl peptidase (Entry 18.06), thereby liberating a dipeptide and mimicking the action of a dipeptidyl peptidase [36].

DPP IV is capable of removing Gly-Pro from the N-terminus of the α-chain of the fibrin monomer produced by the action of thrombin on fibrinogen. Because these modified monomers show a greatly reduced rate of polymerization, it has been suggested that DPP IV lining the inner walls of the blood vessels of most tissues [4] may exert an anticoagulant activity [23].

DPP IV appears to be a useful cytochemical marker for T lymphocyte subpopulations, e.g. Tμ-lymphocytes [24]. Monoclonal lymphocyte lines of "amplifier phenotype" in particular, which would be expected to be Tμ with respect to their Fc receptors, show substantial levels of cell surface DPP IV [6]. The enzyme occurs exclusively on the plasma membrane surface of rat alveolar macrophages [7].

Substance P, an undecapeptide neurotransmitter and potent inflammatory mediator, with the N-terminal sequence Arg-Pro-Lys-Gln-, is an example of a physiological peptide that is degraded by purified dipeptidyl peptidase IV [25,26], and by renal brush border preparations [27].

Gly-Pro-NNapOMe has been utilized for subcellular localization by light and electron microscopy [2,28]. This method has been used in combination with a separate method to demonstrate alkaline phosphatase in the same section, thereby providing a reliable demonstration of the capillary bed in many organs [4]. Activity is readily detectable in human and animal sera [29], and has been shown to be abnormally elevated in patients with hepatobiliary diseases. Abnormally depressed levels have been reported in patients with gastric cancer [30], various blood cancers [31] and rheumatoid arthritis and systemic lupus erythematosus [32]. Multiple electrophoretic forms occur in the sera of patients with hepatitis [33].

An enzyme similar to DPP IV has been detected in flavobacteria [34]. DPP IV acts synergistically with dipeptidyl peptidase I to degrade large polypeptides that cannot be degraded by DPP I alone, in sequencing studies [35].

References

[1] Hopsu-Havu & Ekfors *Histochemie* **17**: 30-37, 1969.
[2] McDonald *et al.* In: *Tissue Proteinases* (Barrett & Dingle, eds) pp. 69-107, North-Holland Publishing Co., Amsterdam, 1971.
[3] Fukasawa *et al.* *J. Histochem. Cytochem.* **29**: 337-343, 1981.

[4]   Lojda *Histochemistry* **59**: 153-166, 1979.
[5]   Feller *Histochem. J.* **14**: 889-895, 1982.
[6]   Hunt *et al.* In: *Non-HLA Antigens in Health, Aging and Malignancy* (Cohen & Singal, eds), pp. 191-195, Alan R. Liss, New York, 1983.
[7]   Sannes *J. Histochem. Cytochem.* **31**: 684-690, 1983.
[8]   Hopsu-Havu *et al.* *Acta Chem. Scand.* **22**: 299-308, 1968.
[9]   Kenney *et al.* *Biochem. J.* **157**: 169-182, 1976.
[10] Fukasawa *et al.* *Biochim. Biophys. Acta* **657**: 179-189, 1981.
[11] Ichinose *et al.* *Biochim. Biophys. Acta* **719**: 527-531, 1982.
[12] Svensson *et al.* *Eur. J. Biochem.* **90**: 489-498, 1978.
[13] Yoshimoto *et al.* *J. Biol. Chem.* **253**: 3708-3716, 1978.
[14] Hopsu-Havu & Sarimo *Hoppe-Seyler's Z. Physiol. Chem.* **348**: 1540-1550, 1967.
[15] Püschel *et al.* *Eur. J. Biochem.* **126**: 359-365, 1982.
[16] Barth *et al.* *Acta Biol. Med. Ger.* **32**: 157-174, 1974.
[17] Kato *et al.* *Biochem. Med.* **19**: 351-359, 1978.
[18] Umezawa & Aoyagi. In: *Proteinase Inhibitors* (Katunuma *et al.* eds), pp. 3-15, Springer-Verlag, New York, 1983.
[19] Booth & Kenny *Biochem. J.* **187**: 31-44, 1980.
[20] Saison *et al.* *Biochem. J.* **216**: 177-183, 1983.
[21] Bella, Jr. *Arch. Biochem. Biophys.* **218**: 156-162, 1982.
[22] Hama *et al.* *Mol. Cell. Biochem.* **43**: 35-42, 1982.
[23] Heymann & Mentlein *Naturwissenschaften* **69**: 189-190, 1982.
[24] Sterry *et al.* *Acta Histochem. Suppl.* **28**: 181-190, 1983.
[25] Heymann & Mentlein *FEBS Lett.* **91**: 360-364, 1978.
[26] Kato *et al.* *Biochim. Biophys. Acta* **525**: 7417-7422, 1978.
[27] Ward & Johnson *Biochem. J.* **171**: 143-148, 1978.
[28] Smith & Van Frank In: *Lysosomes in Biology and Pathology* (Dingle & Dean, eds), vol. 4, pp. 193-249, North-Holland Publishing Co., Amsterdam, 1975.
[29] Nagatsu *et al.* *Enzymologia* **34**: 73-76, 1968.
[30] Hino *et al.* *Clin. Chim. Acta* **62**: 5-11, 1975.
[31] Fujita *et al.* *Clin. Chim. Acta* **81**: 215-217, 1977.
[32] Fujita *et al.* *Clin. Chim. Acta* **88**: 15-20, 1978.
[33] Kasahara *et al.* *Biomed. Res.* **3**: 265-269, 1982.
[34] Yoshimoto & Tsuru *Agric. Biol. Chem.* **44**: 1961-1963, 1980.
[35] Seifert & Caprioli *Biochemistry* **17**: 436-441, 1978.
[36] Tsunasawa *et al.* *J. Biochem.* **93**: 1217-1220, 1983.

Bibliography

1966

Hopsu-Havu, V. K. & Glenner, G. G. A new dipeptide naphthylamidase hydrolyzing glycyl-prolyl-β-naphthylamide. *Histochemie* **7**: 197-201.

1967

Hopsu-Havu, V. K. & Sarimo, S. R. Purification and characterization of an aminopeptidase hydrolyzing glycyl-proline-naphthylamide. *Hoppe-Seyler's Z. Physiol. Chem.* **348**: 1540-1550.

1968

Hopsu-Havu, V. K., Rintola, P. & Glenner, G. G. A hog kidney aminopeptidase liberating N-terminal dipeptides. Partial purification and characteristics. *Acta Chem. Scand.* **22**: 299-308.

Nagatsu, I., Nagatsu, T. & Glenner, G. G. Species differences of serum amino acid β-naphthylamidases. *Enzymologia* **34**: 73-76.

Nagatsu, I., Nagatsu, T. & Yamamoto, T. Hydrolysis of amino acid β-naphthylamides by aminopeptidases in human parotid saliva and human serum. *Experientia* **24**: 347-348.

1969

Hopsu-Havu, V. K. & Ekfors, T. O. Distribution of a dipeptide naphthylamidase in rat tissues and its localization by using diazo coupling and labeled antibody techniques. *Histochemie* **17**: 30-38.

1970

Hopsu-Havu, V. K., Jansén, C. T. & Järvinen, M. A human serum aminopeptidase capable of splitting juxtaterminal bonds involving proline. *Clin. Chim. Acta* **28**: 25-36.

Oya, H., Nagatsu, I., Harada, M. & Nagatsu, T. Hydrolysis of amino acid β-naphthylamides by aminopeptidases in the parotid gland. *Experientia* **26**: 252-253.

1971

McDonald, J. K., Callahan, P. X., Ellis, S. & Smith, R. E. Polypeptide degradation by dipeptidyl aminopeptidase I (cathepsin C) and related peptidases. In: *Tissue Proteinases* (Barrett, A. J. & Dingle, J. T. eds), pp. 69-107 (see pp. 96-99), North-Holland Publishing Co., Amsterdam.

1972

Oya, H., Kato, T., Nagatsu, I. & Nagatsu, T. Arylaminopeptidase activities in bovine dental pulp. *Arch. Oral Biol.* **17**: 1245-1248.

Oya, H., Nagatsu, I. & Nagatsu, T. Purification and properties of glycylprolyl β-naphthylamidase in human submaxillary gland. *Biochim. Biophys. Acta* **258**: 591-599.

1974

Barth, A., Schulz, H. & Neubert, K. Untersuchungen zur Reinigung und Charakterisierung der Dipeptidylaminopeptidase IV. *Acta Biol. Med. Ger.*

32: 157-174.

Neubert, K. & Schulz, H. Neue chromogene Substrate zur Bestimmung von Dipeptidylaminopeptidasen. *Acta Biol. Med. Germ.* **33**: 161-168.

Oya, H., Harada, M. & Nagatsu, T. Peptidase activity of glycylprolyl β-naphthylamidase from human submaxillary gland. *Arch. Oral Biol.* **19**: 489-491.

1975

Caprioli, R. M. & Seifert, W. E. Hydrolysis of polypeptides and proteins utilizing a mixture of dipeptidylaminopeptidases with analysis by GC/MS. *Biochem. Biophys. Res. Commun.* **64**: 295-303.

Hino, M., Nagatsu, T., Kakumu, S., Okuyama, S., Yoshii, Y. & Nagatsu, I. Glycylprolyl β-naphthylamidase activity in human serum. *Clin. Chim. Acta* **62**: 5-11.

Smith, R. E. & Van Frank, R. M. The use of amino acid derivatives of 4-methoxy-β-naphthylamine for assay and subcellular localization of tissue proteinases. In: *Lysosomes in Biology and Pathology* (Dingle, J. T. & Dean, R. T. eds), pp. 193-249, North-Holland Publishing Co., Amsterdam.

1976

Hino, M., Fuyamada, H., Hayakawa, T., Nagatsu, T., Oya, H., Nakagawa, Y., Takemoto, T. & Sakakibara, S. X-prolyl dipeptidyl aminopeptidase activity, with X-proline p-nitroanilides as substrates, in normal and pathological human sera. *Clin. Chim.* **22**: 1256-1261.

Kenny, A. J., Booth, A. G., George, S. G., Ingram, J., Kershaw, D., Wood, E. J. & Young, A. R. Dipeptidyl peptidase IV, a kidney brush-border serine peptidase. *Biochem. J.* **157**: 169-182.

Nagatsu, T., Hino, M., Fuyamada, H., Hayakawa, T., Sakakibara, S., Nakagawa, Y. & Takemoto, T. New chromogenic substrates for X-prolyl dipeptidyl-aminopeptidase. *Anal. Biochem.* **74**: 466-476.

1977

Danielsen, E. M., Sjöström, H., Norén, O. & Dabelsteen, E. Immunoelectrophoretic studies on pig intestinal brush border proteins. *Biochim. Biophys. Acta* **494**: 332-342.

Fujita, K., Hirano, M., Tokunaga, K., Nagatsu, I., Nagatsu, T. & Sakakibara, S. Serum glycylproline p-nitroanilidase activity in blood cancers. *Clin. Chim. Acta* **81**: 215-217.

Kenny, A. J. Proteinases associated with cell membranes. In: *Proteinases in Mammalian Cells and Tissues* (Barrett, A. J. ed), pp. 393-444 (see pp. 417-424), North-Holland Publishing Co., Amsterdam.

Lojda, Z. Studies on glycyl-proline naphthylamidase. I. Lymphocytes. *Histochemie* **54**: 299-309.

McDonald, J. K. & Schwabe, C. Intracellular exopeptidases. In: *Proteinases in Mammalian Cells and Tissues* (Barrett, A. J. ed), pp.

311-391 (see pp. 371-376), North-Holland Publishing Co., Amsterdam.

Yoshimoto, T. & Walter, R. Post-proline dipeptidyl aminopeptidase (dipeptidyl aminopeptidase IV) from lamb kidney. *Biochim. Biophys. Acta* **485**: 391-401.

1978

Fujita, K., Hirano, M., Ochiai, J., Funabashi, M., Nagatsu, I., Nagatsu, T. & Sakakibara, S. Serum glycylproline p-nitroanilidase activity in rheumatoid arthritis and systemic lupus erythematosus. *Clin. Chim. Acta* **88**: 15-20.

Fukasawa, K. M., Fukasawa, K. & Harada, M. Dipeptidyl aminopeptidase IV, a glycoprotein from pig kidney. *Biochim. Biophys. Acta* **535**: 161-166.

Heymann, E. & Mentlein, R. Liver dipeptidyl aminopeptidase IV hydrolyzes substance P. *FEBS Lett.* **91**: 360-364.

Kar, N. C. & Pearson, C. M. Dipeptidyl peptidases in human muscle disease. *Clin. Chim. Acta* **82**: 185-192.

Kato, T., Nagatsu, T., Fukasawa, K., Harada, M., Nagatsu, I. & Sakakibara, S. Successive cleavage of N-terminal Arg^1-Pro^2 and Lys^3-Pro^4 from substance P but no release of Arg^1-Pro^2 from bradykinin, by X-Pro dipeptidyl-aminopeptidase. *Biochim. Biophys. Acta* **525**: 417-422.

Kato, T., Nagatsu, T., Kimura, T. & Sakakibara, S. Fluorescence assay of X-prolyl dipeptidyl-aminopeptidase activity with a new fluorogenic substrate. *Biochem. Med.* **19**: 351-359.

Kato, T., Nagatsu, T., Kimura, T. & Sakakibara, S. Studies on substrate specificity of X-prolyl dipeptidyl-aminopeptidase using new chromogenic substrates, x-y-p-nitroanilides. *Experientia* **34**: 319-320.

Kenny, A. J. & Booth, A. G. Microvilli: their ultrastructure, enzymology and molecular organization. *Essays Biochem.* **14**: 1-44.

Kullertz, G., Fischer, G. & Barth, A. Beiträge zum Katalysemechanismus der Dipeptidyl-peptidase IV. *Acta Biol. Med. Germ.* **37**: 559-567.

Seifert, W. E., Jr. & Caprioli, R. M. Hydrolysis of proteins using dipeptidyl aminopeptidases: analysis of the N-terminal portion of spinach plastocyanin. *Biochemistry* **17**: 436-441.

Svensson, B., Danielsen, M., Staun, M., Jeppesen, L., Norén, O. & Sjöström, H. An amphiphilic form of dipeptidyl peptidase IV from pig small intestinal brush-border membrane: Purification by immunoadsorbent chromatography and some properties. *Eur. J. Biochem.* **90**: 489-498.

Ward, P. E. & Johnson, A. R. Renal inactivation of substance P in the rat. *Biochem. J.* **171**: 143-148.

Wolf, B., Fischer, G. & Barth, A. Kinetische Untersuchungen an der Dipeptidyl-peptidase IV. *Acta Biol. Med. Germ.* **37**: 409-420.

Yoshimoto, T., Fischl, M., Orlovski, R. C. & Walter, R. Post-proline cleaving enzyme and post-proline dipeptidyl aminopeptidase: Comparison of two peptidases with high specificity for proline residues. *J. Biol. Chem.* **253**: 3708-3716.

1979

Booth, A. G., Hubbard, L. M. L. & Kenny, A. J. Proteins of the kidney microvillar membrane. Immunoelectrophoretic analysis of the membrane hydrolases: identification and resolution of the detergent- and proteinase-solubilized forms. *Biochem. J.* **179**: 397-405.

Fukasawa, K. M., Fukasawa, K. & Harada, M. Immunochemical identity of dipeptidyl aminopeptidase IV from pig serum, liver, submaxillary gland and kidney. *Experientia* **35**: 1142-1143.

Gossrau, R. Peptidasen II. Zur Lokalisation der Dipeptidylpeptidase IV (DPP IV). Histochemische und biochemische Untersuchung. *Histochemistry* **60**: 231-248.

Kato, T., Hama, T., Nagatsu, T., Kuzuya, H. & Sakakibara, S. Changes of X-prolyl-dipeptidyl-aminopeptidase activity in developing rat brain. *Experientia* **35**: 1329-1330.

Kato, T., Iwase, K., Nagatsu, T., Sakakibara, S. & Fujita, K. Comparison of X-Prolyl dipeptidyl-aminopeptidase activity in human cerebrospinal fluid with that in serum. *Experientia* **35**: 20-21.

Kato, T., Iwase. K., Nagatsu, T., Masami, H., Takemoto, T. & Sakakibara, S. A new assay of X-prolyl dipeptidyl-aminopeptidase activity in human serum with glycylproline p-phenylazoanilide as substrate. *Mol. Cell. Biochem.* **24**: 9-13.

Kato, T., Nagatsu, T., Shiio, T. & Sakakibara, S. Reduction of serum X-prolyl dipeptidyl-aminopeptidase activity in tumour-bearing mice and reversal of reduced enzyme activity by lentinan, an anti-tumour polysaccharide. *Experientia* **35**: 409-411.

Krutzsch, H. C. & Pisano. J. J. Preparation of dipeptidyl aminopeptidase IV for polypeptide sequencing. *Biochim. Biophys. Acta* **576**: 280-289.

Lojda, Z. Studies on dipeptidyl(amino)peptidase IV (glycyl-proline naphthylamidase). *Histochemistry* **59**: 153-166.

MacNair, R. D. C. & Kenny, A. J. Proteins of the kidney microvillar membrane. The amphipathic form of dipeptidyl peptidase IV. *Biochem. J.* **179**: 379-395.

Nagatsu, T., Iwase, K., Kasahara, Y., Kubono, K., Sakakibara, S., Aoyama, H. & Izawa, Y. Correlation between X-prolyl dipeptidyl-aminopeptidase and serum amine oxidase in serum of patients with post-burn keloids. *Clin. Chem.* **25**: 376-378.

1980

Barth, A., Fischer, G., Neubert, K., Heins, J. & Mager, H. Untersuchungen zum Funktionsmechanismus der Dipeptidyl-Peptidase IV - ein Beitrag zur Chemie der enzymatischen Katalyse. *Wiss. Z. Tech. Hochsch. Leuna-Merseburg* **22**: 352-371.

Barth, A., Mager, H., Fischer, G., Neubert, K. & Schwartz, G. Struktur-Activitäts-Beziehungen bie der enzymatischen Hydrolyse von Dipeptid-Arylamiden durch Dipeptidyl-Peptidase IV. *Acta Biol. Med. Ger.* **39**: 1129-1142.

Barth, A., Mager, H., Fischer, G., Neubert, K. & Schwarz, G. Struktur-Activitäts-Beziehungen bei der enzymatischen Hydrolyse von Dipeptid-Arylamiden durch Dipeptidyl-Peptidase IV. *Acta. Biol. Med. Ger.* **39**: 1129-1142.

Erickson, R. H. & Kin, Y. S. Effect of Triton X-100 on the electrophoretic mobility of solubilized intestinal brush border membrane dipeptidyl peptidase IV. *Biochem. Biophys. Acta* **614**: 210-214.

Iijima, K.-I., Takei, T., Hino, M. & Hayakawa, T. A new fluorescence assay for dipeptidylpeptidase IV using tripeptide L-prolyl-L-prolyl-L-alanine as substrate. *J. Biochem. Biophys. Methods* **3**: 89-96.

Kato, T., Kojima, K., Imai, K. & Nagatsu, T. Changes in the activities of collagenase-like peptidase and dipeptidylaminopeptidase IV and hydroxyproline contents in developing rat salivary glands. *Arch. Oral Biol.* **25**: 181-185.

Kojima, J., Kanatani, M., Nakamura, N., Kashiwagi, T., Tohjoh, F. & Akiyama, M. Serum and liver glycylproline dipeptidyl aminopeptidase activity in rats with experimental hepatic cancer. *Clin. Chim. Acta* **107**: 105-110.

Kojima, K., Hama, T., Kato, T. & Nagatsu, T. Rapid chromatographic purification of dipeptidyl peptidase IV in human submaxillary gland. *J. Chromatog.* **189**: 233-240.

Kreil, G., Haiml, L. & Suchanek, G. Stepwise cleavage of the pro part of promelittin by dipeptidylpeptidase IV. Evidence for a new type of precursor-product conversion. *Eur. J. Biochem.* **111**: 49-58.

Mentlein, R., Heiland, S. & Heymann, E. Simultaneous purification and comparative characterization of six serine hydrolases from rat liver microsomes. *Arch. Biochem. Biophys.* **200**: 547-559.

Nakayama, K., Yamada, M. & Hirayama, C. The effect of ethanol on glycyl-prolyl dipeptidyl-aminopeptidase activity in the rat pancreas and liver. *Biochem. Pharmacol.* **29**: 3210-3211.

1981

Feller, A. C. & Parwaresch, M. R. Specificity and polymorphism of diaminopeptidase IV in normal and neoplastic Tμ lymphocytes. *J. Cancer Res. Clin. Oncol.* **101**: 59-63.

Fukasawa, K. M., Fukasawa, K., Hiraoka, B. Y. & Harada, M. Comparison of dipeptidyl peptidase IV prepared from pig liver and kidney. *Biochim. Biophys. Acta* **657**: 179-189.

Fukasawa, K. M., Fukasawa, K., Sahara, N., Harada, M, Kondo, Y. & Nagatsu, I. Immunohistochemical localization of dipeptidyl aminopeptidase IV in rat kidney, liver, and salivary glands. *J. Histochem. Cytochem.* **29**: 337-343.

Hama, T., Nagatsu, T., Kobayashi, S., Azuma, S., Miyachi, T., Kumagai, Y. & Kato, T. Changes in form of dipeptidyl-aminopeptidase IV in urine from patients with renal disease. *Clin. Chim. Acta* **113**: 217-221.

Hutchinson, D. R., Halliwell, R. P., Lockhart, J. D. F. & Parke, D. V.

Glycylprolyl-p-nitroanilidase in hepatobiliary disease. *Clin. Chim. Acta* **109**: 83-89.

Sahara, N., Fukasawa, K. M., Fukasawa, K., Araki, N. & Suzuki, K. Immunohistochemical localization of dipeptidyl aminopeptidase (DAP) IV in the rat submandibular gland during postnatal development. *Histochemistry* **72**: 229-236.

Sahara, N., Fukasawa, K., Harada, M. & Suzuki, K. Immunohistochemical localization of dipeptidyl aminopeptidase (DAP) IV in the rat endocrine organs. *Acta Histochem. Cytochem.* **14**: 581-587.

Yoshii, Y., Kasugai, T., Kato, T., Nagatsu, T. & Sakakibara, S. Changes in serum dipeptidyl-aminopeptidase IV (glycylprolyl dipeptidyl-aminopeptidase) activity of patients with gastric carcinoma after surgical excision and the enzyme activity in the carcinoma tissue. *Biochem. Med.* **25**: 276-282.

1982

Bella, A. M., Jr., Erickson, R. H. & Kim, Y. S. Rat intestinal brush border membrane dipeptidyl-aminopeptidase IV: kinetic properties and substrate specificities of the purified enzyme. *Arch. Biochem. Biophys.* **218**: 156-162.

Chilosi, M., Pizzolo, G., Menestrina, F., Iannucci, A. M., Bonetti, F. & Fiore-Donati, L. Dipeptidyl(amino)peptidase IV (DAP-IV) histochemistry on normal and pathologic lymphoid tissues. *Am. J. Clin. Pathol.* **77**: 714-719.

Feller, A. C. Cytochemical reactivity of Tμ lymphocytes in human lymphatic tissue for dipeptidylaminopeptidase IV. *Histochem. J.* **14**: 889-895.

Feller, A. C., Heijnen, C. J., Ballieux, R. E. & Parwaresch, M. R. Enzymehistochemical staining of Tμ lymphocytes for glycyl-proline-4-methoxy-beta-naphthylamide-peptidase (DAP IV). *Br. J. Haematol.* **51**: 227-234.

Feller, A. C., Parwaresch, M. R., Bartels, H. & Lennert, K. Enzyme-cytochemical heterogeneity of human chronic T-lymphocytic leukaemia as demonstrated by reactivity to dipeptidylaminopeptidase IV (DAP IV; EC 3.4.14.4). *Leuk. Res.* **6**: 801-808.

Fujiwara, K., Katyal, S. L. & Lombardi, B. Influence of age, sex and cancer on the activities of gamma-glutamyl transpeptidase and of dipeptidyl aminopeptidase IV in rat tissues. *Enzyme* **27**: 114-118.

Hama, T., Okada, M., Kojima, K., Kato, T., Matsuyama, M. & Nagatsu, T. Purification of dipeptidyl-aminopeptidase IV from human kidney by anti dipeptidyl-aminopeptidase IV affinity chromatography. *Mol. Cell. Biochem.* **43**: 35-42.

Heymann, E. & Mentlein, R. A negative blood-clotting factor lining the vessels. *Naturwissenchaften* **69**: 189-190.

Ichinose, M., Maeda, R., Fukuoi, T., Watanabe, B., Ishimaru, T., Izumi, M., Miyake, S. & Takamori, M. Partial purification and characterization of glycylprolyl dipeptidyl aminopeptidase in porcine pancreas. *Biochim.*

Biophys. Acta **719**: 527-531.

Kasahara, Y., Fujii, N., Naka, H. & Nagatsu, T. Multiple forms of glycylprolyl dipeptidyl-aminopeptidase (dipeptidyl peptidase IV) in human sera from patients with hepatitis. *Biomed. Res.* **3**: 265-269.

Mentlein, R. & Heymann, E. Dipeptidyl peptidase IV inhibits the polymerization of fibrin monomers. *Arch. Biochem. Biophys.* **217**: 748-750.

Püschel, G., Mentlein, R. & Heymann, E. Isolation and characterization of dipeptidyl peptidase IV from human placenta. *Eur. J. Biochem.* **126**: 359-365.

Yoshimoto, T., Kita, T., Ichinose, M. & Tsuru, D. Dipeptidyl aminopeptidase IV from porcine pancreas. *J. Biochem.* **92**: 275-282.

1983

Browne, P. & O'Cuinn, G. An evaluation of the role of a pyroglutamyl peptidase, a post-proline cleaving enzyme and a post-proline dipeptidyl amino peptidase, each purified from the soluble fraction of guinea-pig brain, in the degradation of thyroliberin *in vitro*. *Eur. J. Biochem.* **137**: 75-87.

Danielsen, E. M., Sjöström, H. & Norén, O. Biosynthesis of intestinal microvillar proteins. Pulse-chase labelling studies on maltase-glucoamylase, aminopeptidase A and dipeptidyl peptidase IV. *Biochem. J.* **210**: 389-393.

Erickson, R. H. & Kim, Y. S. Interaction of purified brush-border membrane aminopeptidase N and dipeptidyl peptidase IV with lectin-Sepharose derivatives. *Biochim. Biophys. Acta* **743**: 37-42.

Erickson, R. H., Bella, A. M., Jr., Brophy, E. J., Kobata, A. & Kim, Y. S. Purification and molecular characterization of rat intestinal brush border membrane dipeptidyl aminopeptidase IV. *Biochim. Biophys. Acta* **756**: 258-265.

Feller, A. C. & Parwaresch, M. R. Dipeptidyl peptidase IV as a cytochemical marker in T-lymphocyte subpopulations. *Acta Histochem. Suppl.* **28**: 237 only.

Fukasawa, K. M., Fukasawa, K., Hiraoka, B. Y. & Harada, M. Characterization of a soluble form of dipeptidyl peptidase IV from pig liver. *Experientia* **39**: 1005-1007.

Hunt, J. C., Vasta, G., McDonald, J. K. & Marchalonis, J. J. Expression of plasma membrane associated dipeptidyl peptidase IV (DPP IV) activity and immunoglobulin V_H-related determinants by human lymphocytes and in vitro leukemia/lymphoma cell lines. In: *Non-HLA Antigens in Health, Aging and Malignancy* (Cohen, E. & Singal, D. P. eds), p. 191-195, Alan R. Liss, New York.

Miura, S., Song, I.-S., Morita, A., Erickson, R. H. & Kim, Y. S. Distribution and biosynthesis of aminopeptidase N and dipeptidyl aminopeptidase IV in rat small intestine. *Biochim. Biophys. Acta* **761**: 66-75.

Morita, A., Chung, Y.-C., Freeman, H. J, Erickson, R. H., Sleisenger, M.

H. & Kim, Y. S. Intestinal assimilation of a proline-containing tetrapeptide. Role of a brush border membrane postproline dipeptidyl aminopeptidase IV. *J. Clin. Invest.* **72**: 610-616.

Sahara, N., Fukasawa, K., Harada, M. & Suzuki, K. Immunohistochemical localization of dipeptidyl peptidase IV in rat digestive organs. *Acta Histochem. Cytochem.* **16**: 494-501.

Saison, M., Verlinden, J., Van Leuven, F., Cassiman, J.-J. & Van Den Berghe, H. Identification of cell surface dipeptidylpeptidase IV in human fibroblasts. *Biochem. J.* **216**: 177-183.

Sannes, P. L. Subcellular localization of dipeptidyl peptidases II and IV in rat and rabbit alveolar macrophages. *J. Histochem. Cytochem.* **31**: 684-690.

Sterry, W., Jansen, M., Haneberg, F. & Steigleder, G.-K. Acid esterase and dipeptidyl peptidase IV as markers of immunological lymphocyte populations. *Acta Histochem. Suppl.* **28**: 181-190.

Tsunasawa, S., Imanaka, T. & Nakazawa, T. Apparent dipeptidyl peptidase activities of acylamino acid-releasing enzymes. *J. Biochem.* **93**: 1217-1220.

Umezawa, H. & Aoyagi, T. Trends in research of low molecular weight protease inhibitors of microbial origin. In: *Proteinase Inhibitors: Medical and Biological Aspects* (Katunuma, N., Umezawa, H. & Holzer, H. eds), pp. 3-15, Springer-Verlag, New York.

Yoshimoto, T. & Tsuru, D. Substrate specificity of aminopeptidase M: evidence that the commercial preparation is contaminated by dipeptidyl aminopeptidase IV and prolidase. *J. Biochem.* **94**: 619-622.

1984

Chikuma, T., Kato, T., Hiramatsu, M., Kanayama, S. & Kumegawa, M. Effect of epidermal growth factor on dipeptidyl-aminopeptidase and collagenase-like peptidase activities in cloned osteoblastic cells. *J. Biochem.* **95**: 283-286.

Chilosi, M., Pizzolo, G., Semenzato, G., de Rossi, G. & Pandolfi, F. Heterogeneous expression of dipeptidyl-amino-peptidase (DAP IV) in T-cell chronic lymphocytic leukemia. *Acta Haematol.* **71**: 277-281.

Crockard, A. D. Cytochemistry of lymphoid cells: a review of findings in the normal and leukaemic state. *Histochem. J.* **16**: 1027-1050.

Crockard, A. D., MacFarlane, E., Andrews, C., Bridges, J. M. & Catovsky, D. Dipeptidylaminopeptidase IV activity in normal and leukemic T-cell subpopulations. *Am. J. Clin. Pathol.* **82**: 294-299.

Heins, J., Neubert, K., Barth, A., Canizaro, P. C. & Behal, F. J. Kinetic investigation of the hydrolysis of aminoacyl p-nitroanilides by dipeptidyl peptidase IV from human and pig kidney. *Biochim. Biophys. Acta* **785**: 30-35.

Hiraoka, B. Y., Fukasawa, K. M. & Fukasawa, K. Chemical modication of dipeptidyl peptidase IV: involvement of an essential tryptophan residue at the substrate binding site. *Arch. Biochem. Biophys.* **234**: 622-628.

Imai, K. & Kato, T. Dipeptidyl aminopeptidase in neonatal rat brain regions. *Experientia* **40**: 263-264.

Kasahara, Y., Leroux-Roels, G., Nakamura, N. & Chisari, F. Glycylprolyl-diaminopeptidase in human leucocytes: selective occurrence in T lymphocytes and influence on the total serum enzyme ativity. *Clin. Chim. Acta* **139**: 295-302.

Kreisel, W., Reutter, W. & Gerok, W. Modification of the intramolecular turnover of terminal carbohydrates of dipeptidylaminopeptidase IV isolated from rat-liver plasma membrane during liver regeneration. *Eur. J. Biochem.* **138**: 435-438.

Mentlein, R., Heymann, E., Scholz, W., Feller, A. C. & Flad, H.-D. Dipeptidyl peptidase IV as a new surface marker for a subpopulation of human T-lymphocytes. *Cell. Immunol.* **89**: 11-19.

Sahara, N. & Suzuki, K. Ultrastructural localization of dipeptidyl peptidase IV in rat salivary glands by immunocytochemistry. *Cell Tissue Res.* **235**: 427-432.

Schön, E., Demuth, H. U., Barth, A. & Ansorge, S. Dipeptidyl peptidase IV of human lymphocytes. Evidence for specific hydrolysis of glycylproline *p*-nitroanilide in T-lymphocytes. *Biochem. J.* **223**: 225-25.

Umezawa, H., Aoyagi, T., Ogawa, K., Naganawa, H., Hamada, M. & Takeuchi, T. Diprotins A and B, inhibitors of dipeptidyl aminopeptidase IV, produced by bacteria. *J. Antibiot.* **37**: 422-425.

Ward, P. E. Immunoelectrophoretic analysis of vascular, membrane-bound angiotensin I converting enzyme, aminopeptidase M, and dipeptidyl(amino)peptidase IV. *Biochem. Pharmacol.* **33**: 3183-3193.

1985

Gonschor, H. & Schäfer, W. Dipeptidyl-Peptidase IV. Reinigung für die Verwendung zur Peptidsequenzierung. *Biol. Chem. Hoppe-Seyler* **366**: 157-165.

Section 14
TRIPEPTIDYL PEPTIDASES

Tripeptidyl Peptidase I

Summary

EC Number: None, but name assigned by analogy with IUB recommendations for naming of dipeptidyl peptidases.

Earlier names: Tripeptidyl aminopeptidase.

Abbreviation: TPP I.

Distribution: Has a lysosomal localization, and is widely distributed in tissues. In the rat, highest levels are seen in spleen, liver, kidney, anterior pituitary, and thyroid. In the hog, highest levels occur in kidney followed by liver, spleen, ovary (pregnant sow), adrenal, thyroid, lung, uterus and pancreas.

Source: Beef anterior pituitary glands [1], pregnant hog ovaries [2], and hog spleen [4].

Action: Catalyzes the hydrolysis of arylamide bonds in unsubstituted tripeptidyl arylamides, and releases tripeptide fragments sequentially from the unsubstituted N-termini of oligopeptides and proteins.

Beef pituitary TPP I releases Phe-Pro-Ala from its -NNap derivative (K_m 0.84 mM at pH 4, 37°C) and from the N-terminus of beef growth hormone (phenylalanyl monomer; M_r 21,000). Its total action on the growth hormone monomer includes the sequential release of ten additional tripeptides; Met-Ser-Leu, Ser-Gly-Leu, Phe-Ala-Asn, Ala-Val-Leu, Arg-Ala-Gln, His-Leu-His, Gln-Leu-Ala, Ala-Asp-Thr, Phe-Lys-Glu, and Phe-Glu-Arg (a total of 33 residues). No action occurs on the alanyl monomer of beef growth hormone, wherein Ala-Phe-Pro is the N-terminal tripeptide [1], but tripeptides are released from β-corticotropin, glucagon, somatostatin, and angiotensins II and III [3]. K_m values (pH 4 and 37°C) for Phe-Pro-Ala-NNap [1] and Ala-Ala-Phe-NPhNO$_2$ are 0.84 mM [1], and 0.09 mM [3] respectively.

The enzyme purified from hog ovaries catalyzes the release of collagen-like (Gly-Pro-X) triplets from model substrates such as Gly-Pro-Ala-NPhNO$_2$ and (Gly-Pro-Ala)$_n$ (average M_r 10,000) at pH 3-5, but has no action on the -NPhNO$_2$ derivatives of Boc-Gly-Pro-Ala-, Pro-Ala- or Ala-. Relative rates of tripeptide release for other nitroanilides are as follows: Gly-Pro-Met- 100, Gly-Pro-Ala- 33, and Val-Pro-Arg- 22. Gly-Pro-Met-NNap (a fluorogenic model for the N-terminus of the helical regions of both α_1 and α_2 chains of Type I collagen) exhibits a K_m of 0.43 mM at pH 4.0 and 37°C [2]. Gly-Pro-Met-NMec (K_m 0.17 mM at pH 5.0 and 37°C), a fluorogenic substrate offering higher sensitivity, is also readily hydrolyzed, whereas little or no activity is seen on the -NMec derivatives of Suc(MeO)-Gly-Pro-Met-, Pro-Met-, and Met- [4].

Requirements: pH 4 and 37°C in the presence of 0.2% Triton X-100 for beef pituitary TPP I [3], or pH 4.5 and 37°C in the presence of 0.05% Brij 35 and 1 mM EDTA for hog ovary TPP I [4].

Substrate, usual: Ala-Ala-Phe-NPhNO$_2$ [1,3] or Gly-Pro-Met-NNap [2].

Substrate, special: Gly-Pro-Met-NMec [4].

Inhibitors: Dip-F (1 mM) and Gly-Pro-Met-CH$_2$Cl (10 μM) are potent inhibitors of the hog ovary enzyme [2]. Although the beef pituitary enzyme is strongly inhibited by Hg^{2+} (10 μM) and p-chloromercuri-phenylsulfonate (1 mM), it is not considered to belong to the cysteine catalytic class because inhibition is not obtained with MalNEt [1]. The hog enzyme reacts similarly to these thiol-blocking reagents [2,4]. TPP I from both sources is unaffected by pepstatin and leupeptin.

Molecular properties: Beef pituitary TPP I has M_r about 57,000, as judged by chromatography on Sephadex G-75 equilibrated with 0.1 M acetic acid. Totally inactivated by freeze-drying [1], and loses activity rapidly above pH 5.0 [3]. Stable for months in 0.1 M acetic acid at 5°C [1].

   Hog ovary TPP I behaves as an aggregate ($M_r > 250,000$) during chromatography on Ultrogel AcA 34 in 0.1 M NaCl-0.1 M sodium acetate buffer, pH 4.5, but dissociates to an active monomer (M_r 55,000) when urea is incorporated at 3 M [2,4]. The monomer is comprised of two subunits of unequal size [5]. High affinity for concanavalin A-Sepharose indicates that it is a glycoprotein. Shows good stability to freeze-drying and storage at pH 3-6 [5].

References
   [1] Doebber *et al. Endocrinology* **103**: 1794-1804, 1978.
   [2] McDonald & Hoisington *Fed. Proc. Fed. Am. Soc. Exp. Biol.* **42**: 1781 only, 1983.
   [3] Ellis & Divor *Fed. Proc. Fed. Am. Soc. Exp. Biol.* **38**: 832 only, 1979.
   [4] McDonald *et al. Biochem. Biophys. Res. Commun.* **126**: 63-71, 1985.
   [5] McDonald & Hoisington *Unpublished observations.*

Bibliography

1978

Doebber, T. W., Divor, A. R. & Ellis, S. Identification of a tripeptidyl aminopeptidase in the anterior pituitary gland: effect on the chemical and biological properties of rat and bovine growth hormone. *Endocrinology* **103**: 1794-1804.

1979

Ellis, S. & Divor, A. R. Specificity and lysosomal localization of tripeptidyl aminopeptidase. *Fed. Proc. Fed. Am. Soc. Exp. Biol.* **38**: 832 only.

1983

McDonald, J. K. & Hoisington, A. R. Identification of an ovarian tripeptidyl peptidase: a tripeptide-releasing aminopeptidase active on collagen-like sequences. Fed. Proc. Fed. Am. Soc. Exp. Biol. **43**: 1781 only.

1985

McDonald, J. K., Hoisington, A. R. & Eisenhauer, D. A. Partial purification and characterization of an ovarian tripeptidyl peptidase: a lysosomal exopeptidase that sequentially releases collagen-related (Gly-Pro-X) triplets. *Biochem. Biophys. Res. Commun.* **126**: 63-71.

Tripeptidyl Peptidase II

Summary

EC Number: None

Earlier names: Tripeptidyl aminopeptidase.

Abbreviation: TPP II.

Distribution: Studies have thus far been limited to rat liver, where the enzyme was recently discovered in the extralysosomal compartment.

Source: Rat liver microsomes [1].

Action: Catalyzes the sequential release of tripeptides from oligopeptides possessing unsubstituted N-termini. The specificity of the partially-purified enzyme has been characterized on a series of ^{32}P-labelled phosphopeptides. Two tripeptides are readily released, in sequence, from the nonapeptide Gly-Val-Leu-Arg-Arg-Ala-Ser(P)-Val-Ala, but a lack of action on the des-Gly peptide suggests that the Arg-Arg linkage may be somewhat resistant. Leu-Arg-Arg, however, is slowly released from the des-Gly-Val peptide. Whereas Arg-Arg-Ala is easily removed from the hexapeptide Arg-Arg-Ala-Ser(P)-Val-Ala, no action occurs if the free N-terminal α-amino group is eliminated by replacing the N-terminal arginyl residue with a guanidovaleryl residue. And, in the case of a pentapeptide substrate, Lys-Arg-Ala-Ser-Val, activity is lost if the lysyl residue is replaced by 6-aminohexanoic acid. The phosphate group is not essential for activity. Compared to the rate of hydrolysis of the phosphorylated (hexapeptide) assay substrate, the approximate relative rates of tripeptide release are 174 for Arg-Arg-Ala-Ser-Val-Ala, 97 for Arg-Arg-Ala-Ser-Val and 45 for Lys-Arg-Ala-Ser-Val.

Requirements: pH 6.5 at 30°C, in a phosphate buffer containing (2 mM) dithiothreitol for enzyme stabilization (see also Comment).

Substrate, usual: Arg-Arg-Ala-Ser(^{32}P)-Val-Ala. Ser(^{32}P)-Val-Ala, the negatively-charged, ^{32}P-labeled cleavage product is adsorbed onto QAE-Sephadex at pH 4.5, and the amount of bound, labeled peptide measured in a scintillation counter. The rate is linear with time up to 25% hydrolysis [1].

Inhibitors: Val-Leu-Arg-Arg-Ala-Ser(P)-Val-Ala, the octapeptide mentioned under Action as being resistant to hydrolysis, actually exerts potent inhibition in the assay system.

Molecular properties: M_r exceeds 1,000,000, as judged by chromatography on Sepharose CL-4B. TPP II contained in liver microsomal fractions held at neutral pH shows no loss of activity after 3 days at 4°C, or after 3 months at -70°C. Activity in diluted microsomal fractions becomes unstable at 4°C, but losses can be prevented by including glycerol (30%) and dithiothreitol (1 mM) in a neutral phosphate buffer. Recoveries of 10-20% are typical following a 150-fold purification.

Comment: Although liver microsome preparations serve as an enriched source of TPP II, the ease with which the enzyme is solubilized suggests that it is most probably not membrane-bound in the cells of origin.

Even though partially purified TPP II hydrolyzes the assay substrate most rapidly at pH 7.5, pH 6.5 is used in the standard assay because when crude extracts are assayed, it helps to exclude an interfering activity that releases the N-terminal dipeptide, Arg-Arg, from the assay substrate. The responsible enzyme could very possibly be dipeptidyl peptidase III (Entry 13.03), in which case EDTA could prove to be a useful inhibitor.

Based upon the finding that the phosphorylation of rat pyruvate kinase increases the sensitivity of the kinase to proteolytic cleavage at the phosphorylated site [2], a search was undertaken for an endopeptidase with a specificity for proteins phosphorylated by cyclic AMP-dependent protein kinase. For this purpose, a ^{32}P-phosphopeptide with a sequence that mimicked the phosphorylated site on the protein kinase was used as a probe. The use of this unprotected peptide, Gly-Val-Leu-Arg-Arg-Ala-Ser(P)-Val-Ala, resulted in the detection of a protease that was capable of cleaving the Ala-Ser(P) bond [3]. Subsequent studies [1], however, revealed that the unprotected phosphopeptide was actually degraded sequentially from the N-terminus by a previously unrecognized exopeptidase.

References
[1] Bålöw *et al. J. Biol. Chem.* **258**: 11622-11628, 1983.
[2] Bergström *et al. Biochim. Biophys. Acta* **532**: 259-267, 1978.
[3] Zetterqvist & Bålöw *Biochem. Soc. Trans.* **9**: 233P only, 1981.

Bibliography

1978

Bergström, G., Ekman, P., Humble, E. & Engström, L. Proteolytic modification of pig and rat liver pyruvate kinase type L including phosphorylatable site. *Biochim. Biophys. Acta* **532**: 259-267.

1981

Zetterqvist, O. & Bålöw, R.-M.   Evidence of a peptidase with possible specificity for sites phosphorylated by cyclic AMP-dependent protein kinase. *Biochem. Soc. Trans.* **9**: 233P only.

1983

Bålöw, R.-M., Ragnarsson, U. & Zetterqvist, O.   Tripeptidyl aminopeptidase in the extralysosomal fraction of rat liver. *J. Biol. Chem.* **258**: 11622-11628.

Section 15
CARBOXYPEPTIDASES

Tissue Carboxypeptidase A

Summary

EC Number: 3.4.17.1

Earlier names: Carboxypolypeptidase, pancreatic carboxypeptidase A.

Abbreviation: tCPA.

Distribution: Highly concentrated, as procarboxypeptidase A, in zymogen granules located in the exocrine cells of the pancreas of most vertebrates. Lower concentrations of an active form of carboxypeptidase A have been demonstrated in a variety of mammalian cells, notably mast cells.

Source: Beef [1-3], hog [4], and human [5] pancreas; rat peritoneal mast cells [6]; and rat skeletal muscle [7].

Action: Catalyzes the release of C-terminal residues possessing an L configuration and an unsubstituted α-carboxyl group. Most active on hydrophobic (aromatic and branched-chain aliphatic) C-terminal residues. C-Terminal Thr, Gln, His, and Ala are also released at appreciable rates, but Asn, Ser and Lys are released slowly, Gly, Asp, Glu, Cys, and $CySO_3H$ very slowly, and Arg, Pro and Hyp apparently not at all. Unfavorable residues located in a penultimate position may retard the release of terminal hydrophobic residues. Dipeptides are hydrolyzed slowly unless the N-terminus is acylated. K_m values for longer polypeptides are affected by the nature of the side chains on at least the last (C-terminal) 5 residues.

Peptides generated from proteins by cyanogen bromide cleavage of methionyl bonds contain C-terminal homoserine lactone. Digestion with tCPA under alkaline conditions (pH 8.5) results in a partial opening of the lactone ring and subsequent release of homoserine [8].

tCPA also shows esterase activity when the C-terminal residue of the substrate has a hydrophobic side chain and a free carboxyl group, as typified by the substitution of β-phenyllactic acid for phenylalanine. The K_m values for such esters are characteristically 20- to 110-fold lower than those for the peptides [9].

Requirements: pH 7.3-7.5 at 37°C, usually in a Tris-HCl buffer. Ionic strengths of 0.4 or higher are required for maximal rates of hydrolysis of assay substrates. Acidic residues are released more rapidly at pH 5.5-6.0, whereas the release of histidine occurs more rapidly under alkaline conditions.

Substrate, usual: Z-Gly-Phe [3,10], Bz-Gly-Phe [11], and ester analogs such as Bz-Gly-L-β-phenyllactate [3,12].

Rates can be measured colorimetrically with ninhydrin [13] or fluorometrically with o-phthalaldehyde [14] for the determination of liberated phenylalanine, or spectrophotometrically by following A_{224} decrease in the case of Z-Gly-Phe [3,9], or by following A_{254} increase in the case of the Bz-Gly- (hippuryl-) derivatives [4,15].

Substrate, special: N-(2-Furanacryloyl)-Phe-Phe (FAPP) with decrease in absorbance followed at 330 nm [16].

Inhibitors: Metal binding agents such as 1,10-phenanthroline, cysteine, pyrophosphate, citrate, and oxalate inhibit reversibly, as does a newly-developed phosphoramidate, N-phosphorylphenylalanine [17].

α-Hydroxy acids such as L-β-phenyllactic acid and L-mandelic acid, split from the C-termini of ester substrates, also inhibit competitively.

Organic acids that contain a free carboxyl group and an aromatic or heterocyclic function, such as benzylsuccinic acid, 3-phenylpropionic acid (hydrocinnamic acid), and indole acetic acid, are the most effective reversible inhibitors, competitive against a variety of substrates. Benzylsuccinic acid, in addition to being a potent inhibitor of tCPA (K_i 1 μM) is also an effective inhibitor of tissue (pancreatic) carboxypeptidase B. On the other hand, a recently developed inhibitor [18], 2-benzyl-3-mercaptopropanoic acid, is a potent specific inhibitor of only tCPA (K_i 11 nM).

Certain organic acids, e.g. *trans-* or *cis-*cinnamic acid, inhibit the hydrolysis of ester substrates, such as furanacryloylphenyl- lactate, but not peptide substrates, such as Z-Gly-Trp. Such findings have suggested the existence of distinct binding sites for each class of substrate.

Protein inhibitors of special interest have been isolated from potato [19], and *Ascaris lumbricoides* [20]. The former is a polypeptide of 39 residues (M_r 4200). tCPA has been shown to bind non-specifically to plasma α_2-macroglobulin [21].

Molecular properties: tCPA is a Zn^{2+}-metalloprotein comprised of a single polypeptide chain of 307 amino acid residues (α form) and one zinc atom. Alanine and asparagine are the N-terminal and C-terminal residues, respectively. tCPA has a high content of sulfur-containing amino acids. The two cysteine residues, at positions 138 and 161, form a disulfide bond. It was once believed that the "active site" cysteinyl residue at position 161 served as the zinc ligand. The complete amino acid sequence [22] and atomic resolution crystal structure [23] have been established.

The molecule is nearly spherical, having dimensions of 50 x 42 x 38 Å.

tCPA of rat peritoneal mast cells has an M_r of 35,000 [6], and the two components of muscle tCPA have M_r values of approximately 37,800 and 39,300 [7].

Two allelomorphs have been identified. One has Ile, Ala and Val at positions 179, 228 and 305; the other has Val, Glu and Leu at these positions. The former is referred to as $tCPA^{Val}$, and the latter as $tCPA^{Leu}$. $tCPA^{Val}$ has an M_r of 34,489, no carbohydrate, pI 6.0, and $A_{278,1\%}$ 19.4.

Four forms (α, β, γ, and δ) of tCPA arise during purification as a consequence of proteolytic cleavage in the N-terminal region: ALA-Arg-SER-Thr-Asn-Thr-Phe-ASN-. $tCPA_\alpha$ or tCPA (Cox) possesses the complete (307 residue) sequence. Serine is N-terminal in $tCPA_\beta$ (305 residues), whereas asparagine is N-terminal in $tCPA_\gamma$, or tCPA (Anson) and in $tCPA_\delta$ or tCPA (Allan). The latter two forms, $tCPA_\gamma$ and $tCPA_\delta$, both contain 300 residues, but the δ form (prepared from an acetone powder) has an 8-fold greater solubility in 1 M NaCl, approaching 23 mg per ml [24]. All forms of tCPA are insoluble in water, but can be dissolved in cold 1 M NaCl buffered at pH 7.5.

The (commercially available) tCPA used in most published biochemical studies has been the γ form, or CPA (Anson). Crystallographic studies, however, have been conducted with the α-form or tCPA (Cox), because large crystals are more easily produced [25]. tCPA (Allan) is the form of the enzyme least characterized chemically.

The α, β and γ forms of tCPA are stable over at least the range pH 7.0-10.0 at 0°C in a variety of buffers containing 1 M NaCl. tCPA (Allan) is stable between pH 4.6 and 11.3 at 0°C for 62 h in buffered 1 M NaCl. tCPA (Anson) is stable in 4 M urea at 0°C, pH 7.5, for 24 h. Inactivation occurs at concentrations above 5 M, but is completely reversible up to at least 7 M urea.

The catalytic properties of tCPA can be altered by substituting other metals for Zn^{2+}. Whereas the Mn^{2+}, Co^{2+} and Ni^{2+} forms of the enzyme hydrolyze both peptides and esters, the Hg^{2+}, Cd^{2+} and Pb^{2+} forms hydrolyze only esters. Cu^{2+}-containing tCPA is totally inactive.

Comment: The term "tissue carboxypeptidase A" has been introduced for the purpose of accommodating within this entry both the carboxypeptidase A of pancreatic origin and that found in other mammalian cells and tissues such as rat peritoneal mast cells [6] and rat muscle [7]. The information included in this entry, however, except where indicated otherwise, is based on studies of the enzyme from beef pancreas.

Mast cell tCPA is located in secretory granules together with chymase I [6], the chymotrypsin-like activity of mast cells [26], thus allowing for their synergistic action on protein substrates. Various lines of evidence have been cited [7] that suggest that muscle tCPA, like chymase, is derived from the dense granules of mast cells rather than from myocytes. In neither case, however, have zymogen precursors been demonstrable. Because mast cell granules tend to contaminate preparations of nuclei, mitochondria and myofibrils, caution should be exercised when attributing enzyme activities to particular cells or organelles.

Numerous chemical and enzymatic properties of the mast cell, muscle and pancreatic enzymes have been compared. Similarities include kinetic parameters (K_m and k_{cat}) for several substrates; substrate specificity on esters, peptides and proteins; pH optima for peptide and ester hydrolysis; enhancement of esterolytic activity by imidazoles and indoles; effects of pH on activity and stability; influence of metal ions and metal-binding agents; and interaction with a range of inhibitors, including the protein inhibitor from potato. Molecular masses and amino acid compositions also show distinct similarities, in particular for the pancreatic and mast cell enzymes.

Some differences have been noted, however, between mast cell and pancreatic tCPA. For example, the influence of ionic strength on esterase activity is different for the muscle enzyme: whereas the optimum for the pancreatic enzyme is 0.3 M NaCl, muscle tCPA shows an optimum at 1.5 M NaCl. The muscle enzyme is also reported to be more heat sensitive. Mast cell tCPA also possesses at least one free sulfhydryl function.

A "novel SH-type" carboxypeptidase was reported to be located in rat liver mitochondria [27], but later studies provided evidence for a mast cell origin [28]. Mitochondrial fractions were apparently contaminated with the dense granules of mast cells.

A carboxypeptidase A-like activity sensitive to thiol-blocking reagents has also been described in cultured human fibroblasts [29]. Justification for its classification under EC 3.4.17.1 [30] seems unclear since the enzyme acts optimally at about pH 5.0 and may have a lysosomal localization [29].

Beef pancreatic tCPA is stored in zymogen granules as a two-subunit (S5) complex (M_r 72,500) and a three-subunit (S6) complex (M_r 96,000). Both complexes are held together noncovalently, and contain the precursors of two proteases - an exopeptidase and an endopeptidase. Activation involves the conversion by trypsin (in the duodenum) of subunit II to a chymotrypsin-like endopeptidase (M_r about 26,000), which then acts in concert with trypsin to convert procarboxypeptidase contained in subunit I to the active enzyme. The peptide released upon activation contains approximately 60 residues. Subunit III of the trimeric complex is believed to be an inactive form of subunit II. As already explained under "Molecular properties" at least three forms of active tCPA, differing in the N-terminal region, can be generated from the zymogen.

References
[1]   Anson *Gen. Physiol.* **30**: 663-669, 1937.
[2]   Cox *et al. Biochemistry* **3**: 44-47, 1964.
[3]   Petra *Methods Enzymol.* **19**: 460-503, 1970.
[4]   Folk & Schirmer *J. Biol. Chem.* **238**: 3884-3894, 1963.

[5]   Peterson et al. Biochemistry **15**: 2501-2508, 1976.
[6]   Everitt & Neurath FEBS Lett. **110**: 292-296, 1980.
[7]   Bodwell & Meyer Biochemistry **20**: 2767-2777, 1981.
[8]   Ambler Methods Enzymol. **11**: 436-445, 1967.
[9]   Whitaker et al. Biochemistry **5**: 386-392, 1966.
[10] Hofmann & Bergmann. J. Biol. Chem. **134**: 225-235, 1940.
[11] Snoke & Neurath. J. Biol. Chem. **181**: 789-802, 1949.
[12] Schwert & Takenaka. Biochim. Biophys. Acta **16**: 570-575, 1955.
[13] Moore & Stein J. Biol. Chem. **176**: 367-388, 1948.
[14] Roth Anal. Chem. **43**: 880-882, 1971.
[15] McClure et al. Biochemistry **3**: 1897-1901, 1964.
[16] Peterson et al. Anal. Biochem. **125**: 420-426, 1982.
[17] Kam et al. Biochemistry **18**: 3032-3038, 1979.
[18] Ondetti et al. Biochemistry **18**: 1427-1430, 1979.
[19] Ryan et al. J. Biol. Chem. **249**: 5495-5499, 1974.
[20] Homandberg & Peanasky J. Biol. Chem. **251**: 2226-2233, 1976.
[21] Barrett & Starkey Biochem. J. **133**: 709-724, 1973.
[22] Bradshaw et al. Proc. Natl. Acad. Sci. USA **63**: 1389-1394, 1969.
[23] Lipscomb et al. Phil. Trans. R. Soc. London, Ser. B **257**:177-214, 1970.
[24] Allen et al. Biochemistry **3**: 40-43, 1964.
[25] Lipscomb et al. J. Mol. Biol. **19**: 423-441, 1966.
[26] Barrett & McDonald Vol. 1 of this work.
[27] Haas & Heinrich Eur. J. Biochem. **96**: 9-15, 1979.
[28] Haas et al. FEBS Lett. **103**: 168-171, 1979.
[29] Guy & Butterworth Clin. Chim. Acta **87**: 63-69, 1978.
[30] Butterworth & Duncan Clin. Chim. Acta **108**: 143-146, 1980.

Bibliography

1929

Waldschmidth-Leitz, E. & Purr, A. Über Proteinase und Carboxy-Polypeptidase aus Pankreas. (XVII. Mitteilung zur Spezifität tierischer Proteasen.) Berichte **62B**: 2217-2226.

1931

Waldschmidt-Leitz, E. The mode of action and differentation of proteolytic enzymes. Physiol. Rev. **11**: 358-370.

1935

Anson, M. L. Crystalline carboxypolypeptidase. Science **81**: 467-468.

1937

Anson, M. L. Carboxypeptidase. I. The preparation of crystalline carboxypeptidase. *J. Gen. Physiol.* **20**: 663-669.

Anson, M. L. Carboxypeptidase. II. The partial purification of pro-carboxypeptidase. *J. Gen. Physiol.* **20**: 777-780.

Anson, M. L. Carboxypeptidase. III. The estimation of carboxypeptidase and pro-carboxypeptidase. *J. Gen. Physiol.* **20**: 781-786.

Bergmann, M. & Fruton, J. S. On proteolytic enzymes. XII. Regarding the specificity of aminopeptidase and carboxypeptidase. A new type of enzyme in the intestinal tract. *J. Biol. Chem.* **117**: 189-202.

1940

Hofmann, K. & Bergmann, M. The specificity of carboxypetidase. *J. Biol. Chem.* **134**: 225-235.

1942

Bergmann, M. & Fruton, J. S. The influence of substrate structure on the kinetics of carboxypeptidase action. *J. Biol. Chem.* **145**: 247-252.

Johnson, M. J. & Berger, J. The enzymatic properties of peptidases. *Adv. Enzymol.* **2**: 69-92.

1946

Putnam, F. W. & Neurath, H. Chemical and enzymatic properties of crystalline carboxypeptidase. *J. Biol. Chem.* **166**: 603-619.

Stahmann, M. A., Fruton, J. S. & Bergmann, M. The specificity of carboxypeptidase. *J. Biol. Chem.* **164**: 753-760.

1947

Neurath, H., Elkins, E. & Kaufman, S. The antipodal specificity and inhibition of crystalline carboxypeptidase. *J. Biol. Chem.* **170**: 221-226.

1948

Elkins-Kaufman, E. & Neurath, H. Kinetics and inhibition of carboxypeptidase activity. *J. Biol. Chem.* **175**: 893-911.

Hanson, H. T. & Smith, E. L. The application of peptides containing β-alanine to the study of the specificity of various peptidases. *J. Biol. Chem.* **175**: 833-848.

Smith, E. L. Action of carboxypetidase on peptide derivatives of L-trypthophan. *J. Biol. Chem.* **175**: 39-47.

Smith, E. L. & Hanson, H. T. Pancreatic carboxypeptidase; a metal protein. *J. Biol. Chem.* **176**: 997-998.

Snoke, J. E., Schwert, G. W. & Neurath, H. The specific esterase activity of carboxypeptidase. *J. Biol. Chem.* **175**: 7-13.

1949

Dekker, C. A., Taylor, S. P., Jr. & Fruton, J. S. Synthesis of peptides of methionine and their cleavage by proteolytic enzymes. *J. Biol. Chem.* **180**: 155-172.

Elkins-Kaufman, E. & Neurath, H.   Structural requirements for specific inhibitors of carboxypeptidase. *J. Biol. Chem.* **178**: 645-654.

Hanson, H. T. & Smith, E. L.   Papain resolution of DL-tryptophan; optical specificity of carboxypeptidase. *J. Biol. Chem.* **179**: 815-818.

Smith, E. L. & Hanson, H. T.   The chemical nature and mode of action of pancreatic carboxypeptidase. *J. Biol. Chem.* **179**: 803-813.

Smith, E. L., Brown, D. M. & Hanson, H. T.   The sedimentation behaviour and molecular weight of pancreatic carboxypeptidase. *J. Biol. Chem.* **180**: 33-36.

Snoke, J. E. & Neurath, H.   Structural requirements of specific substrates for carboxypeptidase. *J. Biol. Chem.* **181**: 789-802.

1950

Dunn, F. W. & Smith, E. L.   Action of carboxypeptidase on derivatives of unnatural amino acids. *J. Biol. Chem.* **187**: 385-392.

Neurath, H. & De Maria, G.   The effect of anions on the activity of carboxypeptidase. *J. Biol. Chem.* **186**: 653-665.

Neurath, H. & Schwert, G. W.   The mode of action of the crystallne pancreatic proteolytic enzymes. *Chem. Rev.* **46**: 69-153.

1951

Lumry, R., Smith, E. L. & Glantz, R. R.   Kinetics of carboxypeptidase action. I. Effect of various extrinsic factors on kinetic parameters. *J. Am. Chem. Soc.* **73**: 4330-4340.

Ravin, H. A. & Seligman, A. M.   The colorimetric estimation of carboxypeptidase activity. *J. Biol. Chem.* **190**: 391-402.

Smith, E. L.   The specificity of certain peptidases. *Adv. Enzymol.* **12**: 191-257 (see pp. 225-251).

Wolf, G. & Seligman, A. M.   Synthesis of naphthyl carbonate derivatives of amino acids as chromogenic substrates for carboxypeptidase. *J. Am. Chem. Soc.* **73**: 2080-2082.

1953

Thompson, E. O. P.   The N-terminal sequence of carboxypeptidase. *Biochim. Biophys. Acta* **10**: 633-634.

1954

Gorini, L. & Labouesse-Mercouroff, J.   Factors influencing the enzymic activity of carboxypeptidase. *Biochim. Biophys. Acta* **13**: 291-293.

Smith, E. L. & Stockell, A.   Amino acid composition of crystalline carboxypeptidase. *J. Biol. Chem.* **207**: 501-514.

Vallee, B. L. & Neurath, H.   Carboxypeptidase, a zinc metalloprotein. *J. Am. Chem. Soc.* **76**: 5006-5007.

1955

Neurath. H.   Carboxypeptidase and procarboxypeptidase. *Methods Enzymol.*

2: 77-83.

Vallee, B. L. & Neurath, H. Carboxypeptidase, a zinc metalloenzyme. *J. Biol. Chem.* **217**: 253-261.

1956

Keller, P. J., Cohen, E. & Neurath, H. Purification and properties of procarboxypeptidase. *J. Biol. Chem.* **223**: 457-467.

1957

Yanari, S. & Mitz, M. A. The mode of action of pancreatic carboxypeptidase. I. Optical and structural specificity. *J. Biol. Chem.* **79**: 1150-1153.

Yanari, S. & Mitz, M. A. The mode of action of pancreatic carboxypeptidase. II. The affinity of carboxypeptidase for substrates and inhibitors. *J. Am. Chem. Soc.* **79**: 1154-1158.

1958

Keller, P. J., Cohen, E. & Neurath, H. Procarboxypeptidase. II. Chromatographic isolation, further characterization, and activation. *J. Biol. Chem.* **230**: 905-915.

Vallee, B. L., Rupley, J. A., Coombs, T. L. & Neurath, H. The release of zinc from carboxypeptidase and its replacement. *J. Am. Chem. Soc.* **80**: 4750-4751.

1959

Davie, E. W., Newman, C. R. & Wilcox, P. E. Action of carboxypeptidase on β-lactoglobulin. *J. Biol. Chem.* **234**: 2635-2641.

1960

Coleman, J. E. & Vallee, B. L. Metallocarboxypeptidases. *J. Biol. Chem.* **235**: 390-395.

Coleman, J. E., Allan, B. J. & Vallee, B. L. Protein spherulites. *Science* **131**: 350-352.

Neurath. H. Carboxypeptidases A and B. In: *The Enzymes* (Boyer, P. D., Lardy, H. & Myrbäck, K. eds), 2nd edn., vol. 4, part A, pp. 11-36 (see pp. 11-34), Academic Press, New York.

Rupley, J. A. & Neurath, H. The physical properties of metal-free carboxypeptidase. *J. Biol. Chem.* **235**: 609-615.

Vallee, B. L., Coombs, T. L. & Hoch, F. L. The "active site" of bovine pancreatic carboxypeptidase A. *J. Biol. Chem.* **235**: PC45-PC47.

Vallee, B. L., Rupley, J. A., Coombs, T. L. & Neurath, H. The role of zinc in carboxypeptidase. *J. Biol. Chem.* **235**: 64-69.

1961

Brown, J. R., Cox, D. J., Greenshields, R. N., Walsh, K. A., Yamasaki, M. & Neurath, H. The chemical structure and enzymatic functions of bovine procarboxypeptidase A. *Proc. Natl. Acad. Sci. USA* **47**: 1554-1560.

Coleman, J. E. & Vallee, B. L.   Metallocarboxypeptidases: stability constants and enzymatic characteristics. *J. Biol. Chem.* **236**: 2244-2249.

Vallee, B. L., Williams, R. J. P. & Coleman, J. E.   Nitrogen and sulphur at the active centre of carboxypeptidase A. *Nature* **190**: 633-634.

1962

Coleman, J. E. & Vallee, B. L.   Apocarboxypeptidase-substrate complexes. *J. Biol. Chem.* **237**: 3430-3436.

Coleman, J. E. & Vallee, B. L.   Metallocarboxypeptidase-substrate complexes. *Biochemistry* **1**: 1083-1092.

Coombs, T. L., Felber, J.-P. & Vallee, B. L.   Metallocarboxypeptidases: mechanism of inhibition by chelating agents, mercaptans, and metal ions. *Biochemistry* **1**: 899-905.

Felber, J.-P., Coombs, T. L. & Vallee, B. L.   The mechanism of inhibition of carboxypeptidase A by 1,10-phenanthroline. *Biochemistry* **1**: 231-238.

Walsh, K. A., Sampath Kumar, K. S. V., Bargetzi, J.-P. & Neurath, H.   Approaches to the selective chemical labeling of the active site of carboxypeptidase A. *Proc. Natl. Acad. Sci. USA* **48**: 1443-1449.

1963

Bargetzi, J.-P., Sampath Kumar, K. S. V., Cox, D. J., Walsh, K. A. & Neurath, H.   The amino acid composition of bovine pancreatic carboxypeptidase A. *Biochemistry* **2**: 1468-1474.

Brown, J. R., Greenshields, R. N., Yamasaki, M. & Neurath, H.   The subunit structure of bovine procarboxypeptidase A-S6. Chemical properties and enzymatic activities of the products of molecular disaggregation. *Biochemistry* **2**: 867-876.

Brown, J. R., Yamasaki, M. & Neurath, H.   A new form of bovine pancreatic procarboxypeptidase A. *Biochemistry* **2**: 877-886.

Folk, J. E.   The porcine pancreatic carboxypeptidase A system. II. Mechanism of the conversion of carboxypeptidase A_1 to carboxypeptidase A_2. *J. Biol. Chem.* **238**: 3895-3898.

Folk, J. E. & Schirmer, E. W.   The porcine pancreatic carboxypeptidase A system. I. Three forms of the active enzyme. *J. Biol. Chem.* **238**: 3884-3894.

Riordan, J. F., Vallee, B. L. & Saunders, D. M.   Acetylcarboxypeptidase. *Biochemistry* **2**: 1460-1467.

Simpson, R. T., Riordan, J. F. & Vallee, B. L.   Functional tyrosyl residues in the active center of bovine pancreatic carboxypeptidase A. *Biochemistry* **2**: 616-622.

Vallee, B. L., Riordan, J. F. & Coleman, J. E.   Carboxypeptidase A: approaches to the chemical nature of the active center and the mechanisms of action. *Proc. Natl. Acad. Sci. USA* **49**: 109-116.

Yamasaki, M., Brown, J. R., Cox, D. J., Greenshields, R. N., Wade, R. D. & Neurath, H.   Procarboxypeptidase A-S6. Further studies of its isolation and properties. *Biochemistry* 2: 859-866.

1964

Allan, B. J., Keller, P. J. & Neurath, H. Procedures for the isolation of crystalline bovine pancreatic carboxypeptidase A. I. Isolation from acetone powders of pancreas glands. *Biochemistry* 3: 40-43.

Bargetzi, J.-P., Thompson, E. O. P., Sampath Kumar, K. S. V., Walsh, K. A. & Neurath, H. The amino- and carboxyl-terminal residues and the self-digestion of bovine pancreatic carboxypeptidase A. *J. Biol. Chem.* 239: 3767-3774.

Coleman, J. E. & Vallee, B. L. Metallocarboxypeptidase-inhibitor complexes. *Biochemistry* 3: 1874-1879.

Coombs, T. L., Omote, Y. & Vallee, B. L. The zinc-binding groups of carboxypeptidase A. *Biochemistry* 3: 653-662.

Cox, D. J., Bovard, F. C., Bargetzi, J.-P., Walsh, K. A. & Neurath, H. Procedures for the isolation of crystalline bovine pancreatic carboxypeptidase A. II. Isolation of carboxypeptidase A$_\alpha$ from procarboxypeptidase A. *Biochemistry* 3: 44-47.

Kaiser, E. T. & Carson, F. W. Studies on the esterase action of carboxypeptidase A. Kinetics of the hydrolysis of acetyl-L-mandelate. *J. Am. Chem. Soc.* 86: 2922-2926.

McClure, W. O., Neurath, H. & Walsh, K. A. The reaction of carboxypeptidase A with hippuryl-DL-β-phenyllactate. *Biochemistry* 3: 1897-1901.

Riordan, J. F. & Vallee, B. L. Succinylcarboxypeptidase. *Biochemistry* 3: 1768-1774.

Sampath Kumar, K. S. V., Clegg, J. B. & Walsh, K. A. The N-terminal sequence of bovine carboxypeptidase A and its relation to zymogen activation. *Biochemistry* 3: 1728-1732.

1965

Bender, M. L., Whitaker, J. R. & Menger, F. The effect of enzyme acetylation on the kinetics of the carboxypeptidase-A-catalyzed hydrolysis of hippuryl-L-β-phenyllactic acid. *Proc. Natl. Acad. Sci. USA* 53: 711-716.

Bethune, J. L. Induction of polymer formation in solutions of bovine pancreas carboxypeptidase A by aromatic compounds. *Biochemistry* 4: 2698-2704.

Bethune, J. L. The polymerization of carboxypeptidase A in solutions containing sodium chloride. *Biochemistry* 4: 2691-2698.

Kaiser, E. T. & Carson, F. W. Inhibition of the esterase action of carboxypeptidase A. *Biochem. Biophys. Res. Commun.* 18: 457-461.

Kaiser, E. T., Awazu, S. & Carson, F. W. The hydrolysis of O-hippurylglycolate catalyzed by carboxypeptidase A. Evidence for possible allosteric effects. *Biochem. Biophys. Res. Commun.* 21: 444-447.

Vallee, B. L. & Riordan, J. F. Carboxypeptidase catalysis. *Science* 150:

388 only.

1966

Bishop, W. H., Quiocho, F. A. & Richards, F. M. The removal and exchange of metal ions in cross-linked crystals of carboxypeptidase A. *Biochemistry* **5**: 4077-4087.

Carson, F. W. & Kaiser, E. T. pH Dependence of the hydrolysis of O-Acetyl-L-mandelate catalyzed by carboxypeptidase A. A critical examination. *J. Am. Chem. Soc.* **88**: 1212-1223.

Coleman, J. E., Pulido, P. & Vallee, B. L. Organic modifications of metallocarboxypeptidases. *Biochemistry* **5**: 2019-2026.

Coombs, T. L. & Vallee, B. L. The interaction of polypeptides and proteins with apocarboxypeptidase A. *Biochemistry* **5**: 3272-3280.

Hanson, H. Hydrolasen: Peptidasen (Exopeptidasen). In: *Hoppe-Seyler/Thierfelder-Handbuch der physiologisch- und pathologisch-chemischen Analyse* (Lang, K., Lehnartz, E., Hoffmann-Ostenhof, O. & Siebert, G. eds.), 10th edn., vol. 6, part C, pp. 1-229 (see pp. 85-109), Springer-Verlag, Berlin.

Lipscomb, W. N., Coppola, J. C., Hartsuck, J. A., Ludwig, M. L., Muirhead, H., Searl, J. & Steitz, T. A. The structure of carboxtpeptidase A. III. Molecular structure at 6 Å resolution. *J. Mol. Biol.* **19**: 423-441.

McClure, W. O. & Neurath, H. The reaction of carboxypeptidase A with chromophoric substrates. *Biochemistry* **5**: 1425-1438.

Quiocho, F. A. & Richards, F. M. The enzymic behavior of carboxypeptidase-A in the solid state. *Biochemistry* **5**: 4062-4076.

Shulman, R. G., Navon, G., Wyluda, B. J., Douglass, D. C. & Yamane, T. Nuclear magnetic resonance studies of the active site of carboxypeptidase A. *Proc. Natl. Acad. Sci. USA* **56**: 39-44.

Simpson, R. T. & Vallee, B. L. Iodocarboxypeptidase. *Biochemistry* **5**: 1760-1767.

Walsh, K. A., Ericsson, L. H. & Neurath, H. Bovine carboxypeptidase A variants resulting from allelomorphism. *Proc. Natl. Acad. Sci. USA* **56**: 1339-1344.

Whitaker, J. R., Menger, F. & Bender, M. L. The kinetics of some carboxypeptidase A and acetylcarboxypeptidase A catalyzed hydrolyses. *Biochemistry* **5**: 386-392.

1967

Abramowitz, N., Schechter, I. & Berger, A. On the size of the active site in proteases. II. Carboxypeptidase A. *Biochem. Biophys. Res. Commun.* **29**: 862-867.

Ambler, R. P. Carboxypeptidases A and B. *Methods Enzymol.* **11**: 436-445.

Awazu, S., Carson, F. W., Hall, P. L. & Kaiser, E. T. Inhibition in ester hydrolyses catalyzed by carboxypeptidase A. *J. Am. Chem. Soc.* **89**: 3627-3631.

Freisheim, J. H., Walsh, K. A. & Neurath, H. The activation of bovine procarboxypeptidase A. I. Isolation and properties of the succinylated enzyme precursor. *Biochemistry* **6**: 3010-3019.

Freisheim, J. H., Walsh, K. A. & Neurath, H. The activation of bovine procarboxypeptidase A. II. Mechanism of activation of the succinylated enzyme precursor. *Biochemistry* **6**: 3020-3028.

Hall, P. L. & Kaiser, E. T. On the question of transesterification in carboxypeptidase-A catalyzed hydrolyses. *Biochem. Biophys. Res. Commun.* **29**: 205-210.

Ludwig, M. L., Hartsuck, J. A., Steitz, T. A., Muirhead, H., Coppola, J. C., Reeke, G. N. & Lipscomb, W. N. The structure of carboxypeptidase A, IV. Preliminary results at 2.8Å resolution and a substrate complex at 6Å resolution. *Proc. Natl. Acad. Sci. USA* **57**: 511-514.

Piras, R. & Vallee, B. L. Procarboxypeptidase A-carboxypeptidase A interrelationships. Metal and substrate binding. *Biochemistry* **6**: 348-357.

Reeke, G. N., Hartsuck, J. A., Ludwig, M. L., Quiocho, F. A., Steitz, T. A. & Lipscomb, W. N. The structure of carboxypeptidase A, VI. Some results at 2.0-Å resolution, and the complex with glycyl-tyrosine at 2.8-Å resolution. *Proc. Natl. Adac. Sci. USA* **58**: 2220-2226.

Riordan, J. F., Sokolovsky, M. & Vallee, B. L. Environmentally sensitive tyrosyl residues. Nitration with tetranitromethane. *Biochemistry* **6**: 358-361.

Riordan, J. F., Sokolovsky, M. & Vallee, B. L. The functional tyrosyl residues of carboxypeptidase A. Nitration with tetranitromethane. *Biochemistry* **6**: 3609-3617.

Roholt, O. A. & Pressman, D. The sequence around the active-center tyrosyl residue of bovine pancreatic carboxypeptidase A. *Proc. Natl. Acad. Sci. USA* **58**: 280-285.

Sokolovsky, M. & Vallee, B. L. Azocarboxypeptidase: functional consequences of tyrosyl and histidyl modification. *Biochemistry* **6**: 700-708.

Steitz, T. A., Ludwig, M. L., Quiocho, F. A. & Lipscomb, W. N. The structure of carboxypeptidase A. V. Studies of enzyme-substrate and enzyme-inhibitor complexes at 6 A resolution. *J. Biol. Chem.* **242**: 4662-4668.

1968

Davies, R. C., Auld, D. S. & Vallee, B. L. The effect of modifiers on the hydrolysis of esters and peptides by carboxypeptidase A. *Biochem. Biophys. Res. Commun.* **31**: 628-633.

Davies, R. C., Riordan, J. F., Auld, D. S. & Vallee, B. L. Kinetics of carboxypeptidase A. I. Hydrolysis of carbobenzoxyglycyl-L-phenylalanine, benzoylglycyl-L-phenylalanine, and hippuryl-*dl*-β-phenyllactic acid by metal-substituted and acetylated carboxypeptidases. *Biochemistry* **7**: 1090-1099.

Lipscomb, W. N., Hartsuck, J. A., Reeke, G. N., Jr., Quiocho, F. A.,

Bethge, P. H., Ludwig, M. L., Steitz, T. A., Muirhead, H. & Coppola, J. C. The structure of carboxypeptidase A. VII. The 2.0-Å resolution studies of the enzyme and of its complex with glycyltyrosine, and mechanistic deductions. *Brookhaven Symp. Biol.* **21**: 24-90.

Navon, G., Shulman, R. G., Wyluda, B. J. & Yamane, T. Nuclear magnetic resonance studies of the active site of carboxypeptidase A. *Proc. Natl. Acad. Sci. USA* **60**: 86-91.

Neurath, H., Bradshaw, R. A., Ericsson, L. H., Babin, D. R., Petra, P. H. & Walsh, K. A. Current status of the chemical structure of bovine pancreatic carboxypeptidase A. *Brookhaven Symp. Biol.* **21**: 1-23.

Vallee, B. L. & Riordan, J. F. Chemical approaches to the mode of action of carboxypeptidase A. *Brookhaven Symp. Biol.* **21**: 91-119.

Vallee, B. L. & Williams, R. J. P. Metalloenzymes: the entatic nature of their activities. *Biochemistry* **59**: 498-505.

Vallee, B. L., Riordan, J. F., Bethune, J. L., Coombs, T. L., Auld, D. S. & Sokolovsky, M. A model for substrate binding and kinetics of carboxypeptidase A. *Biochemistry* **7**: 3547-3556.

1969

Bradshaw, R. A. The amino acid sequence of bovine carboxypeptidase A. III. Specificity of peptide-bond cleavage by thermolysin and the complete sequence of the cyanogen bromide fragment F_{III}. *Biochemistry* **8**: 3871-3877.

Bradshaw, R. A., Babin, D. R., Nomoto, M., Srinivasin, N. G., Ericsson, L. H., Walsh, K. A. & Neurath, H. The amino acid sequence of bovine carboxypeptidase A. II. Tryptic and chymotryptic peptides of the cyanogen bromide fragment$_{III}$. *Biochemistry* **8**: 3859-3871.

Bradshaw, R. A., Ericsson, L. H., Walsh, K. A. & Neurath, H. The amino acid sequence of bovine carboxypeptidase A. *Proc. Natl. Acad. Sci. USA* **63**: 1389-1394.

Bradshaw, R. A., Neurath, H. & Walsh, K. A. Considerations of the concept of structural homology as applied to bovine carboxypeptidases A and B. *Proc. Natl. Acad. Sci. U.S.A.* **63**: 406-411.

Hall, P. L., Kaiser, B. L. & Kaiser, E. T. pH Dependence and competitive product inhibition of the carboxypeptidase A catalyzed hydrolysis of O-(*trans*-cinnamoyl)-L-β-phenyllactate. *J. Am. Chem. Soc.* **91**: 485-497.

Kagan, H. M. & Vallee, B. L. Environmental sensitivity of azo chromophores in arsanilazocarboxypeptidase. *Biochemistry* **8**: 4223-4231.

Kaiser, B. L. & Kaiser, E. T. Effect of D_2O on the carboxypeptidase-catalyzed hydrolysis of O-(trans-cinnamoyl)-L-β-phenyllactate and N-(N-benzoylglycyl)-L-phenylalanine. *Proc. Natl. Acad. Sci. USA* **64**: 36-41.

Lin, Y., Means, G. E. & Feeney, R. E. An assay for carboxypeptidases A and B on polypeptides from protein. *Anal. Biochem.* **32**: 436-445.

Lipscombe, W. N., Hartsuck, J. A., Quiocho, F. A. & Reeke, G. N., Jr. The structure of carboxypeptidase A, IX. The X-ray diffraction results in the light of the chemical sequence. *Proc. Natl. Acad. Sci. USA* **64**: 28-

35.
Nomoto, M., Srinivasen, N. G., Bradshaw, R. A., Wade, R. D. & Neurath, H. The amino acid sequence of bovine carboxypeptidase A. I. Preparation and properties of the fragments obtained by cyanogen bromide cleavage. *Biochemistry* **8**: 2755-2762.

Pétra, P. H. & Neurath, H. Heterogeneity of bovine carboxypeptidase A. II. Chromatographic purification of carboxypeptidase A (Cox). *Biochemistry* **8**: 5029-5036.

Pétra, P. H. & Neurath, H. The heterogeneity of bovine carboxypeptidase A. I. The chromatographic purification of carboxypeptidase A (Anson). *Biochemistry* **8**: 2466-2475.

Pétra, P. H., Bradshaw, R. A., Walsh, K. A. & Neurath, H. Identification of the amino acid replacements characterizing the allotypic forms of bovine carboxypeptidase A. *Biochemistry* **8**: 2762-2768.

Peanasky, R. J., Gratecos, D., Baratti, J. & Rovery, M. Mode of activation and N-terminal sequence of subunit II in bovine procarboxypeptidase A and of porcine chymotrypsinogen C. *Biochim. Biophys. Acta* **181**: 82-92.

Pitout, M. J. A rapid spectrophotometric method for the assay of carboxypeptidase A. *Biochem. Pharmacol.* **18**: 1829-1836.

Pitout, M. J. & Nel, W. The inhibitory effect of ochratoxin A on bovine carboxypeptidase A in vitro. *Biochem. Pharmacol.* **18**: 1837-1843.

Vallee, B. L., Rupley, J. A., Coombs, T. L. & Neurath, H. The release of zinc from carboxypeptidase and its replacement. *J. Am. Chem. Soc.* **80**: 4750-4751.

1970

Auld, D. S. & Vallee, B. L. Kinetics of carboxypeptidase A. II. Inhibitors of the hydrolysis of oligopeptides. *Biochemistry* **9**: 602-609.

Behnke, W. D., Wade, R. D. & Neurath, H. Interactions of the endopeptidase subunit of bovine procarboxypeptidase A-S6. *Biochemistry* **9**: 4179-4188.

Lipscombe, W. N., Reeke, G. N., Jr., Hartsuck, J. A., Quiocho, F. A. & Bethge, P. H. The structure of carboxypeptidase A. VIII. Atomic interpretation at 0.2 nm resolution, a new study of the complex of glycyl-L-tyrosine with CPA, and mechanistic deductions. *Phil. Trans. Roy. Soc. Lond. B* **257**: 177-214.

Navon, G., Shulman, R. G., Wyluda, B. J. & Yamane, T. Nuclear magnetic resonance study of the binding of fluoride ions to carboxypeptidase A. *J. Mol. Biol.* **51**: 15-30.

Pétra, P. H. Bovine procarboxypeptidase and carboxypeptidase A. *Methods Enzymol.* **19**: 460-503.

Riordan, J. F. & Hayashida, H. Chemical evidence for a functional carboxyl group in carboxypeptidase A. *Biochem. Biophys. Res. Commun.* **41**: 122-127.

Vallee, B. L., Riordan, J. F., Auld, D. S. & Latt, S. A. Chemical approaches to the mode of action of carboxypeptidase A. *Phil. Trans. Roy. Soc. Lond. B* **257**: 215-230.

1971

Fretto, L. & Strickland, E. H. Effect of temperature upon the conformations of carboxypeptidase A (Anson),A_γ^{Leu}, A_γ^{Val}, and A_α , β. *Biochim. Biophys. Acta* **235**: 473-488.

Hartsuck, J. A. & Lipscomb, W. N. Carboxypeptidase A. In: *The Enzymes* (Boyer, P. D. ed.), 3rd edn., vol. 3, pp. 1-56, Academic Press, New York.

Hass, G. M. & Neurath, H. Affinity labeling of bovine carboxypeptidase A_γ^{Leu} by N-bromoacetyl-N-methyl-L-phenylalanine. II. Sites of modification. *Biochemistry* **10**: 3541-3546.

Latt, S. A. & Vallee, B. L. Spectral properties of cobalt carboxypeptidase. The effects of substrates and inhibitors. *Biochemistry* **10**: 4263-4270.

Pétra, P. H. Modification of carboxyl groups in bovine carboxypeptidase A. I. Inactivation of the enzyme by N-ethyl-5-phenylisoxazolium-3'-sulfonate (Woodward Reagent K). *Biochemistry* **10**: 3163-3170.

Pétra, P. H., Hermodson, M. A., Walsh, K. A. & Neurath, H. Characterization of bovine carboxypeptidase A (Allan). *Biochemistry* **10**: 4023-4025.

Quiocho, F. A., Bethge, P. H., Lipscomb, W. N., Studebaker, J. F., Brown, R. D. & Koenig, S. H. X-Ray diffraction and nuclear magnetic resonance dispersion studies on derivatives of carboxypeptidase A. *Cold Spring Harbor Symp. Quant. Biol.* **36**: 561-567.

Reeck, G. R., Walsh, K. A. & Neurath, H. Isolation and characterization of carboxypeptidases A and B from activated pancreatic juice. *Biochemistry* **10**: 4690-4698.

Ryan, C. A. Inhibition of carboxypeptidase A by a naturally occurring polypeptide from potatoes. *Biochem. Biophys. Res. Commun.* **44**: 1265-1270.

Schechter, I. & Zazepizki, E. On the mechanism of acceleration of carboxypeptidase A activity. *Eur. J. Biochem.* **18**: 469-473.

Uren, J. R. Affinity chromatography of proteolytic enzymes. *Biochim. Biophys. Acta* **236**: 67-73.

Zelikson, R., Eilam-Rubin, E. & Kulka, R. G. The chymotrypsinogens and procarboxypeptidases of chick pancreas. *J. Biol. Chem.* **246**: 6115-6120.

1972

Auld, D. S., Latt, S. A. & Vallee, B. L. An approach to inhibition kinetics. Measurement of enzyme-substrate complexes by electronic energy transfer. *Biochemistry* **11**: 4994-4999.

Behnke, W. D. & Vallee, B. L. The spectrum of cobalt bovine procarboxypeptidase A, an index of catalytic function. *Proc. Natl. Acad. Sci. U.S.A.* **69**: 2442-2445.

Bunting, J. W. & Murphy, J. The mechanism of substrate inhibition of the esterase activity of carboxypeptidase A. *Biochem. Biophys. Res. Commun.* **48**: 1316-1322.

Byers, L. D. & Wolfenden, R. A potent reversible inhibitor of carboxypeptidase A. *J. Biol. Chem.* **247**: 606-608.

Glovsky, J., Hall, P. L. & Kaiser, E. T. On the action of carboxypeptidase A on ester substrates in alkaline aolution. *Biochem. Biophys. Res. Commun.* **47**: 244-247.

Jones, S. R. & Hofmann, T. Penicillocarboxypeptidase-S, a nonspecific SH-dependent exopeptidase. *Can. J. Biochem.* **50**: 1297-1310.

Quiocho, F. A., McMurray, C. H. & Lipscomb, W. N. Similarities between the conformation of arsanilazotyrosine 248 of carboxypeptidase A in the crystalline state and in solution. *Proc. Natl. Acad. Sci. U.S.A.* **69**: 2850-2854.

Uren, J. R. & Neurath, H. Mechanism of activation of bovine procarboxypeptidase A S_5. Alterations in primary and quaternary structure. *Biochemistry* **11**: 4483-4492.

1973

Bunting, J. W. & Myers, C. D. Reversible inhibition of the esterase activity of carboxypeptidase A by carboxylate anions. *Can. J. Chem.* **51**: 2639-2649.

Byers, L. D. & Wolfenden, R. Binding of the by-product analog benzylsuccinic acid by carboxypeptidase A. *Biochemistry* **12**: 2070-2078.

Johansen, J. T. & Vallee, B. L. Conformations of arsanilazotyrosine-248 carboxypeptidase A$_\alpha^{+\beta}{}_\gamma^{+}$. Comparison of crystals and solution. *Proc. Natl Acad. Sci. U.S.A.* **70**: 2006-2010.

Naik, V. R. & Horton, H. R. Kinetics and spectral properties of carboxypeptidase A labeled with the 2-hydroxy-5-nitrobenzyl reporter group. *J. Biol. Chem.* **248**: 6709-6717.

Rosenberg, R. C., Root, C. A., Wang, R.-H., Cerdonio, M. & Gray, H. B. The nature of the ground states of cobalt (II) and nickel (II) carboxypeptidase A. *Proc. Natl. Acad. Sci. U.S.A.* **70**: 161-163.

1974

Auld, D. S. & Holmquist, B. Carboxypeptidase A. Differences in the mechanisms of ester and peptide hydrolysis. *Biochemistry* **13**: 4355-4361.

Fridkin, M. & Goren, H. J. Synthesis and catalytic properties of the heptapeptide L-seryl-L-prolyl-L-cysteinyl-L-seryl-α-L-glutamyl-L-threonyl-L-tyrosine. *Eur. J. Biochem.* **41**: 273-283.

Ryan, C. A., Hass, G. M. & Kuhn, R. W. Purification and properties of a carboxypeptidase inhibitor from potatoes. *J. Biol. Chem.* **249**: 5495-5499.

Uren, J. R. & Neurath, H. Intrinsic enzymatic activity of bovine procarboxypeptidase A S_5. *Biochemistry* **13**: 3512-3520.

1975

Hass, G. M., Nau, H., Biemann, K., Grahn, D. T., Ericsson, L. H. & Neurath. H. The amino acid sequence of a carboxypeptidase inhibitor from potatoes. *Biochemistry* **14**: 1334-1342.

Johansen, J. T. & Vallee, B. L. Environment and conformation dependent sensitivity of the arsanilazotyrosine-248 carboxypeptidase A chromophore. *Biochemistry* **14**: 649-660.

Marinkovic, D. V. & Marinkovic, J. N. Studies of human carboxypeptidase A purification and properties from human pancreas. *Biochem. Med.* **14**: 125-134.

1976

Ako, H., Hass, G. M., Grahn, D. T. & Neurath, H. Carboxypeptidase inhibitor from potatoes. Interaction with derivatives of carboxypeptidase A. *Biochemistry* **15**: 2573-2578.

Bazzone, T. J. & Vallee, B. L. Bovine procarboxypeptidase A: kinetics of peptide and ester hydrolysis. *Biochemistry* **15**: 868-875.

Hass, G. M., Ako, H., Grahn, D. T. & Neurath, H. Carboxypeptidase inhibitor from potatoes. The effects of chemical modifications on inhibitory activity. *Biochemistry* **15**: 93-100.

Homandberg, G. A. & Peanasky, R. J. Characterization of proteins from *Ascaris lumbricoides* which bind specifically to carboxypeptidase. *J. Biol. Chem.* **251**: 2226-2233.

Makinen, M. W., Yamamura, K. & Kaiser, E. T. Mechanism of action of carboxypeptidase A in ester hydrolysis. *Proc. Natl. Acad. Sci. U.S.A.* **73**: 3882-3886.

1977

Ager, S. P. & Hass, G. M. Affinity chromatography of pancreatic carboxypeptidases using a carboxypeptidase inhibitor from potatoes as ligand. *Anal. Biochem.* **83**: 285-295.

Kester, W. R. & Matthews, B. W. Comparison of the structures of carboxypeptidase A and thermolysin. *J. Biol. Chem.* **252**: 7704-7710.

1978

Bunting, J. W. & Kabir, S. H. The pH-dependence of the non-specific esterase activity of carboxypeptidase A. *Biochim. Biophys. Acta* **527**: 98-107.

Guy, G. J. & Butterworth, J. Carboxypeptidase A activity of cultured skin fibroblasts and relationship to cystic fibrosis. *Clin. Chim. Acta* **87**: 63-69.

Van Wart, H. E. & Vallee, B. L. Enzymatically inactive, exchange-inert Co(III)-carboxypeptidase A: role of inner sphere coordination in peptide and ester catalysis. *Biochemistry* **17**: 3385-3394.

1979

Bauer, R., Christensen, C., Johansen, J. T., Bethune, J. L. & Vallee, B. L. Perturbed angular correlation γ ray (PAC) spectroscopy of [111]Cd carboxypeptidase Aα. *Biochem. Biophys. Res. Commun.* **90**: 679-685.

Bazzone, T. J., Sokolovsky, M., Cueni, L. B. & Vallee, B. L. Single-step isolation and resolution of pancreatic carboxypeptidases A and B.

Biochemistry **18**: 4362-4366.

Billo, E. J. Kinetics of dissociation of zinc ion from carboxypeptidase A: Catalysis and inhibition by amino acids. *J. Inorg. Biochem.* **11**: 339-347.

Billo, E. J. Kinetics of interaction of ligands with carboxypeptidase A. *J. Inorg. Biochem.* **10**: 331-339.

Haas, R., Heinrich, P. C. & Sasse, D. Proteolytic enzymes of rat-liver mitochondria. Evidence for a mast cell origin. *FEBS Lett.* **103**: 168-171.

Haas, R. & Heinrich, P. C. A novel SH-type carboxypeptidase in the inner membrane of rat-liver mitochondria. *Eur. J. Biochem.* **96**: 9-15.

Kam, C.-M., Nishino, N. & Powers, J. C. Inhibition of thermolysin and carboxypeptidase A by phosphoramidates. *Biochemistry* **18**: 3032-3038.

Leary, T. R., Grahn, D. T., Neurath, H. & Hass, G. M. Structure of potato carboxypeptidase inhibitor: disulfide pairing and exposure of aromatic residues. *Biochemistry* **18**: 2252-2256.

Makinen, M. W., Kuo, L. C., Dymowski, J. J. & Jaffer, S. Catalytic role of the metal ion of carboxypeptidase A in ester hydrolysis. *J. Biol. Chem.* **254**: 356-366.

Marinkovic, D. V. N-Terminal amino acid sequences of human carboxypeptidases A, B1 and B2. *Biochem. Med.* **22**: 11-26.

Nakano, A., Tasumi, M., Fujiwara, K., Fuwa, K. & Miyazawa, T. Nuclear magnetic relaxation study on the interaction of glycyl-L-tyrosine with manganese-carboxypeptidase A in solution. *J. Biochem.* **86**: 1001-1011.

Ondetti, M. A., Condon, M. E., Reid, J., Sabo, E. F., Cheung, H. S. & Cushman, D. W. Design of potent and specific inhibitors of carboxypeptidases A and B. *Biochemistry* **18**: 1427-1430.

Poorman, R., Kuo, M., Johnson, D. I., Lin, S. & Sebastian, J. F. Inhibition of carboxypeptidase A catalyzed peptide hydrolysis by 3-phenylpropanoate at activating and nonactivating substrate concentrations. *Can. J. Biochem.* **57**: 357-365.

Urdea, M. S. & Legg, J. I. A peptidase-inactive derivative of carboxypeptidase A modified specifically at tyrosine 248. Cobalt(III)(ethylenediamine-N,N'-diacetato)(arsanilazotyrosinato 248 carboxypeptidase A). *J. Biol. Chem.* **254**: 11868-11874.

Wu, L. N.-Y. L. & Horton, R. Structure of 2-hydroxy-5-nitrobenzylated carboxypeptidase A. *Biochim. Biophys. Acta* **577**: 22-33.

1980

Butterworth, J. & Duncan, J. J. Chromatography of carboxypeptidase A and B activity of cultured skin fibroblasts: relationship to cystic fibrosis. *Clin. Chim. Acta* **108**: 143-146.

Cueni, L. B., Bazzone, T. J., Riordan, J. F. & Vallee, B. L. Affinity chromatographic sorting of carboxypeptidase A and its chemically modified derivatives. *Anal. Biochem.* **107**: 341-349.

Everitt, M. T. & Neurath, H. Rat peritoneal mast cell carboxypeptidase: localization, purification and enzymatic properties. *FEBS Lett.* **110**:

292-296.

Hass, G. M. & Ryan, C. A.   Cleavage of the carboxypeptidase inhibitor from potatoes by carboxypeptidase A. *Biochem. Biophys. Res. Commun.* **97**: 1481-1486.

Homandberg, G. A., Minor, S. T. & Peanasky, R. J.   Modification of the carboxypeptidase A active site residue Glu-270 prevents interaction with a protein protease inhibitor from Ascaris. *Biochim. Biophys. Acta* **612**: 384-394.

Lipscomb, W. N.   Carboxypeptidase A mechanisms. *Proc. Natl. Acad. Sci. USA.* **77**: 3875-3878.

Rees, D. C. & Lipscomb, W. N.   Structure of potato inhibitor complex of carboxypeptidase A at 5.5-Å resolution. *Proc. Natl. Acad. Sci. USA.* **77**: 277-280.

Rees, D. C. & Lipscomb, W. N.   Structure of the potato inhibitor complex of carboxypeptidase A at 2.5-Å resolution. *Proc. Natl. Acad. Sci. USA* **77**: 4633-4637.

Rees, D. C., Honzatko, R. B. & Lipscomb, W. N.   Structure of an actively exchanging complex between carboxypeptidase A and a substrate analogue. *Proc. Natl. Acad. Sci. USA* **77**: 3288-3291.

Rogers, R. J. & Billo, E. J.   Inhibition of cobalt (II) carboxypeptidase A by ligands. Kinetic evidence for the formation of a ternary enzyme-metal-ligand complex. *J. Inorg. Biochem.* **12**: 335-341.

Scheule, R. K., Van Wart, H. E., Vallee, B. L. & Scheraga, H. A.   Resonance Raman spectroscopy of arsanilazocarboxypeptidase A: conformational equilibria in solution and crystal phases. *Biochemistry* **19**: 759-766.

Thompson, J. S., Gehring, H. & Vallee, B. L.   Structure and function of carboxypeptidase Aα in supercooled water. *Proc. Natl. Acad. Sci. USA.* **77**: 132-136.

1981

Kobayashi, Y., Kobayashi, R. & Hirs, C. H. W.   Identification of zymogen E in a complex with bovine procarboxypeptidase A. *J. Biol. Chem.* **256**: 2466-2470.

Koide, A. & Yoshizawa, M.   Crystallization and identification of a binary complex of a elastase-I and a carboxypeptidase A from porcine pancreas. *Biochem. Biophys. Res. Commun.* **100**: 1091-1098.

Mock, W. L., Chen, J.-T. & Tsang, J. W.   Hydrolysis of a thiopeptide by cadmium carboxypeptidase A. *Biochem. Biophys. Res. Commun.* **102**: 389-396.

Rees, D. C. & Lipscomb, W. N.   Binding of ligands to the active site of carboxypeptidase A. *Proc. Natl. Acad. Sci. USA* **78**: 5455-5459.

Rees, D. C., Lewis, M., Honzatko, R. B., Lipscomb, W. N. & Hardman, K. D.   Zinc environment and *cis* peptide bonds in carboxypeptidase A at 1.75-Å resolution. *Proc. Natl. Acad. Sci. USA* **78**: 3408-3412.

Scheule, R. K., Han, S. L., Van Wart, H. E., Vallee, B. L. & Scheraga, H. A.   Resonance raman spectroscopy of arsanilazocarboxypeptidase A: mode of inhibitor binding and active site topography. *Biochemistry* **20**:

1778-1784.

Wicker, C. & Puigserver, A. Further studies on subunit III of bovine procarboxypeptidase A. Structure and reactivity of the weakly functional active site. *FEBS Lett.* **128**: 13-16.

Woodbury, R. G., Everitt, M. T. & Neurath, H. Mast cell proteases. *Methods Enzymol.* **80**: 588-609.

1982

Avilés, F. X., San Segundo, B., Vilanova, M., Cuchillo, C. M. & Turner, C. The activation segment of procarboxypeptidase A from porcine pancreas constitutes a folded structural domain. *FEBS Lett.* **149**: 257-260.

Bachovchin, W. W., Kanamori, K., Vallee, B. L. & Roberts, J. D. Nitrogen-15 nuclear magnetic resonance of arsanilazotyrosine-248 carboxypeptidase A and its complex with βphenylpropionate. Structure and dynamics in solution. *Biochemistry* **21**: 2885-2892.

Bartlett, P. A., Spear, K. L. & Jacobsen, N. E. A thioamide substrate of carboxypeptidase A. *Biochemistry* **21**: 1608-1611.

Kuo, L. C. & Makinen, M. W. Hydrolysis of esters by carboxypeptidase A requires a penta-coordinate metal ion. *J. Biol. Chem.* **257**: 24-27.

Palmer, A. R., Ellis, P. D. & Wolfenden, R. Extreme state of ionization of benzylsuccinate bound by carboxypeptidase A. *Biochemistry* **21**: 5056-5059.

Peterson, L. M., Holmquist, B. & Bethune, J. L. A unique activity assay for carboxypeptidase A in human serum. *Anal. Biochem.* **125**: 420-426.

Quinto, C., Quiroga, M., Swain, W. F., Nikovits, W. C., Jr., Standring, D. N., Pictet, R. L., Valenzuela, P. & Rutter, W. J. Rat preprocarboxypeptidase A: cDNA sequence and preliminary characterization of the gene. *Proc. Natl. Acad. Sci. USA* **79**: 31-35.

San Segundo, B., Martinez, M. C., Vilanova, M., Cuchillo, C. M. & Avilés, F. X. The severed activation segment of porcine pancreatic procarboxypeptidase A is a powerful inhibitor of the active enzyme. Isolation and characterization of the activation peptide. *Biochim. Biophys. Acta* **707**: 74-80.

Vendrell, J., Aviles, F. X., San Segundo, B. & Cuchillo, C. M. Isolation and re-association of the subunits from the pro-(carboxypeptidase A)-pro-(proteinase E) binary complex from pig pancreas. *Biochem. J.* **205**: 449-452.

1983

Galdes, A., Auld, D. S. & Vallee, B. L. Cryokinetic studies of the intermediates in the mechanism of carboxypeptidase A. *Biochemistry* **22**: 1888-1893.

Geoghegan, K. F., Galdes, A., Martinelli, R. A., Holmquist, B.. Auld, D. S. & Vallee, B. L. Cryospectroscopy of intermediates in the mechanism of carboxypeptidase A. *Biochemistry* **22**: 2255-2262.

Geoghegan, K. F., Holmquist, B., Spilburg, C. A. & Vallee, B. L. Spectral properties of cobalt carboxypeptidase A. Interaction of the metal atom with anions. *Biochemistry* **22**: 1847-1852.

King, S. W. & Fife, T. H. Effects of metal ion substitution on carboxypeptidase A catalyzed hydrolysis of O-*trans*-cinnamoyl-L-β-phenyllactate. *Biochemistry* **22**: 3603-3610.

Martinez, M. C., Nieuwenhuysen, P., Clauwaert, J. & Cuchillo, C. M. Pro-(carboxypeptidases A) from whole pig pancreas. Their mass, size, shape and solvation. *Biochem. J.* **215**: 23-27.

Pearce, G. & Ryan, C. A. A rapid, large-scale method for purification of the metallo-carboxypeptidase inhibitor from potato tubers. *Anal. Biochem.* **130**: 223-225.

Peterson, L. M. & Holmquist, B. Human serum procarboxypeptidase A. *Biochemistry* **22**: 3077-3082.

Rees, D. C. & Lipscomb, W. N. Crystallographic studies on apocarboxypeptidase A and the complex with glycyl-L-tyrosine. *Proc. Natl. Acad. Sci. USA* **80**: 7151-7154.

1984

Auld, D. S., Galdes, A., Geoghegan, K. F., Holmquist, B., Martinelli, R. A. & Vallee, B. L. Cryospectrokinetic characterization of intermediates in biochemical reactions: carboxypeptidase A. *Proc. Natl. Acad. Sci. USA* **81**: 5041-5045.

Dua, R. D. & Gupta, K. K. Role of metal ions in goat carboxypeptidase A-catalysed hydrolysis of acyl peptides. *Biochem. Int.* **9**: 379-389.

Galardy, R. E. & Kortylewicz, Z. P. Inhibition of carboxypeptidase A by aldehyde and ketone substrate analogues. *Biochemistry* **23**: 2083-2087.

Hirose, J. & Wilkins, R. G. Interaction of cobalt(II) complexes with apoarsanilazotyrosine-248 carboxypeptidase A. *Biochemistry* **23**: 3149-3153.

Hook, V. Y. H. & Loh, Y. P. Carboxypeptidase B-like converting enzyme activity in secretory granules of rat pituitary. *Proc. Natl. Acad. Sci. USA* **81**: 2776-2780.

Kerfelec, B., Chapus, C. & Puigserver, A. Two-step dissociation of bovine 6S procarboxypeptidase A by dimethylmaleylation. *Biochem. Biophys. Res. Commun.* **121**: 162-167.

Tanaka, T., Suda, H., Naganawa, H., Hamada, M., Takeuchi, T., Aoyagi, T. & Umezawa, H. Production of (S)-α-benzylmalic acid, inhibitor of carboxypeptidase A by actinomycetes. *J. Antibiot.* **37**: 682-684.

Tissue Carboxypeptidase B

Summary

EC Number: 3.4.17.2.

Earlier names: Protaminase, pancreatic carboxypeptidase B.

Abbreviation: tCPB.

Distribution: Occurs primarily as an inactive zymogen, procarboxypeptidase B, in the pancreatic secretions of most vertebrates. Activity on carboxypeptidase B substrates, at neutral pH, has also been detected in human urine [1], rat kidney cortex [2], and beef adrenal medulla [3].

Source: Hog [4,5], beef [6,7] and human [8] pancreas. Activated extracts of pancreatic acetone powders are commonly used. Affinity chromatography has been used to separate carboxypeptidases A and B [9].

Action: Catalyzes the release of basic amino acids such as lysine, hydroxylysine, arginine, ornithine, and S-(β-aminoethyl)cysteine from the unsubstituted C-termini of polypeptides and N-acylated dipeptides provided the terminal residue possesses an L-configuration and the penultimate residue is not proline. Histidine is not released by tCPB. As in the case of tissue carboxypeptidase A, the rate of removal of C-terminal residues is greatly affected by the structure of the penultimate residue. Thus, the relative rates of hydrolysis of hippuryl (Bz-Gly-) derivatives by beef tCPB, at pH 7.6, are as follows: Hip-Lys 100, Hip-Arg 71, Hip-Orn 30, and Hip-Pro-Lys 0.5. Non-acylated dipeptides such as Gly-Lys are hydrolyzed poorly [10].

tCPB also hydrolyzes esters of α-hydroxy acids that contain an unsubstituted terminal carboxyl group, such as hippurylphenyllactic acid. Beef tCPB shows a substantial amount of intrinsic activity on carboxypeptidase A substrates such as Z-Gly-Phe, but the hog enzyme shows little or none [4], so hog tCPB is generally used in biochemical studies requiring near-absolute specificity.

Requirements: pH 7.6-8.0 at 37°C, usually in a Tris-HCl buffer with ionic strength maintained above 0.1.

Substrate, usual: Bz-Gly-Arg. The rate of arginine release is usually determined spectrophotometrically by following the increase in A_{254} [5], but may also be determined colorimetrically [10].

Substrate, special: Bz-Ala-Lys and Bz-Ala-Arg, which may also be used as spectrophotometric substrates, generally exhibit lower K_m values and higher k_{cat} values than the corresponding hippuryl derivatives. This is especially true for the human pancreatic enzyme [11].

Inhibitors: Benzylsuccinic acid, a "bi-product analogue" that structurally resembles the products of hydrolysis of C-terminal phenylalanine peptides, reversibly inhibits both carboxypeptidases A and B [12]. On the other hand, inhibitors that mimic the side chains of arginine or lysine, as do guanidinoethylmercaptosuccinic acid (GEMSA; K_i 4 μM) and aminopropylmercaptosuccinic acid (K_i 8 μM), are potent competitive inhibitors that selectively inhibit tCPB [13]. An even more potent and specific inhibitor, 2-mercaptomethyl-5-guanidinopentanoic acid (K_i 0.4 nM), contains a sulfhydryl group that is believed to bind to the catalytically important zinc atom of tCPB [14]. 4-Acetamidobutylguanidine, a reagent that alkylates a tyrosine residue located at the active center, produces rapid and irreversible inhibition [14a].

Various amino acids and their derivatives act as competitive inhibitors of tCPB. Typically, they contain a free carboxyl group and a basic side chain. Examples include 6-aminohexanoic acid (K_i 1.1 mM), L-arginine (K_i 0.5 mM) and N-benzoylarginine (K_i 40 μM). Agmatine and cadaverine are not inhibitory.

Metal-chelating agents such as 1,10-phenanthroline, 8-hydroxyquinoline-5-sulfonic acid, and 2,2'-dipyridyl show reversible inhibition [4]. Other metal binding agents such as phosphate and citrate below pH 7.5, as well as pyrophosphate and borate above pH 8.0, are also inhibitory [15].

Various alcohols have a pronounced effect on tCPB. For example, 0.3 M butan-1-ol greatly enhances the rate of Hip-Arg hydrolysis, whereas rates on ester substrates are greatly reduced [16].

Molecular properties: tCPB from beef and hog pancreas is a metalloprotein comprised of one zinc atom and a single polypeptide chain. The enzyme from beef pancreas contains 308 amino acid residues, the sequence of which is homologous (49% identity) with beef tCPA [17]. Optimal alignment of the two sequences is based on the assumption that tCPB lacks the first three residues corresponding to tCPA, extends one residue beyond the C-terminus, and has one additional residue inserted between positions 186 and 190 [17]. M_r 34,606, pI about 6.0, and $A_{280,1\%}$ 21.0. tCPB however, contains 7 cysteine residues, 5 more than tCPA, six of these form disulfide bonds, and one (Cys-290) remains free and is reactive with sulfhydryl reagents, but only after removal of the zinc atom [18].

X-ray diffraction studies, with 2.8 Å resolution, have revealed remarkable similarities in the three-dimensional structures of tCPA and tCPB from beef pancreas [19]. Differences in folding are limited to chain termini and external loops. The active sites of both enzymes are also alike, with the zinc atom and its ligands (His-69, Glu-72 and His-196), Arg-145 (the site of substrate carboxyl group binding), and Glu-270 (whose side chain is involved in peptide bond cleavage) showing a similar orientation in the two molecules. The difference in specificity of tCPB is attributed to the presence of Asp-255 in the center of its binding pocket, as compared to Ile-255 in tCPA.

The single zinc atom in the molecule is easily exchanged for other metals, such as cobalt or cadmium, by simply incubating the enzyme with a solution of the appropriate metal salt at pH 7.75 [20]. Both the cobalt and cadmium enzymes retain good activity on ester substates, such as Hip-argininic acid, but only the cobalt enzyme acts on peptide substrates such as Hip-Arg and Hip-Lys.

Concentrated solutions of tCPB may be stored at -10°C in Tris-HCl buffer, pH 7.5, for periods up to one year without loss of activity.

Comment: tCPB, particularly the hog enzyme, has found wide use in end group analysis of proteins, and, when used in conjunction with tCPA, has proved valuable for sequence determination and protein modification [21]. tCPB has been used successfully in digests containing 0.5% SDS as a denaturing and solubilizing agent [22].

Procarboxypeptidase B occurs in pancreatic secretions released into the duodenum, where it undergoes activation by trypsin. The beef proenzyme contains 505 amino acids residues, M_r about 57,000. Its conversion to the active carboxypeptidase (M_r 34,606) involves more than one hydrolytic event besides the cleavage of the Arg-Thr linkage that gives rise to the N-terminal threonine of tCPB.

References
[1] Innerfield *et al. Proc. Soc. Exp. Biol. Med.* **116**: 573-575, 1964.
[2] Innerfield *et al. Life Sci.* **3**: 267-275, 1964.
[3] Wallace *et al. Life Sci.* **31**: 1793-1796, 1982.
[4] Folk *et al. J. Biol. Chem.* **235**: 2272-2277, 1960.
[5] Folk *Methods Enzymol.* **19**: 504-508, 1970.
[6] Wintersberger *et al. Biochemistry* **1**: 1069-1078, 1962.
[7] Kycia *et al. Arch. Biochem. Biophys.* **123**: 336-342, 1968.
[8] Marinkovic *et al. Biochem. J.* **163**: 253-260, 1977.
[9] Bazzone *et al. Biochemistry* **18**: 4362-4366, 1979.
[10] Folk & Gladner *J. Biol. Chem.* **231**: 379-391, 1958.
[11] McKay *et al. Arch. Biochem. Biophys.* **197**: 487-492, 1979.
[12] Zisapel & Sokolovsky *Biochem. Biophys. Res. Commun.* **58**: 951-959, 1974.
[13] McKay & Plummer *Biochemistry* **17**: 401-405, 1978.
[14] Ondetti *et al. Biochemistry* **18**: 1427-1430, 1979.
[14a]Plummer *et al. J. Biol. Chem.* **241**: 1648-1650, 1966.

[15] Wolff *et al.* *J. Biol. Chem.* **237**: 3094-3099, 1962.
[16] Folk *et al.* *J. Biol. Chem.* **237**: 3105-3109, 1962.
[17] Titani *et al.* *Proc. Natl. Acad. Sci. USA* **72**: 1666-1670, 1975.
[18] Wintersberger *et al.* *Biochemistry* **4**: 1526-1532, 1965.
[19] Schmid & Herriott *J. Mol. Biol.* **103**: 175-190, 1976.
[20] Folk & Gladner *Biochim. Biophys. Acta* **48**: 139-147, 1961.
[21] Ambler *Methods Enzymol.* **11**: 436-445, 1967.
[22] Koorajian & Zabin *Biochem. Biophys. Res. Commun.* **18**: 384-388, 1965.

Bibliography

1931

Waldschmidt-Leitz, E., Ziegler, F., Schäffner, A. & Weil, L. Uber die Struktur der Protamine I. Protaminase und die Produkte ihrer Einwirkung auf Clupein und Salmin. *Hoppe-Seyler's Z. Physiol. Chem.* **197**: 219-236.

1933

Calvery, H. O. Crystalline egg albumin. The hydrolysis of crystalline egg albumin by pepsin, papain-hydrocyanic acid, and pancreatic proteinase and the subsequent action of some other enzymes on the hydrolysis products produced by these enzymes. *J. Biol. Chem.* **102**: 73-88.

1947

Portis, R. A. & Altman, K. I. The action of proteolytic enzymes and protaminase on salmine sulfate. *J. Biol. Chem.* **169**: 203-209.

1956

Folk, J. E. A new pancreatic carboxypeptidase. *J. Am. Chem. Soc.* **78**: 3541-3542.
Folk, J. E. The influence of the lysine-glucose reaction on enzymatic digestion. *Arch. Biochem. Biophys.* **64**: 6-18.
Keller, P. J., Cohen, E. & Neurath, H. Purification and properties of procarboxypeptidase. *J. Biol. Chem.* **223**: 457-467.

1958

Folk, J. E. & Gladner, J. A. Carboxypeptidase B. I. Purification of the zymogen and specificity of the enzyme. *J. Biol. Chem.* **231**: 379-391.
Gladner, J. A. & Folk, J. E. Carboxypeptidase B. II. Mode of action on protein substrates and its application to carboxyl terminal group analysis. *J. Biol. Chem.* **231**: 393-401.

1959

Weil, L., Seibles, T. S. & Telka, M. Studies on the specificity of protaminase. *Arch. Biochem. Biophys.* **79**: 44-54.

1960

Folk, J. E., Piez, K. A., Carroll, W. R. & Gladner, J. A. Carboxypeptidase B. IV. Purification and characterization of the porcine enzyme. *J. Biol. Chem.* **235**: 2272-2277.

Neurath, H. Carboxypeptidases A and B. In: *The Enzymes* (Boyer, P. D., Lardy, H. & Myrbäck, K. eds), 2nd edn., vol. 4, part A, pp. 11-36, (see pp. 34-36), Academic Press, New York.

1961

Folk, J. E. & Gladner, J. A. Influence of cobalt and cadmium on the peptidase and esterase activities of carboxypeptidase B. *Biochim. Biophys. Acta* **48**: 139-147.

Folk, J. E., Braunberg, R. C. & Gladner, J. A. Carboxypeptidase B. V. Amino- and carboxyl-terminal sequences. *Biochim. Biophys. Acta* **47**: 595-596.

1962

Cox, D. J., Wintersberger, E. & Neurath, H. Bovine pancreatic procarboxypeptidase B. II. Mechanism of activation. *Biochemistry* **1**: 1078-1082.

Erdös, E. G. & Sloane, E. M. An enzyme in human blood plasma that inactivates bradykinin and kallidins. *Biochem. Pharmacol.* **11**: 585-592.

Folk, J. E., Wolff, E. C., Schirmer, E. W. & Cornfield, J. The kinetics of carboxypeptidase B activity. III. Effects of alcohol on the peptidase and esterase activities; kinetic models. *J. Biol. Chem.* **237**: 3105-3109.

Wintersberger, E., Cox, D. J. & Neurath, H. Bovine pancreatic procarboxypeptidase B. I. Isolation, properties and activation. *Biochemistry* **1**: 1069-1078.

Wolff, E. C., Schirmer, E. W. & Folk, J. E. The kinetics of carboxypeptidase B activity. I. Kinetic parameters. *J. Biol. Chem.* **237**: 3094-3099.

1963

Bargetzi, J.-P., Sampath-Kumar, K. S. V., Cox, D. J., Walsh, K. A. & Neurath, H. The amino acid composition of bovine pancreatic carboxypeptidase A. *Biochemistry* **2**: 1468-1474.

1964

Erdös, E. G., Sloane, E. M. & Wohler, I. M. Carboxypeptidase in blood and other fluids. I. Properties, distribution and partial purification of the enzyme. *Biochem. Pharmol.* **13**: 893-905.

Innerfield, I., Gimble, F. S. & Blincoe, E. Tissue peptidase activity following orally given proteases. *Life Sci.* **3**: 267-275.

Innerfield, I., Harvey, R., Luongo, F. & Blincoe, E. Urine peptidase activity following single and multiple oral doses of streptokinase. *Proc. Soc. Exp. Biol. Med.* **116**: 573-575.

1965

Avrameas, S. & Uriel, J. Systematic fractionation of swine pancreatic hydrolases. II. Fractionation of enzymes insoluble in ammonium sulfate solution at 0.40 saturation. *Biochemistry* **4**: 1750-1757.

Barrett, J. T. An antibody to porcine carboxypeptidase B. *Int. Arch. Allergy Appl. Immunol.* **26**: 158-166.

Koorajian, S. & Zabin, I. Carboxypeptidase studies on β galactosidase: detection of one C-terminal lysine per monomer. *Biochem. Biophys. Res. Comm.* **18**: 384-388.

Wintersberger, E. Isolation and structure of an active-center peptide of bovine carboxypeptidase B containing the zinc-binding sulfhydryl group. *Biochemistry* **4**: 1533-1536.

Wintersberger, E., Neurath, H., Coombs, T. L. & Vallee, B. L. A zinc-binding thiol group in the active center of bovine carboxypeptidase B. *Biochemistry* **4**: 1526-1532.

1966

Hanson, H. Hydrolasen: Peptidasen (Exopeptidasen). In: *Hoppe-Seyler/Thierfelder Handbuch der physiologisch- und pathologisch-chemischen Analyse*. (Lang, K., Lehnartz, E., Hoffmann-Ostenhof, O. & Siebert, G. eds), 10th edn., vol. 6, part C, pp. 1-229 (see pp. 109-118), Springer Verlag, Berlin.

Pascale, J., Avrameas, S. & Uriel, J. The characterization of rat pancreatic zymogens and their active forms by gel diffusion techniques. *J. Biol. Chem.* **241**: 3023-3027.

Prahl, J. W. & Neurath, H. Pancreatic enzymes of the spiny Pacific dogfish. II. Procarboxypeptidase B and carboxypeptidase B. *Biochemistry* **5**: 4137-4145.

1967

Ambler, R. P. Carboxypeptidases A and B. *Methods Enzymol.* **11**: 436-445.

Erdös, E. G., Yang, H. Y. T., Tague, L. L. & Manning, N. Carboxypeptidase in blood and other fluids. III. The esterase activity of the enzyme. *Biochem. Pharmacol.* **16**: 1287-1297.

1968

Elzinga, M. & Hirs, C. H. W. Primary structure of bovine carboxypeptidase B. II. Tryptic peptides from the reduced, aminoethylated protein. *Arch. Biochem. Biophys.* **123**: 343-352.

Elzinga, M. & Hirs, C. H. W. Primary structure of bovine carboxypeptidase B. IV. Amino acid sequence of a disulfide-containing loop. *Arch. Biochem. Biophys.* **123**: 361-367.

Elzinga, M., Lai, C. Y. & Hirs, C. H. W. Primary structure of bovine carboxypeptidase B. III. The carboxyl-terminal sequence. *Arch. Biochem. Biophys.* **123**: 353-360.

Jones, D. D. & Miller, W. G. Studies on the action of carboxypeptidase B on polylysine. *Biochim. Biophys. Acta* **159**: 411-413.

Kycia, J. H., Elzinga, M., Alonzo, N. & Hirs, C. H. W. Primary structure of bovine carboxypeptidase B. I. Preparation of enzyme from pancreatic juice. *Arch. Biochem. Biophys.* **123**: 336-342.

1969

Bradshaw, R. A., Neurath, H. & Walsh, K. A. Considerations of the concept of structural homology as applied to bovine carboxypeptidases A and B. *Proc. Natl. Acad. Sci. U.S.A.* **63**: 406-411.

Plummer, T. H., Jr. Isolation and sequence of peptides at the active center of bovine carboxypeptidase B. *J. Biol. Chem.* **244**: 5246-5253.

1970

Folk, J. E. Carboxypeptidase B (porcine pancreas). *Methods Enzymol.* **19**: 504-508.

1971

Akanuma, H., Kasuga, A., Akanuma, T. & Yamasaki, M. The effective use of affinity chromatography for the study of complex formation of bovine carboxypeptidase B with basic and aromatic amino acid analogues. *Biochem. Biophys. Res. Commun.* **45**: 27-33.

Folk, J. E. Carboxypeptidase B. In: *The Enzymes* (Boyer, P. D. ed.), 3rd edn., vol. 3, pp. 57-79, Academic Press, New York.

Kemmler, W., Peterson, J. D. & Steiner, D. F. Studies on the conversion of proinsulin to insulin. I. Conversion in vitro with trypsin and carboxypeptidase B. *J. Biol. Chem.* **246**: 6786-6791.

Plummer, T. H., Jr. Evidence for a carboxyl group at the active center of bovine carboxypeptidase B. *J. Biol. Chem.* **246**: 2930-2935.

Reeck, G. R., Walsh, K. A. & Neurath, H. Isolation and characterization of carboxypeptidases A and B from activated pancreatic juice. *Biochemistry* **10**: 4690-4698.

Reeck, G. R., Walsh, K. A., Hermodson, M. A. & Neurath, H. New forms of bovine carboxypeptidase B and their homologous relationships to carboxypeptidase A. *Proc. Natl. Acad. Sci. USA* **68**: 1226-1230.

Sokolovsky, M. & Zisapel, N. Porcine carboxypeptidase B. I. Affinity chromatography and specificity. *Biochim. Biophys. Acta* **250**: 203-206.

Wünsch, E., Jaeger, E. & Schönsteiner-Altmann, G. Chromophor-Substrate. VII. Zur Spezifität der Carboxypeptidase B. *Hoppe-Seyler's Z. Physiol. Chem.* **352**: 1580-1583.

1972

Hass, G. M., Govier, M. A., Grahn, D. T. & Neurath, H. Modification of bovine carboxypeptidase B with N-Bromoacetyl-N-methyl-L-phenylalanine. *Biochemistry* **11**: 3787-3791.

Kimmel, M. T. & Plummer, T. H., Jr. Identification of a glutamic acid at the active center of bovine carboxypeptidase B. *J. Biol. Chem.* **247**:

7864-7869.

Moore, G. J. & Benoiton, N. L. Positive cooperativity in the porcine carboxypeptidase B-catalyzed hydrolysis of neutral peptide substrates. *Biochem. Biophys. Res. Commun.* **47**: 581-587.

Sokolovsky, M. Porcine carboxypeptidase B. Nitration of the functional tyrosyl residue with tetranitromethane. *Eur. J. Biochem.* **25**: 267-273.

Zisapel, N. & Sokolovsky, M. Porcine carboxypeptidase B: multiple substrates binding modes. *Biochem. Biophys. Res. Commun.* **46**: 357-363.

1973

Zisapel, N. & Sokolovsky, M. Metallocarboxypeptidases: a cadmium-carboxypeptidase B with peptidase activity. *Biochem. Biophys. Res. Commun.* **53**: 722-729.

Zisapel, N., Kurn-Abramowitz, N. & Sokolovsky, M. Basic and non-basic substrates of carboxypeptidase B. *Eur. J. Biochem.* **35**: 507-511.

Zisapel, N., Kurn-Abramowitz, N. & Sokolovsky, M. Peptide inhibitors and activators of carboxypeptidase B. *Eur. J. Biochem.* **35**: 512-516.

1974

Geokas, M. C., Wollesen, F. & Rinderknecht, H. Radioimmunassay for pancreatic carboxypeptidase B in human serum. *J. Lab. Clin. Med.* **84**: 574-583.

Oshima, G., Kato, J. & Erdös, E. G. Subunits of human plasma carboxypeptidase N (kininase I, anaphylatoxin inactivator). *Biochim. Biophys. Acta* **365**: 344-348.

Schmid, M. F., Herriott, J. R. & Lattman, E. E. The structure of bovine carboxypeptidase B: results at 5.5 Ångström resolution. *J. Mol. Biol.* **84**: 97-101.

Schmidt, J. J. & Hirs, C. H. W. Primary structure of bovine carboxypeptidase B. Inferences from the locations of the half-cystines and identification of the active site arginine. *J. Biol. Chem.* **249**: 3756-3764.

Sokolovsky, M. & Zisapel, N. Chemical approaches to the mode of action of carboxypeptidase B. *Isr. J. Chem.* **12**: 631-641.

Zisapel, N. & Sokolovsky, M. Affinity labelling of carboxypeptidase B: modification of a methionyl residue. *Biochem. Biophys. Res. Commun.* **58**: 951-959.

1975

Geokas, M. C., Largman, C., Brodrick, J. W., Raeburn, S. & Rinderknecht, H. Human pancreatic carboxypeptidase B. I. Isolation, purification and characterization of fraction II. *Biochim. Biophys. Acta* **391**: 396-402.

Moore, G. J. & Benoiton, N. L. Effect of modifiers on the hydrolysis of basic and neutral peptides by carboxypeptidase B. *Can. J. Biochem.* **53**: 747-757.

Titani, K., Ericsson, L. H., Walsh, K. A. & Neurath, H. Amino-acid sequence of bovine carboxypeptidase B. *Proc. Natl. Acad. Sci. USA* **72**: 1666-1670.

Zisapel, N. & Sokolovsky, M. On the interaction of esters and peptides with carboxypeptidase B. *Eur. J. Biochem.* **54**: 541-547.

1976

Brodrick, J. W., Geokas, M. C. & Largman, C. Human carboxypeptidase B. II. Purification of the enzyme from pancreatic tissue and comparison with the enzymes present in pancreatic secretion. *Biochim. Biophys. Acta* **452**: 468-481.

Corbin, N. C., Hugli, T. E. & Müller-Eberhard, H. J. Serum carboxypeptidase B: a spectrophotometric assay using protamine as substrate. *Anal. Biochem.* **73**: 41-51.

Schmid, M. F. & Herriott, J. R. Structure of carboxypeptidase B at 2.8 Å resolution. *J. Mol. Biol.* **103**: 175-190.

1977

Marinkovic, D. V., Marinkovic, J. N., Erdös, E. G. & Robinson, C. J. G. Purification of carboxypeptidase B from human pancreas. *Biochem. J.* **163**: 253-260.

1978

McKay, T. J. & Plummer, T. H., Jr. By-product analogues for bovine carboxypeptidase B. *Biochemistry* **17**: 401-405.

1979

Akanuma, H. & Yamasaki, M. Ligand bindings of bovine carboxypeptidase B. I. Kinetic studies using dipeptide substrates and their carboxyl-terminal amino acid analogs. *J. Biochem.* **85**: 775-783.

Bazzone, T. J., Sokolovsky, M., Cueni, L. B. & Vallee, B. L. Single-step isolation and resolution of pancreatic carboxypeptidases A & B. *Biochemistry* **18**: 4362-4366.

Butterworth, J. & Duncan, J. J. Carboxypeptidase B activity of cultured skin fibroblasts and relationship to cystic fibrosis. *Clin. Chim. Acta* **97**: 39-43.

Danner, J., Somerville, J. E., Turner, J. & Dunn, B. M. Multiple binding sites of carboxypeptidase B: the evaluation of dissociation constants by quantitative affinity chromatography. *Biochemistry* **18**: 3039-3045.

Marinkovic, D. V. N-Terminal amino acid sequences of human carboxypeptidases A, B_1, and B_2. *Biochem. Med.* **22**: 11-26.

Marinkovic, D. V., Marinkovic, J. N. & Hammon. K. Studies of human carboxypeptidase B purification and properties from human small intestine. *Biochem. Med.* **22**: 1-10.

McKay, T. J., Phelan, A. W. & Plummer, T. H., Jr. Comparative studies on human carboxypeptidases B and N. *Arch. Biochem. Biophys.* **197**: 487-492.

Ondetti, M. A., Condon, M. E., Reid, J., Sabo, E. F., Cheung, H. S. &

Cushman, D. W. Design of potent and specific inhibitors of carboxypeptidases A and B. *Biochemistry* **18**: 1427-1430.

1980

Kuroda, K., Akanuma, H., Sukenaga, Y., Sugihara, H. & Yamasaki, M. Ligand bindings of bovine carboxypeptidase B. III. Hydrophobic activators in dipeptide hydrolysis. *J. Biochem.* **87**: 1681-1689.

Sukenaga, Y., Akanuma, H. & Yamasaki, M. Ligand bindings of bovine carboxypeptidase B. IV. Oligopeptide substrates and extended active center. *J. Biochem.* **87**: 1691-1701.

Sukenaga, Y., Akanuma, H., Suekane, C. & Yamasaki, M. Ligand bindings of bovine carboxypeptidase B. II. Affinity chromatography and cooperative ligations. *J. Biochem.* **87**: 695-707.

1981

Wu, G. Y., Pereyra, B. & Seifter, S. Specificity of trypsin and carboxypeptidase B for hydroxylysine residues in denatured collagens. *Biochemistry* **20**: 4321-4324.

1982

Wallace, E. F., Evans, C. J., Jurik, S. M., Mefford, I. N. & Barchas, J. D. Carboxypeptidase B activity from adrenal medulla. Is it involved in the processing of proenkephalin? *Life Sci.* **31**: 1793-1796.

Zisapel, N. & Sokolovsky, M. Mechanistic implications of cyanide binding to carboxypeptidase B. *Int. J. Peptide Protein Res.* **19**: 470-479.

Zisapel, N., Mallul, Y. & Sokolovsky, M. Tyrosyl interactions at the active site of carboxypeptidase B. *Int. J. Peptide Protein Res.* **19**: 480-486.

1983

Zisapel, N., Blank, T. & Sokolovsky, M. Metal ion effects on target sites of modification in metallocarboxypeptidase B. *J. Inorg. Biochem.* **18**: 253-262.

1984

Umezawa, H., Aoyagi, T., Ogawa, K., Iinuma, H., Naganawa, H., Hamada, M. & Takeuchi, T. Histargin, a new inhibitor of carboxypeptidase B, produced by actinomycetes. *J. Antibiot.* **37**: 1088-1090.

Lysosomal Carboxypeptidase A

Summary

EC Number: 3.4.16.1

Earlier names: Cathepsin I, cathepsin A, and catheptic carboxypeptidase A.

Abbreviation: lCPA.

Distribution: Has a lysosomal localization and a wide species and tissue distribution. Activity attributed to lCPA has also been identified in cerebral grey and white matter of human brain, and cartilage of rabbit ear.

Source: Beef spleen [1-3], rat liver [4,5] and brain [6], hog kidney [7,8], and chicken skeletal muscle [9].

Action: Catalyzes the release of C-terminal hydrophobic residues having an unsubstituted α-carboxyl group. Residues of other types including proline, but not arginine or lysine, may also be released, provided the penultimate residue is hydrophobic.

The approximate relative rates at which acylated dipeptides are hydrolyzed (at pH 5.2) by the hog kidney enzyme are reported [8] to be: Z-Glu-Tyr 100, Z-Phe-Ala 1850, Z-Phe-Phe 648, Z-Phe-Pro 333, Z-Gly-Phe 70, Z-Gly-Leu 57, Z-Gly-Ala 35, Z-Gly-Tyr 26, and Z-Gly-Pro 6. The rate on Z-Glu-Tyr is generally twice that observed on Z-Glu-Phe. Bz-Gly-Arg and Bz-Gly-Lys are not attacked.

Non-acylated dipeptides and tripeptides are not attacked, but limited endopeptidase activity may occur at hydrophobic sites in small protected peptides, as for example at the Glu-Tyr bond in Z-Glu-Tyr-OEt and at the Val-Tyr bond in Z-Val-Tyr-OMe [3]. As was first demonstrated with glucagon [2], action on larger peptides is restricted to the sequential removal of amino acids from the C-termini. Angiotensin II and Met-enkephalin-Arg[6]-Phe[7] are similarly attacked [6].

No action occurs on proteins such as hemoglobin, serum albumin, or ribonuclease [10]. Earlier accounts of protein hydrolysis are now attributed to contaminating cathepsin D.

Requirements: pH 5.2-5.5 at 37°C. Some preparations of lCPA are markedly stabilized by sucrose (0.5 M) or KCl (0.1 M).

Substrate, usual: Z-Glu-Tyr [11], with rates measured colorimetrically with ninhydrin [12] or fluorometrically with o-phthalaldehyde [13] for the determination of liberated tyrosine.

Substrate, special: Z-Phe-Ala may offer greater sensitivity since it is reported to be hydrolyzed about 20 times faster than Z-Glu-Tyr by hog kidney *l*CPA [14].

N-(2-Furanacryloyl)-Phe-Phe, which was identified as a special substrate for tissue carboxypeptidase A of pancreatic origin (Entry 15.01), may also serve as a useful substrate in a spectrophotometric assay for *l*CPA. This seems especially likely in view of its exceptional rate on Z-Phe-Phe, as noted above.

Inhibitors: *l*CPA is classified as a serine protease on the basis of its sensitivity to Dip-F and Pms-F. A sensitivity to certain thiol reagents (4-chloromercuribenzoate) and heavy metals (Hg^{2+} and Ag^{2+}) has also been noted. Tos-PheCH$_2$Cl shows partial inhibition, whereas EDTA and IAcOH are ineffective. *l*CPA is particularly sensitive to the potato inhibitor that is specific for serine proteases [15], but is unaffected by the potato inhibitor active against metallo-carboxypeptidases such as tCPA and tCPB.

Pepstatin, leupeptin and antipain have no effect, whereas chymostatin is weakly inhibitory [16].

Molecular properties: *l*CPA possesses a serine active center. As purified from most sources it shows a pronounced tendency to form stable aggregates that dissociate in the presence of urea [17]. The M_r 35,000 monomer from beef spleen contains N-terminal Asn, and the pI is 5.0-5.2 [17]. Small and large aggregate forms of the hog kidney enzyme have been designated "A,S" (M_r 100,000) and "A,L" (M_r 400,000); pI is 5.8 [7]. Both forms contain three structural subunits with M_r values of 20,000, 25,000 and 55,000. The M_r 25,000 subunit contains the active serine. The native A,S form of *l*CPB contains the subunits in a 1:1:1 ratio [18].

Preparations of *l*CPA from rat liver lysosomes contain a major (A1) component (M_r 200,000) according to one study [19]. Another study [16] of rat liver *l*CPA reported three components: A1 (M_r 100,000; pI 4.7), A2 (M_r 200,000; pI 4.8), and A3 (M_r 420,000; pI 4.9).

*l*CPA from most sources is a heat-labile enzyme, but the enzyme from chicken skeletal muscle is a notable exception. The hog kidney enzyme is markedly stabilized by the presence of sucrose (0.5 M) and KCl (0.1 M). *l*CPA is most stable at pH 5.0, and stores well in the frozen state; it becomes labile at pH 7.0, and is extremely unstable at alkaline pH [19a].

Comment: It appears that early reports of low levels of *l*CPA in beef spleen [20] and its absence from brain [21] may now be attributed to the use of alkaline extraction conditions [19a].

An enzyme has been described under the name "carboxamidopeptidase" (peptidyl aminoacylamidase, Entry 18.07) that catalyzes the release of amino acid amides from the C-termini of peptides such as oxytocin, vasopressin and substance P, at neutral pH. Recent evidence has shown, however, that peptidyl aminoacylamidase is a *l*CPA-like enzyme capable of releasing C-terminal amino acids optimally at pH 5.2 [22]. Conversely, *l*CPA has been shown to possess peptidyl aminoacylamidase activity at neutral pH [23].

As is explained in Entry 15.06, tyrosine carboxypeptidase (EC 3.4.16.3) of the hog thyroid shares many properties with *l*CPA and may yet prove to be the the same enzyme.

Caution should be exercised when assaying impure preparaions on Z-Glu-Tyr, because lysosomal carboxypeptidase B (Entry 15.04) also acts on this substrate. It should be possible to exclude interference by this sulfhydryl-dependent enzyme by the incorporation of IAcOH in assay mixtures. It may also help to use the special substrate Z-Phe-Ala, which should be less susceptible to attack by *l*CPB.

Significant elevations of *l*CPA have been noted in the skeletal muscle of rabbits made nutritionally dystrophic with a vitamin E-deficient diet. In chickens with hereditary muscular dystrophy, *l*CPA was the first of the lysosomal cathepsins to be significantly elevated during the development of the disease [24].

References

[1]   Fruton & Bergmann *J. Biol. Chem.* **130**: 19-27, 1939.
[2]   Iodice *Arch. Biochem. Biophys.* **121**: 241-242, 1967.
[3]   Logunov & Orekhovich *Biochem. Biophys. Res. Commun.* **46**: 1161-1168, 1972.
[4]   Taylor & Tappel *Biochim. Biophys. Acta* **341**: 99-111, 1974.
[5]   Matsuda & Misaka *J. Biochem.* **76**: 639-649, 1974.
[6]   Marks *et al. Peptides* **2**: 159-164, 1981.
[7]   Doi *et al. J. Biochem.* **75**: 889-894, 1974.
[8]   Kawamura *et al. J. Biochem.* **76**: 915-924, 1974.
[9]   Iodice *et al. Arch. Biochem. Biophys.* **117**: 477-486, 1966.
[10] Logunov & Orekhovich *Biochemistry USSR* **37**: 1061-1066, 1972.
[11] Taylor & Tappel *Biochim. Biophys. Acta* **341**: 112-119, 1974.
[12] Moore & Stein *J. Biol. Chem.* **176**: 367-388, 1948.
[13] Roth *Anal. Chem.* **43**: 880-882, 1971.
[14] Kawamura *et al. J. Biochem.* **76**: 915-924, 1974.
[15] Worowski *Experientia* **31**: 637-638, 1975.
[16] Matsuda & Misaka *J. Biochem.* **78**: 31-39, 1975.
[17] Logunov & Orekhovich *Biochemistry USSR* **37**: 715-721, 1972.
[18] Kawamura *et al. J. Biochem.* **88**: 1559-1561, 1980.
[19] Taylor & Tappel *Biochim. Biophys. Acta* **341**: 112-119, 1974.
[19a] Bowen & Davison *Biochem. J.* **131**: 417-419, 1973.
[20] Press *et al. Biochem. J.* **74**: 501-514, 1960.
[21] Marks & Lajtha *Biochem. J.* **97**: 74-83, 1965.
[22] Simmons & Walter In:*Neurohypophyseal Peptide Hormones and Other*

Biologically Active Peptides (Schlesinger, D. H. ed.), pp. 151-165, Elsevier/North Holland, 1981.

[23] Matsuda *J. Biochem.* **80**: 659-669, 1976.

[24] Iodice *et al. Arch. Biochem. Biophys.* **152**: 166-174, 1972.

Bibliography

1939

Fruton, J. S. & Bergmann, M. On the proteolytic enzymes of animal tissues. I. Beef spleen. *J. Biol. Chem.* **130**: 19-27.

1941

Fruton, J. S., Irving, G. W., Jr. & Bergmann, M. On the proteolytic enzymes of animal tissues. II. The composite nature of beef spleen cathepsin. *J. Biol. Chem.* **138**: 249-262.

Fruton, J. S., Irving, G. W., Jr. & Bergmann, M. On the proteolytic enzymes of animal tissues. III. The proteolytic enzymes of beef spleen, beef kidney and swine kidney. Classification of the cathepsins. *J. Biol. Chem.* **141**: 763-774.

1960

Greenbaum, L. M. Cathepsins and kinin-forming and -destroying enzymes. In: *The Enzymes* (Boyer, P. D. ed.), 3rd edn., vol. 3, pp. 475-483 (see pp. 481-482), Academic Press, New York.

Lichtenstein, N. & Fruton, J. S. Studies on beef spleen cathepsin A. *Proc. Natl. Acad. Sci. U.S.A.* **46**: 787-791.

1965

Iodice, A. A. & Weinstock, I. M. Cathepsin A in nutritional and hereditary muscular dystrophy. *Nature* **207**: 1102 only.

Shibko, S. & Tappel, A. L. Rat-kidney lysosomes: isolation and properties. *Biochem. J.* **95**: 731-741.

1966

Iodice, A. A., Leong, V. & Weinstock, I. M. Separation of cathepsins A and D of skeletal muscle. *Arch. Biochem. Biophys.* **117**: 477-486.

1967

Iodice, A. A. The carboxypeptidase nature of cathepsin A. *Arch. Biochem. Biophys.* **121**: 241-242.

1969

Ali, S. Y. & Evans, L. Studies on the cathepsins in elastic cartilage. *Biochem. J.* **112**: 427-433.

1971

Misaka, E. & Tappel, A. L.   Inhibition studies of cathepsins A, B, C and D from rat liver lysosomes. *Comp. Biochem. Physiol.* **38B**: 651-662.

1972

Iodice, A. A., Chin, J., Perker, S. & Weinstock, I. M.   Cathepsins A, B, C, D and autolysis during development of breast muscle of normal and dystrophic chickens. *Arch. Biochem. Biophys.* **152**: 166-174.

Logunov, A. I. & Orekhovich, V. N.   Isolation and properties of bovine spleen highly purified cathepsin A. *Biochemistry U.S.S.R. (Engl. Transl.)* **37**: 715-721.

Logunov, A. I. & Orekhovich, V. N.   Isolation and some properties of cathepsin A from bovine spleen. *Biochem. Biophys. Res. Commun.* **46**: 1161-1168.

Logunov, A. I. & Orekhovich, V. N.   Substrate specificity of cathepsin A and its behaviour toward inhibitors. *Biochemistry U.S.S.R. (Engl. Transl.)* **37**: 888-892.

1973

Bowen, D. M. & Davison, A. N.   Cathepsin A in human brain and spleen. *Biochem. J.* **131**: 417-419.

1974

Bowen, D. M. & Davison, A. N.   Macrophages and cathepsin A activity in multiple sclerosis brain. *J. Neurol. Sci.* **21**: 227-231.

Doi, E.   Stabilization of pig kidney cathepsin A by sucrose and chloride ion, and inhibition of the enzyme activity by diisopropyl fluorophosphate and sulfhydryl reagents. *J. Biochem.* **75**: 881-887.

Doi, E., Kawamura, Y., Matoba, T. & Hata, T.   Cathepsin A of two different molecular sizes in pig kidney. *J. Biochem.* **75**: 889-894.

Kawamura, Y., Matoba, T., Hata, T. & Doi, E.   Purification and some properties of cathepsin A of large molecular size from pig kidney. *J. Biochem.* **76**: 915-924.

Matsuda, K. & Misaka, E.   Studies on cathepsins of rat liver lysosomes. I. Purification and multiple forms. *J. Biochem.* **76**: 639-649.

Taylor, S. L. & Tappel, A. L.   Characterization of rat liver lysosomal cathepsin A1. *Biochim. Biophys. Acta* **341**: 112-119.

Taylor, S. L. & Tappel, A. L.   Identification and separation of lysosomal carboxypeptidases. *Biochim. Biophys. Acta* **341**: 99-111.

1975

Kawamura, Y., Matoba, T., Hata, T. & Doi, E.   Purification and some properties of cathepsin A of small molecular size from pig kidney. *J. Biochem.* **77**: 729-737.

Matsuda, K. & Misaka, E.   Studies on cathepsins of rat liver lysosomes. II. Comparative studies on multiple forms of cathepsin A. *J. Biochem.*

78: 31-39.

Worowski, K. Inhibition of cathepsin A activity by the potato protease inhibitor. *Experientia* **31**: 637-638.

1976

Grynbaum, A. & Marks, N. Characterization of a rat brain catheptic carboxypeptidase (cathepsin A) inactivating angiotensin-II. *J. Neurochem.* **26**: 313-318.

Kar, N. C. & Pearson, C. M. Arylamidase and cathepsin-A activity of normal and dystrophic human muscle. *Proc. Soc. Exp. Biol. Med.* **151**: 583-586.

Marks, N., Grynbaum, A. & Benuck, M. On the sequential cleavage of myelin basic protein by cathepsins A and D. *J. Neurochem.* **27**: 765-768.

Matsuda, K. Studies on cathepsins of rat liver lysosomes. III. Hydrolysis of peptides, and inactivation of angiotensin and bradykinin by cathepsin A. *J. Biochem.* **80**: 659-669.

1977

Kawamura, Y., Matoba, T., Hata, T. & Doi, E. Substrate specificities of cathepsin A,L and A,S from pig kidney. *J. Biochem.* **81**: 435-441.

McDonald, J. K. & Schwabe, C. Intracellular exopeptidases. In: *Proteinases in Mammalian Cells and Tissues* (Barrett, A. J. ed.), pp. 311-391 (see pp. 329-335), North Holland Publishing Co., Amsterdam.

1980

Kawamura, Y., Matoba, T. & Doi, E. Subunit structure of pig kidney cathepsin A. *J. Biochem.* **88**: 1559-1561.

Obled, C., Arnal, M. & Valin, C. Variations through the day of hepatic and muscular cathepsin A (carboxypeptidase A; EC.3.4.12. 2), C (dipeptidyl peptidase; EC 3.4.14.1) and D (endopeptidase D; EC 3.4.23.5) activities and free amino acids of blood in rats: influence of feeding schedule. *Br. J. Nutr.* **44**: 61-69.

Simmons, W. H. & Walter, R. Carboxamidopeptidase: purification and characterization of a neurohypophyseal hormone inactivating peptidase from toad skin. *Biochemistry* **19**: 39-48.

1981

Marks, N., Sachs, L. & Stern, F. Conversion of Met-enkephalin-Arg[6]-Phe[7] by a purified brain carboxypeptidase (cathepsin A). *Peptides* **2**: 159-164.

Simmons, W. H. & Walter, R. Enzyme inactivation of oxytocin: properties of carboxamidopeptidase. In: *Neurohypophyseal Peptide Hormones and Other Biologically Active Peptides* (Schlesinger, D. H. ed.), pp. 151-165, Elsevier/North Holland Publ. Co., New York.

Lysosomal Carboxypeptidase B

Summary

EC Number: 3.4.18.1

Earlier names: Cathepsin IV, cathepsin B2, catheptic carboxypeptidases A, B and G.

Abbreviation: *l*CPB.

Distribution: Generally distributed in the lysosomes of mammalian cells.

Source: Rat liver lysosomes [1], beef spleen [2-4], and rabbit lung [5].

Action: Catalyzes the release of C-terminal residues from N-acylated dipeptides and polypeptides possessing a free α-carboxyl group. Dipeptides are not hydrolyzed, but many tripeptides are. Tri-lysine is not hydrolyzed [1], but tetra-lysine is [2]. Like cathepsin B, *l*CPB catalyzes the deamidation of Bz-Arg-NH$_2$, but unlike cathepsin B, it exhibits no action on arylamide derivatives, e.g. Bz-Arg-NPhNO$_2$ [6], Bz-Arg-NNap [7]. It releases arginine from Z-Arg-Arg, but has no action on Z-Arg-Arg-NNap [2].

The substrate specificity of *l*CPB is relatively broad. Except for proline, amino acids of all classes are released at appreciable rates from N-acylated dipeptides. Approximate relative rates observed at pH 5.0 for beef spleen *l*CPB are as follows: Z-Gly-Met 100, Z-Gly-Phe 93, Z-Gly-Ser 85, Z-Gly-Asp 70, Z-Gly-Arg 58, Z-Gly-Leu 47, and Z-Gly-Gly 22 [2].

Polypeptides also serve as substrates for *l*CPB. The rat liver enzyme acts on Leu-Trp-Met-Arg-Phe-Ala (releasing four residues) and glucagon [1]. The beef spleen enzyme acts on Val-Leu-Ser-Glu-Gly (releasing three residues) and glucagon [2]. It also acts on the hormone relaxin (M_r 5521) to release Ser from the C-terminus of the B chain, followed by Trp, Val and Gly [8]. Leu, Arg and Ala are released from the A chain [9]. The rabbit lung enzyme catalyzed the release of Ala from Thr-Pro-Lys-Ala, and Leu and His from angiotensin [5]. Further action on these peptides is apparently precluded by the presence of a prolyl residue in the penultimate position.

The rabbit lung enzyme may possess a degree of endopeptidase activity, as indicated by its apparent ability to cleave the leucyl bond at position 15 of the B chain of oxidized beef insulin. This bond, however, is known to be especially sensitive to cleavage by cathepsins B [10] and N [11]. Thus, the possibility of trace contamination by such endopeptidases needs to be considered, particularly in light of the report that rat liver *l*CPB has no action on the A and B chains of insulin [1]. Peptides with an amidated C-terminus, such as substance P, are not attacked [5].

Requirements: pH 5.0 and 37°C; preincubation with 10 mM ETDA and 10 mM dithiothreitol.

Substrate, usual: Bz-Arg-NH$_2$, with liberated ammonia being determined colorimetrically by microdiffusion and reaction with Nessler's reagent [12], or fluorometrically by reaction with phthalaldehyde [13]. Cathepsin B is also detected by this assay.

Substrate, special: Bz-Gly-Arg, with liberated arginine detected colorimetrically with ninhydrin [14], or fluorometrically with fluorescamine [15]. This assay excludes cathepsin B activity.

Inhibitors: Sulfhydryl-blocking agents (4-chloromercuribenzoate, iodoacetate) and heavy metals (Hg^{2+}, Pb^{2+}) are typical inhibitors. Leupeptin (1 μM), and to a lesser extent antipain (1 μM), were recently reported to be effective inhibitors of rabbit lung *l*CPB. Pepstatin, bestatin, puromycin, Pms-F and MalNEt had little or no effect [5].

Molecular properties: *l*CPB is a cysteine protease, M_r about 50,000 [4,10]. The native enzyme consists of 2 subunits of approximately equal size (M_r 25,000), and contains two thiol groups [3], and about 10% carbohydrate [5]. *l*CPB shows maximum stability between pH 5.0 and 6.0, and is very unstable above pH 7. The pI is approximately 5.0 [4]. *l*CPB is more stable in storage with the active-site thiol blocked than free [3].

Comment: Activities previously referred to as catheptic carboxypeptidase B and catheptic caboxypeptidase G [16], as well as histone hydrolase [17], are now attributed to *l*CPB.

A thiol-dependent carboxypeptidase with a lysosomal distribution has been detected in cultured human skin fibroblasts. The acidic carboxypeptidase found in cells taken from matched pairs of control and cystic fibrosis patients exhibits many properties of *l*CPB [18]. Although a lowered plasma carboxypeptidase B-type activity has been reported in cystic fibrosis patients [19], no consistent differences were detected when the *l*CPB activities of skin fibroblasts were compared [18].

T-cell activating factor (TAF), previously known as LAF, produced by macrophages in a serum-free medium contains *l*CPB activity. The activity of TAF is inhibited by *l*CPB inhibitors and can be mimicked by commercial hog pancreatic CPB [20].

References
[1] Ninjoor *et al. Biochim. Biophys. Acta* **370**: 308-321, 1974.

[2] McDonald & Ellis *Life Sci.* **17**: 1269-1276, 1975.
[3] Otto & Riesenkönig *Biochim. Biophys. Acta* **379**: 462-475, 1975.
[4] McDonald & Schwabe In: *Proteinases in Mammalian Cells and Tissues* (Barrett, A. J. ed.), pp. 335-341, North Holland, 1977.
[5] Lones *et al. Arch. Biochem. Biophys.* **221**: 64-78, 1983.
[6] Otto *Hoppe-Seyler's Z. Physiol. Chem.* **348**: 1449-1460, 1967.
[7] McDonald *et al. Nature* **225**: 1048-1049, 1970.
[8] Schwabe *et al. Biochem. Biophys. Res. Commun.* **75**: 503-510, 1977.
[9] Schwabe *et al. Recent Prog. Horm. Res.* **34**: 123-211, 1978.
[10] Otto In: *Tissue Proteinases* (Barrett, A. J. & Dingle, J. T. eds.), pp. 1-25, North-Holland, 1971.
[11] Evans & Etherington *FEBS Lett.* **99**: 55-58, 1979.
[12] Seligson & Seligson. *J. Lab. Clin. Med.* **38**: 324-330, 1951.
[13] Taylor *et al. Anal. Biochem.* **60**: 153-162, 1974.
[14] Moore & Stein *J. Biol. Chem.* **176**: 367-388, 1948.
[15] Udenfriend *et al. Science* **178**: 871-872, 1972.
[16] Taylor & Tappel *Biochim. Biophys. Acta* **341**: 99-111, 1974.
[17] DeLumen & Tappel *Biochim. Biophys. Acta* **293**: 217-225, 1973.
[18] Butterworth & Duncan *Clin. Chim. Acta* **97**: 39-43, 1979.
[19] Koheil & Forstner *Biochim. Biophys. Acta* **524**: 151-161, 1978.
[20] Dessaint *et al. J. Immunopharmacol.* **1**: 399-414, 1979.

Bibliography

1939

Fruton, J. S. & Bergmann, M. On the proteolytic enzymes of animal tissues. I. Beef spleen. *J. Biol. Chem.* **130**: 19-27.

1941

Fruton, J. S., Irving, G. W., Jr. & Bergmann, M. On the proteolytic enzymes of animal tissues. III. The proteolytic enzymes of beef spleen, beef kidney, and swine kidney. Classification of the cathepsins. *J. Biol. Chem.* **141**: 763-774.

1952

Tallan, H. H., Jones, M. E. & Fruton, J. S. On the proteolytic enzymes of animal tissues. X. Beef spleen cathepsin C. *J. Biol. Chem.* **194**: 793-805.

1960

Greenbaum, L. M. Cathepsins and kinin-forming and -destroying enzymes. In: *The Enzymes* (Boyer, P. D. ed.), 3rd edn., vol. 3, pp. 475-483 (see pp. 480-481), Academic Press, New York.

1962

Greenbaum, L. M. & Sherman, R. Studies on catheptic carboxypeptidase. *J. Biol. Chem.* **237**: 1082-1085.

1965

Greenbaum, L. M. & Yamafuji, K. Catheptic degradation of bradykinin. *Life Sci.* **4**: 657-663.

1966

Greenbaum, L. M. & Yamafuji, K. The role of cathepsins in the inactivation of plasma kinins. In: *Hypotensive Peptides* (Erdös, E. G., Back, N., Sicuteri, F. & Wilde, A. F. eds), pp. 252-262, Springer-Verlag, New York.

1967

Otto, K. Über ein neues Kathepsin. Reiningung aus Rindermilz, Eigenschaften, sowie Vergleich mit Kathepsin B. *Hoppe-Seyler's Z. Physiol. Chem.* **348**: 1449-1460.

Otto, K. & Schepers, P. Über die katheptische Inaktivierung einiger Enzyme der Rattenleber, insbesondere der Glucokinase. *Hoppe-Seyler's Z. Physiol. Chem.* **348**: 482-490.

1970

McDonald, J. K., Zeitman, B. & Ellis, S. Leucine naphthylamide: an inappropriate substrate for the histochemical detection of cathepsins B and B'. *Nature* **225**: 1048-1049 (see erratum **226**: 90).

1971

Mellors, A. A thiol-dependent cathepsin IV in rat liver lysosomes. *Arch. Biochem. Biophys.* **144**: 281-285.

1972

Distelmaier, P., Hüber, H. & Otto, K. Cathepsins B1 and B2 in various organs of the rat. *Enzymologia* **42**: 363-375.

1973

De Lumen, B. O. & Tappel, A. L. Histone hydrolase activity of rat liver lysosomal cathepsin B2. *Biochim. Biophys. Acta* **293**: 217-225.

1974

Ninjoor, V., Taylor, S. L. & Tappel, A. L. Purification and characterization of rat liver lysosomal cathepsin B2. *Biochim. Biophys. Acta* **370**: 308-321.

Taylor, A., Ninjoor, V., Dowd, D. M. & Tappel, A. L. Cathepsin B2 measurement by sensitive fluorometric ammonia analysis. *Anal. Biochem.*

60: 153-162.

Taylor, S. L. & Tappel, A. L.   Identification and separation of lysosomal carboxypeptidases. *Biochim. Biophys. Acta* **341**: 99-111.

1975

McDonald, J. K. & Ellis, S.   On the substrate specificity of cathepsins B1 and B2 including a new fluorogenic substrate for cathepsin B1. *Life Sci.* **17**: 1269-1276.

Otto, K. & Riesenkönig, H.   Improved purification of cathepsin B1 and cathepsin B2. *Biochim. Biophys. Acta* **379**: 462-475.

1976

Afroz, H., Otto, K., Müller, R. & Fuhge, P.   On the specificity of bovine spleen cathepsin B2. *Biochim. Biophys. Acta* **452**: 503-509.

1977

McDonald, J. K. & Schwabe, C.   Intracellular exopeptidases.   In: *Proteinases in Mammalian Cells and Tissues* (Barrett, A. J. ed.), pp. 311-391 (see pp. 335-341), North-Holland Publishing Co., Amsterdam.

Schwabe, C., McDonald, J. K. & Steinetz, B. G.   Primary structure of the B-chain of porcine relaxin. *Biochem. Biophys. Res. Commun.* **75**: 503-510.

1978

Guy, G. J. & Butterworth, J.   Carboxypeptidase A activity of cultured skin fibroblasts and relationship to cystic fibrosis. *Clin. Chim. Acta* **87**: 63-69.

Schwabe, C., Steinetz, B., Weiss, G., Segaloff, A., McDonald, J. K., O'Byrne, E., Hochman, J., Carriere, B. & Goldsmith, L.   Relaxin. *Recent Prog. Horm. Res.* **34**: 123-211.

1979

Butterworth, J. & Duncan, J. J.   Carboxypeptidase B activity of cultured skin fibroblasts and relationship to cystic fibrosis. *Clin. Chim. Acta* **97**: 39-43.

Dessaint, J.-P., Katz, S. P. & Waksman, B. H.   Catheptic carboxypeptidase B as a major component in "T-cell activating factor" of macrophages. *J. Immunopharmacol.* **1**: 399-414.

Kar, N. C. & Pearson, C. M.   Activity of some proteolytic enzymes in normal and dystrophic human muscle. *Clin. Biochem.* **12**: 37-39.

1980

Butterworth, J. & Duncan, J. J.   Chromatography of carboxypeptidase A and B activity of cultured skin fibroblasts: relationship to cystic fibrosis. *Clin. Chim. Acta* **108**: 143-146.

1983

Lones, M., Chatterjee, R., Singh, H. & Kalnitsky, G. Lysosomal carboxypeptidase B from rabbit lung. Purification and characterization. *Arch. Biochem. Biophys.* **221**: 64-78.

Entry 15.05

Lysosomal Prolyl Carboxypeptidase

Summary

EC Number: 3.4.16.2

Earlier names: Angiotensinase C, lysosomal carboxypeptidase C, lysosomal angiotensinase, acid angiotensinase, and proline carboxypeptidase.

Abbreviation: *l*PCP.

Distribution: Widely distributed in mammalian tissues, especially lysosome-rich tissues such as kidney, spleen and liver. Has also been detected in biological fluids such as urine and synovial fluid from arthritic joints [1], and in cells such as polymorphonuclear neutrophils, macrophages and lymphocytes isolated from human blood [1]. Human endothelial cells isolated from umbilical arteries and veins are especially high in *l*PCP activity [2]. Human foreskin fibroblasts and pulmonary artery endothelial cells are also relatively high in activity [2]. No activity is detectable in human plasma or red cells.

Source: Lysosomal fractions from rat kidney [3], and homogenates of hog kidney [4,5], human kidney [6] and beef spleen [7].

Action: Catalyzes the hydrolysis of C-terminal, penultimate prolyl bonds to release C-terminal, preferentially hydrophobic, amino acids from N-acylated dipeptides and polypeptides possessing a free α-carboxyl group. Dipeptides of the Pro-X type are not hydrolyzed. Approximate relative rates observed at pH 5.5 for rat liver *l*PCP are as follows: Z-Pro-Phe 100, Z-Pro-Tyr 83, Z-Pro-Met 75, Z-Pro-Val 67, Z-Pro-Ala 13, and Z-Pro-Ser 5 [7]. Relative rates for the hog kidney enzyme have been reported to be: Z-Pro-Phe 100, Z-Pro-Val 110, Z-Pro-Tyr 83, and Z-Pro-Trp 19 [8]; and for human kidney *l*PCP: Z-Pro-Phe 100, Z-Pro-Ala 420, Z-Pro-Val 360, Z-Pro-Leu 160, Z-Pro-Ser 95, Z-Pro-Tyr 71, and Z-Pro-Gly 6 [6]. No action occurs on Z-Pro-Phe-NH$_2$, Z-Pro-Pro, and Z-Phe-Pro. Z-Glu-Tyr and Bz-Gly-Arg substrates used for the assay of lysosomal carboxypeptidases A and B are not hydolyzed. Phenylalanine is rapidly released from the C-termini of angiotensin II, angiotensin III, and des-Arg[9]-bradykinin. K$_m$ values are approximately 1 mM, 2 mM and 0.77 mM for Z-Pro-Phe, angiotensin II, and angiotensin III, respectively [6]. *l*PCP also inactivates saralasin (Sar^1-Ala^8-angiotensin II), an antagonist of angiotensin II that has considerable clinical interest [9].

At pH 7.0, these polypeptide substrates are hydrolyzed at rates that are at least 50% of the rates observed at pH 5.0 [6]. This is also true for dipeptidyl peptidase I, another lysosomal exopeptidase possessing pronounced angiotensinase activity [10].

Requirements: pH 5.0-5.5 at 37°C.

Substrate, usual: Z-Pro-Phe [4,8], rates being measured colorimetrically with ninhydrin [11] or fluorometrically with phthalaldehyde [12].

Substrate, special: Z-Pro-[^{3}H]Ala, in conjunction with a scintillation method to determine the amount of free tritium-labeled alanine released into the reaction mixture. This step is preceded by an extraction of the reaction mixture to remove unreacted substrate [2].

Inhibitors: *l*PCP exhibits the characteristics of a serine protease in that it is completely inhibited by 1 mM Dip-F, but is unaffected by sulfhydryl-blocking reagents, chelating agents and sulfhydryl compounds. Pepstatin at relatively high concentrations (0.3 mM) is inhibitory, but aprotinin (Trasylol) is not [6].

Molecular properties: *l*PCP has a serine active center, M_r about 210,000 in the hog [5] and about 115,000 in the human [6]. The former is reported to be comprised of at least 8 subunits (M_r about 25,000) whereas the latter is said to consist of 2 subunits, one large (M_r 66,500) and one small (M_r 45,000).

*l*PCP from all sources is a remarkably stable enzyme: it readily tolerates the autolysis (22 h at pH 3.5 and 37°C) and heat treatment (40 min at 65°C at pH 5.0) steps used in the purification of dipeptidyl peptidase 1 [7].

Comment: Preparations of *l*PCP may be contaminated by dipeptidyl peptidase I, and *vice versa*, as a consequence of shared physicochemical properties. However, DPP I can be selectively inactivated with p-chloromercuriphenylsulfonate. Alternatively, *l*PCP can be selectively inactivated with Dip-F [10].

It has been postulated that renal *l*PCP plays an important part in regulating angiotensin levels in the kidney and elsewhere [6]; however, its distribution in diverse cells such as macrophages, lymphocytes, endothelial cells, and fibroblasts suggests a broader role in intracellular protein degradation.

References
    [1]   Kumamoto *et al*. *J. Clin. Invest.* **67**: 210-215, 1981.
    [2]   Skidgel *et al*. *Anal. Biochem.* **118**: 113-119, 1981.
    [3]   Matsunaga *et al*. *Japn. Circ. J.* **33**: 545-551, 1969.
    [4]   Yang *et al*. *Biochem. Pharmacol.* **19**: 1201-1211, 1970.
    [5]   Kakimoto *et al*. *Biochim. Biophys. Acta* **302**: 178-182, 1973.
    [6]   Odya *et al*. *J. Biol. Chem.* **253**: 5927-5931, 1978.

[7]   McDonald *et al. Biochem. Biophys. Res. Commun.* **46**: 62-70, 1972.
[8]   Yang *et al. Nature* **218**: 1224-1226, 1968.
[9]   Hollenberg *et al. J. Clin. Invest.* **57**: 39-46, 1976.
[10] McDonald *et al. J. Biol. Chem.* **249**: 234-240, 1974.
[11] Moore & Stein. *J. Biol. Chem.* **211**: 907-913, 1954.
[12] Roth *Anal. Chem.* **43**: 880-882, 1971.

Bibliography

1968

Johnson, D. C. & Ryan, J. W.   Degradation of antiotensin II by a carboxypeptidase of rabbit liver. *Biochim. Biophys. Acta* **160**: 196-203.
Yang, H. Y. T., Erdös, E. G. & Chiang, T. S.   New enzymatic route for the inactivation of angiotensin. *Nature* **218**: 1224-1226.

1969

Matsunaga, M., Saito, N., Kira, J., Ogino, K. & Takayasu, M.   Acid angiotensinase as a lysosomal enzyme. *Jpn. Circ. J.* **33**: 545-551.

1970

Yang, H. Y. T. & Erdös, E. G.   Prolylcarboxypeptidase: a recently described lysosomal enzyme. In: *Immunopathology of Inflammation* (Forscher, B. K. & Houck, J. C. eds), pp. 146-148, Excerpta Medica, Amsterdam.
Yang, H. Y. T., Erdös, E. G., Chiang, T. S., Jenssen, T. A. & Rodgers, J. G.   Characteristics of an enzyme that inactivates angiotensin II (angiotensinase C). *Biochem. Pharmacol.* **19**: 1201-1211.

1971

Canonico, P. G., Zweiten, M. Van. & Pinkerton, R. S.   Lysosomal processing of exogenous proteins. *Fed. Proc. Fed. Am. Soc. Exp. Biol.* **30**: 600 only.
Matsunaga, M.   Nature of lysosomal angiotensinase activity. *Jpn. Circ. J.* **35**: 333-338.
Sorrells, K. & Erdös, E. G.   Prolylcarboxypeptidase (angiotensinase C) in subcellular particles of leucocytes. *Fed. Proc. Fed. Am. Soc. Exp. Biol.* **30**: 600 only.
Sorrells, K. & Erdös, E. G.   Prolylcarboxypeptidase in biological fluids. *Adv. Exp. Med. Biol.* **23**: 393-397.

1972

McDonald, J. K., Zeitman, B. B. & Ellis, S.   Detection of a lysosomal carboxypeptidase and a lysosomal dipeptidase in highly-purified dipeptidyl aminopeptidase I (cathepsin C) and the elimination of their activities from preparations used to sequence peptides. *Biochem.*

Biophys. Res. Commun. **46**: 62-70.

1973

Kamimoto, T., Oshima, G., Yeh, H. S. J. & Erdös, E. G. Purification of lysosomal prolylcarboxypeptidase angiotensinase C. *Biochim. Biophys. Acta* **302**: 178-182.

1977

McDonald, J. K. & Schwabe, C. Intracellular exopeptidases. In: *Proteinases in Mammalian Cells and Tissues* (Barrett, A. J. ed), pp. 311-391, (see pp. 341-344), North-Holland Publishing Co., Amsterdam.

1978

Odya, C. E., Marinkovic, D. V., Hammon, K. J., Stewart, T. A. & Erdös, E. G. Purification and properties of prolylcarboxypeptidase (Angiotensinase C) from human kidney. *J. Biol. Chem.* **253**: 5927-5931.

1981

Kumamoto, K., Stewart, T. A., Johnson, A. R. & Erdös, E. G. Prolylcarboxypeptidase (angiotensinase C) in human lung and cultured cells. *J. Clin. Invest.* **67**: 210-215.

Odya, C. E. & Erdös, E. G. Human prolylcarboxypeptidase. *Methods Enzymol.* **80**: 460-466.

Skidgel, R. A., Wickstrom, E., Kumamoto, K. & Erdös, E. G. Rapid radioassay for prolylcarboxypeptidase (angiotensinase C). *Anal. Biochem.* **118**: 113-119.

Lysosomal Tyrosine Carboxypeptidase

Summary

EC Number: 3.4.16.3 (but see Comment).

Earlier names: Acetylphenylalanyltyrosine (APAT) hydrolase, thyroid peptide carboxypeptidase, and thyroid peptidase.

Abbreviation: *T*CP.

Distribution: *T*CP activity has been demonstrated in the thyroid glands of several mammalian species [1]. Among rat tissues, activity in thyroid gland extracts is exceeded only by that in kidney extracts [1]. However, the action of kidney extracts on Ac-Phe-Tyr tends to be exaggerated as a result of the release of free phenylalanine by the high levels of aminoacylase in that organ. In addition, lysosomal carboxypeptidase A is known to exert some action on Ac-Phe-Tyr at pH 4.1 [2]. In the hog thyroid, *T*CP is localized predominantly in lysosome-like organelles [3].

Source: Hog thyroid gland [1,4].

Action: Catalyzes the release of tyrosine, preferentially, from the unsubstituted C-termini of N-acylated dipeptides and of polypeptides, provided the penultimate residue is aromatic or hydrophobic. Comparable rates have been reported for action on Ac-Phe-Tyr, Gly-Leu-Tyr, Gly-Phe-Phe, and Trp-Leu at pH 4.1 [1,4]. (The possibility exists, however, that the action on Trp-Leu was due to contamination of the enzyme preparation used by lysosomal dipeptidase I, Entry 17.10.) Ac-Phe-3,5-diiodotyrosine also is hydrolyzed [1], but no action occurs on either Z-Glu-Tyr or Z-Gly-Phe. No endopeptidase activity is detectable on the A and B chains of oxidized insulin, nor on intact thyroglobulin [5].

　　*T*CP releases mono-iodotyrosine, di-iodotyrosine and thyroxine from intermediate peptides produced by the action of thyroid aspartic proteinase (Volume 1, Entry 3.12) on [^{131}I]thyroglobulin, resulting in the liberation of 25-30% of the total iodinated amino acids [6].

Requirements: pH 4.1 and 37°C for the hog enzyme [4], and pH 5.6 and 37°C for the rat enzyme [1].

Substrate, usual: Ac-Phe-Tyr. Z-Phe-Tyr gives comparable rates of hydrolysis, but lacks adequate solubility.

Inhibitors: p-Chloromercuribenzoate (10 μM) and Tos-Phe-CH$_2$Cl (0.25 mM) give complete inhibition, whereas IAcOH and MalNEt give only slight inhibition. Dip-F (1 mM) has no effect [1].

Molecular properties: Its sensitivity to certain thiol reagents suggests that *I*TCP is a cysteine protease, but it does not show a requirement for sulfhydryl activation. Although the enzyme has been purified more than 1200-fold, little physico-chemical characterization has been reported.

Comment: We consider lysosomal tyrosine carboxypeptidase to be distinct from lysosomal carboxypeptidase A (cathepsin A), because of its insensitivity to Dip-F and its inactivity on Z-Glu-Tyr. The Enzyme Nomenclature Committee has classified *I*TCP as a serine protease [7], but we can find no evidence in support of this.

It appears that the synergistic action of thyroid aspartic proteinase and *I*TCP observed *in vitro* most probably occurs *in vivo* as well. Both proteases reside within lysosome-like organelles of the thyroid gland [3] which have been reported to migrate under the influence of thyrotropin from the basal part of secretory epithelial cells of the thyroid follicle to the apical part, where they appear to fuse with colloid (thyroglobulin) droplets entering from the lumen [8].

References
[1] Dunn & McQuillan *Biochim. Biophys. Acta* **235**: 149-158, 1971.
[2] Taylor & Tappel *Biochim. Biophys. Acta* **341**: 112-119, 1974.
[3] Jablonski & McQuillan *Biochim. Biophys. Acta* **132**: 454-471, 1967.
[4] Menzies & McQuillan *Biochim. Biophys. Acta* **132**: 444-453, 1967.
[5] McQuillan *et al.* *Nature* **192**: 333-336, 1961.
[6] Dopheide *et al.* *Biochim. Biophys. Acta* **181**: 105-115, 1969.
[7] IUB-IUPAC Nomenclature Committee *Enzyme Nomenclature 1984.* Academic Press, 1984.
[8] Wetzel *et al.* *J. Cell Biol.* **25**: 593-618, 1965.

Bibliography

1961

McQuillan, M. T., Mathews, J. D. & Trikojus, V. M. Proteolysis of thyroglobulin by thyroid enzymes. *Nature* **192**: 333-336.

1967

Jablonski, P. & McQuillan, M. T. The distribution of proteolytic enzymes in the thyroid gland. *Biochim. Biophys. Acta* **132**: 454-471.
Menzies, C. A. & McQuillan, M. T. Partial purification and properties of

a peptidase from thyroid glands. *Biochim. Biophys. Acta* **132**: 444-453.

1969

Dopheide, T. A. A., Menzies, C. A., McQuillan, M. T. & Trikojus, V. M. Studies with purified pig thyroglobulin and thyroid enzymes. *Biochim. Biophys. Acta* **181**: 105-115.

1971

Dunn, N. W. & McQuillan, M. T. Purification and properties of a peptidase from thyroid glands. *Biochim. Biophys. Acta* **235**: 149-158.

1977

McDonald, J. K. & Schwabe, C. Intracellular exopeptidases. In: *Proteinases in Mammalian Cells and Tissues* (Barrett, A. J. ed.),pp. 311-391 (see pp. 334-335), North-Holland Publishing Co., Amsterdam.

Microsomal Prolyl Carboxypeptidase

Summary

EC Number: 3.4.17.-

Earlier names: Carboxypeptidase P, microsomal carboxypeptidase.

Abbreviation: mPCP.

Distribution: Characterized as a membrane-bound enzyme in studies with microsomal fractions from hog kidney homogenates. The tissue distribution of mPCP has not been reported.

Source: Hog kidney [1].

Action: Catalyzes the hydrolysis of penultimate prolyl bonds to release C-terminal amino acids from N-acylated dipeptides and polypeptides possessing a free α-carboxyl group. Dipeptides of the Pro-X type are also hydrolyzed, but at significantly lower rates. Approximate relative rates observed at pH 7.75 for hog kidney mPCP are as follows: Z-Pro-Ala 100, Z-Gly-Pro-Ala 100, Z-Leu-Gly-Pro-Ala 100, Z-Gly-Pro-Gly-Gly-Pro-Ala 100, Z-Pro-Gly 20, Pro-Ala 12, and Pro-Phe 11 [1]. Some substrates lacking a penultimate prolyl residue are also hydrolyzed, as for example, on the same relative scale: Z-Ala-Tyr 32, Z-Gly-Tyr 13, Z-Gly-Phe 12, and Z-Gly-Leu 5. Little or no action occurs on Z-Pro-Pro, Z-Hyp-Gly, Z-Leu-Pro, Z-Gly-Gly, Z-Leu-Tyr, and Z-Leu-Ala. No chain length dependence has been noted.

Requirements: pH 7.75 and 1 mM $MnCl_2$ at 37°C.

Substrate, usual: Z-Pro-Ala with rates measured colormetrically using a modified ninhydrin reagent [1].

Substrate, special: None reported, but it would appear that Z-Pro-[^{3}H]Ala could be utilized in the manner described for lysosomal prolyl carboxypeptidase (see Entry 15.05).

Inhibitors: None reported, but inhibition should be possible with chelating agents and the protein inhibitor from potatoes, which is active against metallocarboxypeptidases [2].

Molecular properties: Hog kidney mPCP is a M_r 240,000 membrane-bound protein. As judged by the 5- to 6-fold activation obtained with Mn^{2+}, mCPC is most probably a metallocarboxypeptidase, but its final classification awaits more extensive characterization.

Comment: mPCP has not yet been obtained completely free of microsomal alanyl aminopeptidase (aminopeptidase M). The same toluene-trypsin extraction procedure developed for the solubilization of the latter enzyme [3] was employed for the solubilization of mPCP [1]. Both exopeptidases are plentiful in kidney microsome fractions, but similarities in molecular weight and behaviour in ion exchange chromatography have made them difficult to separate.

References

[1] Dehm & Nordwig *Eur. J. Biochem.* **17**: 372-377, 1970.
[2] Ryan *et al. J. Biol. Chem.* **249**: 5495-5499, 1974.
[3] Wachsmuth *et al. Biochemistry* **5**: 169-174, 1966

Bibliography

1968

Nordwig, A. & Dehm, P. Breakdown of protein: cleavage of peptides of the X-Pro-Y type by kidney peptidases. *Biochim. Biophys. Acta* **160**: 293-295.

1970

Dehm, P. & Nordwig, A. The cleavage of prolyl peptides by kidney peptidases. Isolation of a microsomal carboxypeptidase from swine kidney. *Eur. J. Biochem.* **17**: 372-377.

1977

Kenny, A. J. Proteinases associated with cell membranes. In: *Proteinases in Mammalian Cells and Tissues* (Barrett, A. J. ed.), pp. 393-444 (see p. 425), North Holland Publishing Co., Amsterdam.

1978

Kenny, A. J. & Booth, A. G. Microvilli: Their ultrastructure, enzymology and molecular organization. *Essays Biochem.* **14**: 1-44.

1979

Booth, A. G., Hubbard, L. M. L. & Kenny, A. J. Proteins of the kidney microvillar membrane. Immunoelectrophoretic analysis of the membrane hydrolases: identification and resolution of the detergent- and proteinase-solubilized forms. *Biochem. J.* **179**: 397-405.

Tubulinyl Tyrosine Carboxypeptidase

Summary

EC Number: 3.4.-.-

Earlier names: Carboxypeptidase-tubulin, soluble carboxypeptidase.

Abbreviation: tTCP.

Distribution: In brain (where tubulin accounts for 10-40% of the soluble protein) tTCP is a microtubule-associated protein that remains associated with microtubules during successive cycles of assembly-disassembly [1]. It is present in soluble (100,000 **g**) supernatant fractions prepared from brain homogenates.

Source: Rat [2], beef [3], and chicken [4] brain.

Action: Catalyzes the hydrolysis of a penultimate α-glutamyl bond to release C-terminal tyrosine from tubulinyl tyrosine without further digestion of tubulin [5], and without evidence of endopeptidase activity [1]. Hydrolysis rates are higher on tubulin from disassembled microtubules as compared to nonassembled tubulin [6]. The products of the reaction are free tyrosine and de-tyrosinated tubulin [5].

  tTCP exhibits a notable degree of specificity for native tubulin. Its rate on reduced and alkylated tubulin is only about 20% of that observed on native dimeric tubulin [3]. tTCP also exhibits a 380-fold greater preference than pancreatic carboxypeptidase A for tubulin over aldolase, another protein with C-terminal tyrosine [3].

Requirements: pH 7.0 and 2 mM Mg^{2+} at 37°C [2].

Substrate, usual: Tubulinyl-[^{14}C]tyrosine [2], the reaction being followed as increasing radioactivity in the trichloroacetic acid supernatant [3] or decreasing radioactivity in the precipitate [2].

Inhibitors: IAcOH (2 mM), $CuCl_2$ (0.1 mM), and $ZnCl_2$ (0.1 mM) are strong inhibitors [2,4]. EDTA (5 mM) and phenylacetate (10 mM), inhibitors of tissue (pancreatic) carboxypeptidase A show either little effect [2] or enhancement of activity [4].

  Z-Glu-Tyr, a model of the C-terminus of tubulinyl tyrosine, is not a substrate, but inhibits strongly [4]. Z-Gly-Tyr is a weaker inhibitor.

Inhibition by an endogenous proteoglycan from rat brain has been traced to a constituent glycosaminoglycan that contains equimolar uronic acid, hexosamine and sulfate, and is detroyed by chondroitinase ABC [7].

Polyamines such as spermine and spermidine (which are present in relatively high concentrations in mammalian nervous tissue) are stimulatory *in vitro* at concentrations considered to be within the physiological range, but are inhibitory at elevated levels. Spermine, for example, is stimulatory at 60 μM, but inhibitory at 4 mM [8].

Molecular properties: Despite reports of strong inhibition of rat brain tTCP by IAcOH, this is probably a metallo-carboxypeptidase. Sheep brain tTCP is inhibited by EDTA, o-phenanthroline and β-phenylpropionate [9]. On the other hand, the inhibitor derived from potato tubers, which is generally effective against metallo-carboxypeptidases, is not inhibitory [9].

The K_m for tubulinyl-[^{14}C]tyrosine is approximately 0.5 μM, or 50 μg per ml, based on studies with sheep brain tTCP [9].

The chicken brain enzyme of M_r 90,000 is unstable in partially (250-fold) purified form. Storage at 4°C for 24 h results in a total loss of activity, so complete purifiction has not been possible [4]. Beef brain tTCP, which has been purified 400-fold [3], and sheep brain tTCP, which has been purified 500-fold [9], are apparently more stable enzymes [3].

Comment: Tubulin undergoes a post-translational addition of tyrosine to the C-terminal glutamic acid of the α-subunit [10]. This reaction is catalyzed by an ATP-dependent enzyme, tubulin-tyrosine ligase [10]. The reaction is reversible in the presence of rather high concentrations of ADP and inorganic phosphate, but detyrosination is more readily accomplished by the exopeptidase, tubulinyl tyrosine carboxypeptidase.

Unassembled tubulin is the preferred substrate for the ligase [6], but tTCP acts preferentially on intact microtubules [6,11]. Consistent with this finding, it has been shown that the C-terminal tyrosine of α-tubulin exchanges rapidly with free tyrosine in cultured muscle cells, but only in the presence of intact microtubules [12]. Although the half-life of tubulin itself is greater than 48 h, that of the C-terminal tyrosine residue is 37 min [12].

Microtubules add dimers at one end and lose them at the other end, in a "treadmill metabolism" that is interrupted by colchicine. Detyrosination occurs late in the transit of the tyrosinated dimers through the microtubule. It has been suggested that loss of dimers and detyrosination are linked [11].

References
[1]   Arce & Barra *FEBS Lett* **157**: 75-78, 1983.
[2]   Argaraña *et al. Mol. Cell. Biochem.* **19**: 17-21, 1978.
[3]   Kumar & Flavin *J. Biol. Chem.* **256**: 7678-7686, 1981.
[4]   Argaraña *et al. J. Neurochem.* **34**: 114-118, 1980.

[5]   Hallak *et al. FEBS Lett.* **73**: 147-150, 1977.
[6]   Arce *et al. J. Neurochem.* **31**: 205-210, 1978.
[7]   Argaraña *et al. J. Biol. Chem.* **256**: 827-830, 1981.
[8]   Barra & Argaraña *Biochem. Biophys. Res. Commun.* **108**: 654-657, 1982.
[9]   Martensen & Flavin *Miami Winter Symp.* **16**: 612 only, 1979.
[10] Arce *et al. Eur. J. Biochem.* **59**: 145-149, 1975.
[11] Deanin *et al. Eur. J. Biochem.* **109**: 207-216, 1980.
[12] Thompson *et al. Proc. Natl. Acad. Sci. USA* **76**: 1318-1322, 1979.

Bibliography

1977

Hallak, M. E., Rodriguez, J. A., Barra, H. S. & Caputto, R.   Release of tyrosine from tyrosinated tubulin. Some common factors that affect this process and the assembly of tubulin. *FEBS Lett.* **73**: 147-150.

1978

Arce, C. A., Hallak, M. E., Rodriguez, J. A., Barra, H. S. & Caputto, R. Capability of tubulin and microtubules to incorporate and to release tyrosine and phenylalanine and the effect of the incorporation of these amino acids on tubulin assembly. *J. Neurochem.* **31**: 205-210.
Argaraña, C., Barra, H. S. & Caputto, R.   Release of [^{14}C]tyrosine from tubulinyl-[^{14}C]tyrosine by brain extract. Separation of a carboxypeptidase from tubulin-tyrosine ligase. *Mol. Cell. Biochem.* **19**: 17-21.

1979

Martensen, T. M. & Flavin, M.   Reversal of post translational tyrosylation of tubulin. *Miami Winter Symp.* **16**: 612 only.
Thompson, W. C., Deanin, G. G. & Gordon, M. W.   Intact microtubules are required for rapid turnover of carboxyl-terminal tyrosine of α-tubulin in cell cultures. *Proc. Natl. Acad. Sci. USA* **76**: 1318-1322.

1980

Argaraña, C. E., Barra, H. S. & Caputto, R.   Tubulinyl-tyrosine carboxypeptidase from chicken brain: properties and partial purification. *J. Neurochem.* **34**: 114-118.
Barra, H. S., Arce, C. A. & Caputto, R.   Total tubulin and its aminoacylated and non-aminoacylated forms during the development of rat brain. *Eur. J. Biochem.* **109**: 439-446.
Deanin, G. G., Preston, S. F., Hanson, R. K. & Gordon, M. W.   On the mechanism of turnover of the carboxy-terminal tyrosine of the alpha chain of tubulin. *Eur. J. Biochem.* **109**: 207-216.

1981

Argaraña, C. E., Barra, H. S. & Caputto, R. Inhibition of tubulinyl-tyrosine carboxypeptidase by brain soluble RNA and proteoglycan. *J. Biol. Chem.* **256**: 827-830.

Kumar, N. & Flavin, M. Preferential action of a brain detyrosinolating carboxypeptidase on polymerized tubulin. *J. Biol. Chem.* **256**: 7678-7686.

1982

Barra, H. S. & Argaraña, C. E. Activation of tubulinyl-tyrosine carboxypeptidase by spermine, spermidine and putrescine. *Biochem. Biophys. Res. Commun.* **108**: 654-657.

Barra, H. S., Argaraña, C. E. & Caputto, R. Enzymatic detyrosination of tubulin tyrosinated in rat brain slices and extracts. *J. Neurochem.* **38**: 112-115.

1983

Arce, C. A. & Barra, H. S. Association of tubulinyl-tyrosine carboxypeptidase with microtubules. *FEBS Lett.* **157**: 75-78.

Plasma Carboxypeptidase B

Summary

EC Number: 3.4.17.3

Earlier names: Carboxypeptidase N (CPN), kininase I, bradykininase I, arginine carboxypeptidase, anaphylatoxin inactivator (AI) and serum carboxypeptidase N (SCPB).

Abbreviation: pCPB.

Distribution: Circulates as an α-globulin in the blood plasma of many mammalian species. It is also present in serum and lymph. pCPB-like activity has been detected in platelets, but not in erythrocytes [1].

Source: Outdated human blood plasma [2-4] and pig serum [5].

Action: Catalyzes the release of unsubstituted basic amino acids, lysine and arginine, from the C-termini of peptides and proteins. Relative rates are: Bz-Gly-Lys 100, Bz-Gly-Arg 23, Pro-Phe-Lys 190, Phe-Ser-Pro-Phe-Arg 20, Z-Gly-Arg 5, Z-Ala-Arg 160, Z-Gly-Orn 1, and Bz-Gly-argininic acid 500 [6]. (Bz-Gly-argininic acid is O-[benzoylglycyl]-α-L-hydroxy-δ-guanidino-*n*-valeric acid.) In contrast to (pancreatic) tCPB, from both hog [6] and human sources [7], human pCPB releases C-terminal lysine about 4 times faster than arginine, but like tCPB, pCPB does not act on peptides containing a penultimate prolyl residue. On the other hand, the rate of hydrolysis is enhanced by the presence of a penultimate alanyl residue. Bz-Ala-Lys, for example, is hydrolyzed 22 times faster than Bz-Gly-Lys [7]. Furthermore, the K_m value for the former substrate (0.35 mM) is about one fourth that for the latter [7]. The K_m for Bz-Gly-argininic acid is 0.1 mM.

pCPB, although originally detected as a kininase active on bradykinin (K_m 0.4 μM) and kallidin [8,9], is now generally believed to contribute to the regulation of the blood levels of other physiologically-active peptides as well. For example, fibrinopeptides A and B [10], the complement-derived anaphylatoxins C3a, C4a and C5a [11,12] and the plasmin-generated fragments of fibrin [13] all contain a C-terminal arginine or lysine that is essential for activity. The more rapid rate of inactivation of C3a compared to C5a has been attributed to the presence of a penultimate alanyl residue in the former.

Salmine, a protamine from salmon sperm that contains a series of C-terminal arginyl residues, is rapidly degraded by pCPB, and has been adapted to a spectrophotometric assay [14].

Requirements: pH 7.6 for the Bz-substrates and pH 7.75 for FA-Ala-Lys, both at 37°C [4]. Co^{2+} enhances the rate of hydrolysis of these peptide substrates [11].

Substrate, usual: Bz-Ala-Lys [4,7] for assay of peptidase activity, or Bz-Gly-argininic acid for assay of esterase activity [2,4]. Rates are determined spectrophotometrically by monitoring the increase in A_{254}. A $\Delta\epsilon$ of 360 $M^{-1}cm^{-1}$ may be used to convert rates to units for the hydrolysis of Bz-substrates [15].

Substrate, special: N-(2-Furanacryloyl)-Ala-Lys (FA-Ala-Lys) with decrease in absorbance followed at 336 nm [4,16]. With an initial substrate concentration of 0.5 mM, units of activity may be calculated from $\Delta\epsilon$ - 1300 $M^{-1}cm^{-1}$ [16]. This assay is unaffected by protein concentration and is therefor the method of choice for the assay of pCPB in plasma and crude preparations.

Inhibitors: Bi-product analogs, compounds that structurally resemble the product of hydrolysis of C-terminal lysine or arginine peptides, are among the most potent competitive inhibitors of pCPB, especially dicarboxylic acid analogs of arginine [7]. The sulfur-containing derivative, guanidinoethylmercaptosuccinic acid (GEMSA), and the all-carbon derivative, guanidinopropylsuccinic acid (GPSA) were equally effective, with K_i values of 1.0 μM. Replacement of a carboxyl group with a sulfhydryl group, however, dramatically increases the efficiency of these bi-product analog inhibitors. For example, 2-mercaptomethyl-3-guanidinoethylthiopropanoic acid exhibits a K_i of 2.0 nM with human pCPB [17]. Metal binding agents such as EDTA, 1,10-phenanthroline and 2-mercaptoethanol are inhibitory, whereas Zn-EDTA enhances activity [2].

6-Aminohexanoic acid and 4-(aminomethyl)cyclohexanecarboxylic acid, both of which are used as drugs, are good inhibitors of pCPB [18].

Molecular properties: Human pCPB is a zinc-containing sialoglycoprotein [3]. The native enzyme, M_r 280,000-310,000 [6,11], consists of two heavy subunits, M_r 83,000-90,000 [3,15,18a], and two light subunits, M_r 48,000-55,000 [3,18a]. Mild proteolysis by trypsin or plasmin reduces the former subunit to one of about M_r 70,000, and the latter to about M_r 49,000 [3]. The native enzyme is readily dissociated by urea or SDS [6]. The separated subunits retain some enzymatic activity, but it is short lived. An unstable, M_r 45,000 form of the enzyme found in hog liver may be identical to the light subunit of pCPB [6].

An M_r 315,000 form of the enzyme purified from hog serum is composed of subunits of M_r 90,000, 50,000 and 30,000. Carbohydrate was detected only in the largest subunit [5]. Carbohydrate amounts to 17% by weight for native human pCPB; it is associated with the heavy subunit, and contains significant amounts of glucosamine, mannose, galactose, fucose and sialic acid [3]. Two active forms of the enzyme exhibiting pI values of 3.8 and 4.3 are fully converted by neuraminidase treatment to an active desialylated enzyme with a pI of about 5.3 [19].

pCPB is unusually rich in hydrophobic and aromatic amino acids, leucine alone accounting for 16% of the total residues, and 20% of those in the heavy chain [3]. The hydrophobic regions of pCPB are believed to facilitate aggregation and thereby stabilize the protein as a high M_r aggregate. pCPB is stable in blood plasma stored for several months. $A_{280.1\%}$ - 11.9 [3].

Comment: Early reports of depressed blood levels of pCPB in cystic fibrosis patients have not been confirmed in more recent studies with better methods [16]. Levels are depressed by cirrhosis of the liver, but are unaffected by diseases of the pancreas [6]. The familial syndrome of hyperbradykininism appears to be associated with depressed levels of circulating pCPB [20].

Elevated blood levels of pCPB have been observed in Hodgkin's disease, pregnancy and dextran shock.

It was recently proposed [18] that hypotension associated with the use of drugs such as 6-aminohexanoic acid and 4-(aminomethyl)-cyclohexane-carboxylic acid may be the consequence of elevated levels of bradykinin. This explanation is consistent with the fact that these drugs are good inhibitors of pCPB.

References

[1]   Erdös *et al. Biochem. Pharmacol.* **13**: 893-905, 1964.
[2]   Erdös *et al. Biochem. Pharmacol.* **16**: 1287-1297, 1967.
[3]   Plummer & Hurwitz *J. Biol. Chem.* **253**: 3907-3912, 1978.
[4]   Plummer & Erdös *Methods Enzymol.* **80**: 442-449, 1981.
[5]   Jeanneret *et al. Hoppe-Seyler's Z. Physiol. Chem.* **357**: 867-872, 1976.
[6]   Oshima *et al. Arch. Biochem. Biophys.* **170**: 132-138, 1975.
[7]   McKay *et al. Arch. Biochem. Biophys.* **197**: 487-492, 1979.
[8]   Erdös & Sloane *Biochem. Pharmacol.* **11**: 585-592, 1962.
[9]   Zacest *et al. Aust. J. Exp. Biol. Med. Sci.* **52**: 601-606, 1974.
[10] Teger-Nilsson *Acta Chem. Scand.* **22**: 3171-3182, 1968.
[11] Bokisch & Müller-Eberhard *J. Clin. Invest.* **49**: 2427-2436, 1970.
[12] Gorski *et al. Proc. Natl. Acad. Sci. USA* **76**: 5299-5302, 1979.
[13] Belew *et al. Biochim. Biophys. Acta* **621**: 169-178, 1980.
[14] Corbin *et al. Anal. Biochem.* **73**: 41-51, 1976.
[15] Folk *et al. J. Biol. Chem.* **235**: 2272-2277, 1960.
[16] Plummer & Kimmel *Anal. Biochem.* **108**: 348-353, 1980.
[17] Plummer & Ryan *Biochem. Biophys. Res. Commun.* **98**: 448-454,1981.
[18] Juillerat-Jeanneret *et al. Hoppe-Seyler's Z. Physiol. Chem.*

363: 51-58, 1982.
[18a]Levin *et al. Proc. Natl. Acad. Sci. USA* **79**: 4618-4622, 1982.
[19] Koheil & Forstner *Biochim. Biophys. Acta* **524**: 156-161, 1978.
[20] Streeten *et al. Lancet* **2**: 1048-1053, 1972.

Bibliography

1962

Erdös, E. G. & Sloane, E. M. An enzyme in human blood plasma that inactivates bradykinin and kallidins. *Biochem. Pharmacol.* **11**: 585-592.

1963

Erdös, E. G., Renfrew, A. G., Sloane, E. M. & Wohler, J. R. Enzymatic studies on bradykinin and similar peptides. *Ann. N. Y. Acad. Sci.* **104**: 222-235.

1964

Erdös, E. G., Sloane, E. M. & Wohler, I. M. Carboxypeptidase in blood and other fluids. I. Properties, distribution, and partial purification of the enzyme. *Biochem. Pharmacol.* **13**: 893-905.

1965

Erdös, E. G., Wohler, I. M., Levine, M. I. & Westerman, M. P. Carboxypeptidase in blood and other fluids. Values in human blood in normal and pathological conditions. *Clin. Chim. Acta* **11**: 39-43.

1967

Erdös, E. G., Yang, H. Y. T., Tague, L. L. & Manning, N. Carboxypeptidase in blood and other fluids. III. The esterase activity of the enzyme. *Biochem. Pharmacol.* **16**: 1287-1297.

1968

Teger-Nilsson, A. C. Degradation of human fibrinopeptides A and B in blood serum in vitro. *Acta Chem. Scand.* **22**: 3171-3182.

1970

Bokisch, V. A. & Müller-Eberhard, H. J. Anaphylatoxin inactivator of human plasma: its isolation and characcretization as a carboxypeptidase. *J. Clin. Invest.* **49**: 2427-2436.
Erdós, E. G. & Yang, H. Y. T. Kininases. In: *Handbook of Experimental Pharmacology* (Erdós, E. G. ed), vol. 25, pp. 289-323, (see pp. 290-294), Springer-Verlag, New York.

1972

Petáková, M., Simonianová, E. & Rybák, M. Carboxypeptidases N (kininases

I) in rat serum, lungs, liver and spleen and the inactivation of kinins (bradykinin). *Physiol. Bohemoslov.* **21**: 287-293.

Streeten, D. H. P., Kerr, L. P., Kerr, C. B., Prior, J. C. & Dalakos, T. G. Hyperbradykininism: a new orthostatic syndrome. *Lancet* **2**: 1048-1053.

1973

Conover, J. H., Conod, E. J. & Hirschhorn, K. Complement components in cystic fibrosis. *Lancet* **2**: 1501 only.

1974

Conover, J. H., Conod, E. J. & Hirschhorn, K. Studies on ciliary dyskinesia factor in cystic fibrosis. IV. Its possible identification as anaphylatoxin (C3a)-IgG complex. *Life Sci.* **14**: 253-266.

Oshima, G., Kato, J. & Erdös, E. G. Subunits of human plasma carboxypeptidase N (kininase I; anaphylatoxin inactivator). *Biochim. Biophys. Acta* **365**: 344-348.

Zacest, R., Oparil, S. & Talamo, R. C. Studies of plasma bradykininases using radiolabelled substrates. *Aust. J. Exp. Biol. Med. Sci.* **52**: 601-606.

1975

Hugli, T. E. Serum anaphylatoxins: formation, characterization and control. In: *Proteases and Biological Control* (Reich, E., Rifkin, D. B. & Shaw, E. eds), pp. 273-290, (see pp. 276-277), Cold Spring Harbor Laboratory.

Lieberman, J. Carboxypeptidase B-like activity and C3 in cystic fibrosis. *Am. Rev. Respir. Dis.* **111**: 100-102.

Oshima, G., Kato, J. & Erdós, E. G. Plasma carboxypeptidase N, subunits and characteristics. *Arch. Biochem. Biophys.* **170**: 132-138.

1976

Corbin, N. C., Hugli, T. E. & Müller-Eberhard, H. J. Serum carboxypeptidase B: A spectrophotometric assay using protamine as substrate. *Anal. Biochem.* **73**: 41-51.

Jeanneret, L., Roth, M. & Bargetzi, J.-P. Carboxypeptidase N from pig serum. *Hoppe-Seyler's Z. Physiol. Chem.* **357**: 867-872.

1978

Koheil, A. & Forstner, G. Isoelectric focusing of carboxypeptidase N. *Biochim. Biophys. Acta* **524**: 156-161.

Plummer, T. H. Jr. & Hurwitz, M. Y. Human plasma carboxypeptidase N. Isolation and characterization. *J. Biol. Chem.* **253**: 3907-3912.

1979

Gorski, J. P., Hugli, T. E. & Müller-Eberhard, H. J. C4a: the third anaphylatoxin of the human complement system. *Proc. Natl. Acad. Sci.*

USA **76**: 5299-5302.
McKay, T. J., Phelan, A. W. & Plummer, T. H. Jr. Comparative studies on human carboxypeptidases B and N. *Arch. Biochem. Biophys.* **197**: 487-492.
Simonianová, E. & Petáková, M. Isolation of carboxypepotidase N by affinity chromatography on column of CNBr-activated sepharose with immobilized antibody. *Coll. Czech. Chem. Commun.* **44**: 626-630.

1980

Belew, M., Gerdin, B., Lindeberg, G., Porath, J., Saldeen, T. & Wallin, R. Structure-activity relationships of vasoactive peptides derived from fibrin or fibrinogen degraded by plasmin. *Biochim. Biophys. Acta* **621**: 169-178.
Mathews, K. P., Pan, P. M., Gardner, N. J. & Hugli, T. E. Familial carboxypeptidase N deficiency. *Ann. Intern. Med.* **93**: 443-445.
Plummer, T. H., Jr. & Kimmel, M. T. An improved spectrophotometric assay for human plasma carboxypeptidase N. *Anal. Biochem.* **108**: 348-353.

1981

Plummer, T. H., Jr. & Erdös, E. G. Human plasma carboxypeptidase N. *Methods Enzymol.* **80**: 442-449.
Plummer, T. H., Jr. & Ryan, T. J. A potent mercapto bi-product analogue inhibitor for human carboxypeptidase N. *Biochem. Biophys. Res. Commun.* **98**: 448-454.

1982

Juillerat-Jeanneret, L., Roth, M. & Bargetzi, J.-P. Some properties of porcine carboxypeptidase N. *Hoppe-Seyler's Z. Physiol. Chem.* **363**: 51-58.
Levin, Y., Skidgel, R. A. & Erdös, E. G. Isolation and characterization of the subunits of human plasma carboxypeptidase N (kininase I). *Proc. Natl. Acad. Sci. USA* **79**: 4618-4622.

1983

Marceau, F., Drumheller, A., Gendreau, M., Lussier, A. & St.-Pierre, S. Rapid assay of human plasma carboxypeptidase N by high-performance liquid chromatographic separation of hippuryl-lysine and its product. *J. Chromatogr.* **266**: 173-177.

Crino Carboxypeptidase B

Summary

EC Number: 3.4.17.10

Earlier names: Enkephalin convertase, enkephalin-forming carboxypeptidase, cobalt-stimulated chromaffin granule carboxypeptidase, and insulin granule-associated carboxypeptidase.

Abbreviation: cCPB.

Distribution: Adrenal medulla (chromaffin granules), pituitary gland, and brain, with negligible activity in most peripheral tissues [1]. Soluble cCPB activity is 20-fold higher in the anterior than the posterior pituitary [1]. In the anterior lobe (which lacks enkephalin), cCPB is concentrated in secretory granules [2].

   Considerable regional variation occurs in the rat brain [3]. Levels are highest in the thalamus-hypothalamus, corpus striatum, hippocampus, and midbrain. Lower levels ($<$10%) occur in the cerebellum and brain stem, while intermediate levels are seen in the cerebral cortex. A membrane-bound form of cCPB is more uniformly distributed throughout the brain [4]. Recent studies show that cCPB distribution in the rat brain is closely associated with the distribution of enkephalinergic neurons [4a].

   cCPB is also present in the secretory granules of insulin-producing islet cells [5].

Source: Soluble extracts of beef brain, pituitary, and adrenal glands [3,6]; beef adrenal chromaffin granule and pituitary membranes [4]; and a transplantable rat insulinoma that consists almost entirely of β cells [5].

Action: In a comparison of rates of arginine release from a series of dansylated tripeptides (Dns-Phe-X-Arg), V_{max} is highest when alanine is the penultimate residue [6]. The rate of arginine release is 3-5-fold greater than for lysine. Penultimate residues, as shown for alanine, glycine, leucine, and isoleucine, are not released. Dansylated dipeptides such as Dns-Ala-Arg are poor substrates [6].

   The enkephalin pentapeptides, Met-enkephalin and Leu-enkephalin, are formed from hexapeptide precursors such as [Met^5]enkephalin-Arg^6, [Met^5]enkephalin-Lys^6, [Leu^5]enkephalin-Arg^6, [Leu^6]enkephalin-Lys^6, as well as from the heptapeptide precursor [Met^5]enkephalin-Arg^6-Arg^7 [3].

Requirements: pH 5.5 and 37°C with 1 mM CoCl$_2$. Activity attributable to cCPB is usually obtained by subtracting the carboxypeptidase activity seen in the absence of Co^{2+} [6].

Substrate, usual: Dansyl-Phe-Leu-Arg, with liberated Dns-Phe-Leu being determined fluorometrically after extraction into chloroform [8]. Activity attributable to cCPB is usually obtained by subtracting either the carboxypeptidase activity seen in the absence of Co^{2+} [6] or that which is not inhibited by GEMSA (see Inhibitors).

Substrate, special: [^{3}H]Benzoyl-Phe-Ala-Arg has been used in a similar assay for radiometric determination [7]. A ligand binding assay that utilizes [^{3}H]GEMSA (see Inhibitors) has also been reported [7a].

Inhibitors: Chelating agents (1 mM) such as 1,10-phenanthroline and EDTA show pronounced inhibition [3,5]. Thiol-blocking reagents such as (10 μM) p-chloromercuriphenyl sulfonate and (1 μM) HgCl$_2$ show moderate and variable degrees of inhibition (see Comment), whereas 1 mM concentrations of IAcOH, IAcNH$_2$, and MalNEt have little or no effect [1,5]. Metal ions (0.1 mM) such as Cd^{2+} and Cu^{2+} show strong inhibition [3], in contrast to the activating effects of Co^{2+} and Ni^{2+} [1,5].

Analogs of arginine, such as guanidinopropylsuccinic acid (GPSA) and guanidinoethylmercaptosuccinic acid (GEMSA), are potent competitive inhibitors at nanomolar concentrations, whereas bromoacetyl-D-arginine is a potent irreversible inhibitor of cCPB [8].

Dipeptides and tripeptides containing C-terminal arginine or lysine are strong competitive inhibitors. Free arginine is a weak inhibitor [3]. Pms-F, leupeptin, pepstatin, Tos-Lys-CH$_2$Cl, and dithiothreitol have no effect [1,3,5].

Molecular properties: A M_r 50,000 glycoprotein consisting of a single polypeptide chain [6]. The membrane-bound form of the enzyme is somewhat larger (M_r 52,000), but shows identical substrate and inhibitor specificities [4]. cCPB shows a remarkable (11-fold) enhancement of activity by 1 mM Co^{2+}; Ni^{2+} produces a smaller (2-fold) enhancement [3].

Comment: In the storage granules of many secretory cells, limited proteolysis (partial crinophagy) is responsible for the release of biologically active fragments from inactive precursors such as proenkephalin A and B, pro-opiomelanocortin, and proinsulin. This process is apparently initiated by a cathepsin B-like proteinase [9] that cleaves on the carboxyl side of basic amino acid pairs that characteristically flank active sequences contained within the larger precursors. A carboxypeptidase B-like enzyme, here termed "crino carboxypeptidase B", is believed to mediate the final stage of prohormone and neurotransmitter peptide activation by removing the arginine or lysine residues that remain attached to the newly-formed C-terminus. Such a role for cCPB is suggested by its regional distribution in the brain, which parallels enkephalin concentration [1], and by its presence in pituitary [2] and insulin [5] secretory granules. Additionally, its pH 5.5 optimum is similar to the predicted internal pH of the insulin secretory granule [10].

Although lysosomal carboxypeptidase B (Entry 15.04) is capable of releasing C-terminal arginine residues exposed by the action of cathepsin B [11], the broad specificity of the lysosomal enzyme would appear to preclude it from involvement in hormone bioactivation.

Preparations of crino carboxypeptidase B tend to be contaminated with the lysosomal cysteine carboxypeptidase; this might possibly account for discrepant reports concerning the effects of thiol reagents and metal chelators [1,5,12].

Although the Co^{2+} activation exhibited by plasma carboxypeptidase B (Entry 15.09) and the low molecular mass (M_r 50,000) of its active subunit bear a notable resemblance to cCPB, inhibitor studies have clearly distinguished these two enzymes [8].

A Co^{2+}-activated carboxypeptidase (M_r 73,000) recently detected in human urine has been shown to be distinct from both tissue (Entry 15.02) and plasma (Entry 15.09) carboxypeptidase B [13]. The urinary enzyme, which is believed to originate in the kidney, bears an interesting resemblance to crino carboxypeptidase B, in that both exhibit arginine- and lysine-cleaving activity that is enhanced by Co^{2+} and inhibited by Cd^{2+} and 1,10-phenanthroline. It seems likely that cCPB (M_r 50,000), which is probably released into the blood by secretory cells [14], could be filtered intact through the glomerulus, and thus appear in the urine. Neverthelee, cCPB is readily distinguished from urinary carboxypeptidase B (Entry 15.11) by its lower pH optimum.

It remains to be established whether the carboxypeptidase B-like activity present in different secretory cells is attributable to the same enzyme. It has been suggested that a family of carboxypeptidase genes may exist that are selectively expressed to meet the special processing needs of specific secretory cells [5].

References
[1]   Fricker *et al. Life Sci.* **31**: 1841-1844, 1982.
[2]   Hook & Loh *Proc. Natl. Acad. Sci. USA* **81**: 2776-2780, 1984.
[3]   Fricker & Snyder *Proc. Natl. Acad. Sci. USA* **79**: 3886-3890, 1982.
[4]   Supattapone *et al. J. Neurochem.* **42**: 1017-1023, 1984.
[4a] Lynch *et al. Proc. Natl. Acad. Sci. USA* **81**: 6543-6547, 1984.
[5]   Docherty & Hutton *FEBS Lett.* **162**: 137-140, 1983.
[6]   Fricker & Snyder *J. Biol. Chem.* **258**: 10950-10955, 1983.
[7]   Stack *et al. Life Sci.* **34**: 113-121, 1984.
[7a] Strittmatter *et al. J. Biol. Chem.* **259**: 11812-11817, 1984.
[8]   Fricker *et al. Biochem. Biophys. Res. Commun.* **111**: 994-1000,

1983.

[9] Docherty *et al. Proc. Natl. Acad. Sci. USA* **80**: 3245-3249, 1983.

[10] Hutton *Biochem. J.* **204**: 171-178, 1982.

[11] McDonald & Ellis *Life Sci.* **17**: 1269-1276, 1975.

[12] Hook *et al. Nature* **295**: 341-342, 1982.

[13] Skidgel *et al. Anal. Biochem.* **140**: 520-531, 1984.

[14] Mains & Eipper *Endocrinology* **115**: 1683-1690, 1984.

Bibliography

1973

Kemmler, W., Steiner, D. F. & Borg, J. Studies on the conversion of proinsulin to insulin. III. Studies *in vitro* with a crude secretion granule fraction isolated from rat islets of Langerhans. *J. Biol. Chem.* **248**: 4544-4551.

1982

Fricker, L. D. & Snyder, S. H. Enkephalin convertase: purification and characterization of a specific enkephalin-synthesizing carboxypeptidase localized to adrenal chromaffin granules. *Proc. Natl. Acad. Sci. USA* **79**: 3886-3890.

Fricker, L. D., Supattapone, S. & Snyder, S. H. Enkephalin convertase: a specific enkephalin synthesizing carboxypeptidase in adrenal chromaffin granules, brain and pituitary gland. *Life Sci.* **31**: 1841-1844.

Hook, V. Y. H., Eiden, L. E. & Brownstein, M. J. A carboxypeptidase processing enzyme for enkephalin precursors. *Nature* **295**: 341-342.

Hutton, J. C. The internal pH and membrane potential of the insulin-secretory granule. *Biochem. J.* **204**: 171-178.

1983

Docherty, K. & Hutton, J. C. Carboxypeptidase activity in the insulin secretory granule. *FEBS Lett.* **162**: 137-141.

Fricker, L. D. & Snyder, S. H. Purification and characterization of enkephalin convertase, an enkephalin-synthesizing carboxypeptidase. *J. Biol. Chem.* **258**: 10950-10955.

Fricker, L. D., Plummer, T. H., Jr. & Snyder, S. H. Enkephalin convertase: potent, selective, and irreversible inhibitors. *Biochem. Biophys. Res. Commun.* **111**: 994-1000.

1984

Hook, V. Y. & Loh, Y. P. Carboxypeptidase B-like converting enzyme activity in secretory granules of rat pituitary. *Proc. Natl. Acad. Sci.* **81**: 2776-2780.

Lynch, D. R., Strittmatter, S. M. & Snyder, S. H. Enkephalin convertase localization by [³H]guanidinoethylmercaptosuccinic acid autoradiography: selective association with enkephalin-containing neurons. *Proc. Natl. Acad. Sci. USA* **81**: 6543-6547.

Mains, R. E. & Eipper, B. A. Secretion and regulation of two biosynthetic enzyme activities, peptidyl-glycine α-amidating monooxygenase and a carboxypeptidase, by mouse pituitary corticotropic tumor cells. *Endocrinology* **115**: 1683-1690.

Stack, G., Fricker, L. D. & Snyder, S. H. A sensitive radiometric assay for enkephalin convertase and other carboxypeptidase B-like enzymes. *Life Sci.* **34**: 113-121.

Steiner, D. F., Docherty, K. & Carroll, R. Golgi/granule processing of peptide hormone and neuropeptide precursors: a minireview. *J. Cell. Biochem.* **24**: 121-130.

Strittmatter, S. M., Lynch, D. R. & Snyder, S. H. [³H]Guanidinoethylmercaptosuccinic acid binding to tissue homogenates. Selective labeling of enkephalin convertase. *J. Biol. Chem.* **259**: 11812-11817.

Supattapone, S., Fricker, L. D. & Snyder, S. H. Purification and characterization of a membrane-bound enkephalin-forming carboxypeptidase, "enkephalin convertase". *J. Neurochem.* **42**: 1017-1023.

1985

Strittmatter, S. M., Lynch, D. R., De Souza, E. B. & Snyder, S. H. Enkephalin convertase visualized in the pituitary and adrenal gland by [³H]guanidinoethylmercaptosuccinic acid (GEMSA) autoradiography: dehydration decreases neurohypophyseal levels. *Endocrinology, in the press.*

Entry 15.11

Urinary Carboxypeptidase B

Summary

EC Number: 3.4.17.-

Earlier names: Renal carboxypeptidase.

Abbreviation: uCPB.

Distribution: The tissue of origin has not yet been established, but uCPB is believed to originate in the brush border of the proximal renal tubules [1] and is thought to be responsible for a carboxypeptidase activity that was detected earlier in a particulate fraction from human cadaver kidneys [2].

Source: Human urine [1].

Action: Releases unsubstituted basic amino acids, lysine and arginine, from the C-termini of αN-protected dipeptides and peptide esters, and from polypeptides. Bz-Gly-Lys is hydrolyzed about 1.6-fold faster than Bz-Gly-Arg. The ester substrate, Bz-Gly-argininic acid, is hydrolyzed faster than either of the benzoylated dipeptides. Rates are enhanced 5-fold if enzyme is preincubated with (1 mM) $CoCl_2$. No carboxypeptidase A-type activity is seen on Bz-Gly-Phe [1].

   The rate of release of C-terminal arginine from bradykinin at pH 7.0 is also enhanced by Co^{2+}: k_{cat} increases to 47 min^{-1}, and K_m decreases to 1 μM, with the result that the catalytic coefficient, k_{cat}/K_m, is increased 1.5-fold.

Requirements: A pH 7.0 HEPES buffer for peptide substrates, or a pH 8.5 Tris buffer for ester substrates; 37°C. Enzyme was activated by preincubation for 2 h at 4°C in buffer containing 1 mM $CoCl_2$ [1].

Substrate, usual: Bz-Ala-Lys in a continuous spectrophotometric assay in which rate of increase in A_{254} is followed [3,4].

Substrate, special: Bz-Gly-argininic acid can be used in a similar assay to measure esterase activity spectrophotometrically.

Inhibitors: In the absence of Co^{2+}, the peptidase activity of uCPB is inhibited by (0.1 mM) Cd^{2+}, (1 mM) 1,10-phenanthroline, and (10 μM)

222

2-mercaptomethyl-3-guanidinoethyl-thiopropanoic acid (MGTA). Aprotinin, p-chloromercuriphenyl sulfonate and Pms-F have little or no effect [1].

Molecular properties: A single chain protein (M_r 73,000) that is resistent to fragmentation and inactivation by trypsin [1]. The smaller (M_r 40,000) form of the enzyme described earlier [2] is now thought to have resulted from partial degradation during purification. uCPB shows no cross reactivity with antisera raised to tissue (pancreatic) carboxypeptidase B [2] or to plasma carboxypeptidase B [1].

The activity of purified uCPB is stable for at least 2 h at pH 8.0 (37°C), but is unstable at room temperature below pH 4.5.

As is true for several metallo-carboxypeptidases, Co^{2+} (1 mM) enhances cCPB activity: k_{cat} is increased about 5-fold [1].

Comment: The separate classification of uCPB is based on differences in relative substrate specificity, effects of pH, stability and lack of immunological cross reactivity. In addition, the urinary enzyme is distinguished from the plasma and pancreatic metallo-carboxypeptidase on the basis of their responses to Cd^{2+} treatment: whereas the esterase activity (measured on Bz-Gly-argininic acid) of the tissue (pancreatic) enzyme (Entry 15.02) is enhanced 100% by treatment with 0.1 mM Cd^{2+} [5] and that of the plasma enzyme (Entry 15.09) is inhibited by 93% [6], the activity of the urinary enzyme is essentially unaffected [1].

uCPB is believed to participate in the urine, and possibly the kidney as well, in the further processing or inactivation of Lys^6- and Arg^6- enkephalins, kinins, and anaphylatoxins [1]. Although uCPB resembles crino carboxypeptidase (Entry 15.10) in specificity, their pH-activity profiles and molecular masses are different. Species lines are crossed, however, in this comparison.

References
[1] Skidgel *et al. Anal. Biochem.* **140**: 520-531, 1984.
[2] Marinkovic *et al. Proc. Soc. Exp. Biol. Med.* **165**: 6-12, 1980.
[3] Wolff *et al. J. Biol. Chem.* **237**: 3094-3099, 1962.
[4] Folk *Methods Enzymol.* **19**: 504-508, 1970.
[5] Folk & Gladner *Biochim. Biophys. Acta* **48**: 139-147, 1961.
[6] Erdös *et al. Biochem. Pharmacol.* **16**: 1287-1297, 1967.

Bibliography

1980

Marinkovic, D. V., Ward, P. E., Erdös, E. G. & Mills, I. H.
Carboxypeptidase-type kininase of human kidney and urine. *Proc. Soc. Exp. Biol. Med.* **165**: 6-12.

1984

Skidgel, R. A., Davis, R. M. & Erdös, E. G. Purification of a human urinary carboxypeptidase (kininase) distinct from carboxypeptidases A, E or N. *Anal. Biochem.* **140**: 520-531.

¨s, E. G. & Mills, I. H.

Carboxypeptidase-type kininase of human kidney and urine. *Proc. Soc. Exp. Biol. Med*

Section 16

PEPTIDYL DIPEPTIDASES

Entry 16.01

Peptidyl Dipeptidase A

Summary

EC Number: 3.4.15.1

Earlier names: Angiotensin I-converting enzyme (ACE), hypertensin-converting enzyme, kininase II, dipeptidyl carboxypeptidase (DCP), peptidase P, carboxycathepsin, dipeptide hydrolase (DH).

Abbreviation: PDP A.

Distribution: Although first detected in horse plasma [1], PDP A is widely distributed in the tissues. It is found primarily on the vascular endothelium [2] and on the brush border epithelial cells of renal tubules [3] and intestine [4]. Levels are generally highest in the lung where it is a membrane-bound component of the vascular endothelium, and is in direct apposition to the pulmonary circulation [5].

In the rat, lung is 8-fold richer in PDP A activity than small intestine, and 100-fold richer than liver [6]. In a survey of 25 rat tissues, the epididymis and testis were found to contain levels of activity that exceeded lung. Activity was associated with seminal plasma, not spermatozoa [7]. In the rabbit, testis is also the richest source, followed by kidney, lung, and small intestine. In the dog and cat, lung is a richer source than testis [8]. In the animal and human brain, the highest levels occur in the posterior lobe of the pituitary gland [9], with a highly variable regional distribution seen for the rest of the brain [10]. PDP A occurs in human urine, where levels seem to correlate with concentrations of Na^+ [11].

Source: Hog [12-14], beef [15] and human [14,16,17] kidney; rabbit [18,19], rat [20], guinea pig [21], hog [22-24], calf [25], beef [26,26a] and human [14,27-29a] lung; hog [30], horse [31] and human [26a,32] blood plasma, and rat brain [33]. Antiserum raised against hog kidney PDP A gives similar inhibition of the enzyme in hog kidney, lung and plasma [13].

Action: Removes C-terminal dipeptides from a wide range of N-acylated peptides and polypeptides provided the C-termini are unsubstituted and a prolyl residue does not reside in the penultimate position or a glutamic

227

or aspartic residue in the terminal position [15,33a]. Z-Pro-Phe-His-Leu, Z-Gly-Phe-Phe-Tyr, Z-Phe-Gly-Pro-Phe-Arg, Bz-Gly-Phe-Phe, Bz-Gly-Gly-Pro, and Bz-Gly-Phe-Pro are attacked by beef kidney PDP A, but Bz-Gly-Phe-Phe-NH$_2$ and Bz-Gly-Gly-Pro-NH$_2$ are not [15]. The human enzyme hydrolyzes Z-Phe-His-Leu about 10 times faster than angiotensin I [34]. Rabbit lung PDP A releases the C-terminal dipeptide (Phe-Arg) from bradykinin almost as rapidly as it does the C-terminal dipeptide (His-Leu) from angiotensin I (K$_m$ 0.07 mM). Corresponding k$_{cat}$ values are 546 and 792 min^{-1} [19]. Bz-Gly-His-Leu (K$_m$ 2.3 mM), a common assay substrate [18,25], is hydrolyzed much more rapidly (k$_{cat}$ 15,430 min^{-1}) [19]. At the optimum (10 mM) Cl$^-$ concentration for bradykinin hydrolysis by the hog lung enzyme, K$_m$ is lowest (0.85 mM) and V$_{max}$ is highest (1.4 μmoles/min/mg protein) [35]. Two dipeptides, Phe-Arg and Ser-Pro, are released sequentially from the C-terminus of bradykinin [35]. Lys-Ala is released by hog PDP A from the C-terminus (-Pro^{28}-Lys^{29}-Ala^{30}) of the B chain of oxidized hog insulin [24].

Human lung PDP A releases two dipeptides consecutively from the C-termini of both Leu-enkephalin (Tyr-Gly-Gly-Phe-Leu) and Met-enkephalin (Tyr-Gly-Gly-Phe-Met) [36]. In addition, this enzyme releases Ser-Glu from the eosinophilotactic peptide (N-f-Val-Gly-Ser-Glu), 4 dipeptides from each of 3 known physiological nonapeptides (the thymus factor, the delta sleep inducing peptide, and the experimental allergic encephalitogenic peptide), 3 dipeptides from bradykinin, 7 dipeptides from fibrinopeptide A, 9 dipeptides from the ribonuclease S-peptide and 14 dipeptides from glucagon. The action of PDP A on the latter 3 polyeptides constitutes total degradation.

Requirements: pH 7-8 and 37°C in a Tris buffer. Phosphate is inhibitory and borate and barbital buffers may cause quenching in fluorometric procedures [36a]. Cl$^-$ (0.01-0.3 M) is required for the hydrolysis of most substrates, such as Bz-Gly-His-Leu, Bz-Gly-Gly-Gly, FA-Phe-Gly-Gly and angiotensin I. The latter is hydrolyzed optimally at 0.1 M Cl$^-$ [35]. Bradykinin, on the other hand, is hydrolyzed by (hog lung) PDP A at a substantial rate (30-50% of maximum) in the absence of Cl$^-$, but maximal rates require 10 mM Cl$^-$ at pH 7.0 [35]. Chloride ion decreases the K$_m$ and increases the V$_{max}$ values for these physiological substrates [35]. Bz-Gly-His-Leu, a common assay substrate, is hydrolyzed most rapidly by rabbit lung PDP A in the presence of 0.3 M Cl$^-$ at its pH 8.2 optimum [18]. Activation of human plasma PDP A by anions (compared at 0.17 M) is maximal for Cl$^-$, followed by NO$_3^-$, Br$^-$, F$^-$, and I$^-$ [34].

Substrate, usual: N-Blocked tripeptides such as Z-Phe-His-Leu, Bz-Gly-His-Leu, and Bz-Gly-Gly-Gly, with rates of dipeptide release measured indirectly with o-phthalaldehyde in a fluorometric procedure [34], or with ninhydrin [37] or 2,4,5-trinitrobenzene sulfonic acid [38] in colorimetric procedures. Corrections may be necessary for dipeptidase activities that cleave His-Leu or Gly-Gly. Liberated hippuric acid (Bz-Gly) can also be measured spectrophotometrically [18], or colorimetrically [39] by use of p-dimethylaminobenzaldehyde.

The chromophoric nitro group contained in t-Boc-Phe(NO_2)-Phe-Gly [40] and Z-Phe(NO_2)-His-Leu [25] permits the direct ultraviolet spectrophotometric assay of PDP A by increase in absorbance at 310 nm.

Substrate, special: Furanacryloyl-Phe-Gly-Gly in a continuous (blue-shift) spectrophotometric assay [14,42], with rate of hydrolysis seen as decreasing absorbance at 328 nm (K_m 0.3 mM, k_{cat} 19,000 min^{-1}). Additionally, p-nitrobenzyloxycarbonyl-Gly-Trp-Gly [36a] and *o*-aminobenzoyl-Gly-Phe(NO_2)-Pro [43] are used in continuous fluorometric assays that are based on the relief of intramolecular quenching by a nitro group of fluorophores such as trytophan or *o*-aminobenzoylglycine.

Inhibitors: Metal-binding agents such as EDTA, 1,10-phenanthroline, and 8-hydroxyquinoline. Phosphate and HEPES buffers are inhibitory [44]. Many compounds that coordinate reversibly with zinc are also inhibitory; these include sulfhydryl compounds such as captopril (SQ 14,225) [45], mercaptoacyl derivatives of thiazolidine carboxylic acid (SA 291, SA 445 and YS 980) [46], and D-Cys-L-Pro (K_i 6 nM) [46a]; carboxylate compounds such as enalapril (MK 421) [47], phosphoramidates [48], phosphoric and phosphonic amides [49], and dipeptide hydroxamates, e.g. Glu(NHOH)-Pro (K_i 3.4 μM) [50].

Chlorambucil-proline inhibits irreversibly and stoichiometrically by alkylating (esterifying) an aspartic or glutamic acid side chain at the active site [51].

Peptide inhibitors include the synthetic peptides originally identified as "bradykinin-potentiating peptides" (BPP) in snake (*Bothrops jararaca*) venom, e.g. pGlu-Trp-Pro-Arg-Pro-Gln-Ile-Pro-Pro-NH_2 (SQ 30,881, teprotide, BPP_{9a}) [52], glutathione, insulin, the B chain of insulin [24], substance P [53], and a variety of small peptides [60].

Oligopeptide inhibitors of microbial origin include ancovenin, a 16 residue peptide from a strain of *Streptomyces* [54], aspergillo-marasmines from *Aspergillus oryzae* [55], and the muraceins (muramyl peptides) from *Nocardia orientalis* [56].

Pseudo-peptide (competitive) inhibitors developed as orally-active antihypertensive drugs effective at nanomolar levels include D-3-mercapto-2-methylpropanoyl-L-proline (SQ 14,225, captopril), which was the prototype for the mercaptoalkanoyl amino acid class of inhibitors [45]. N-[(S)-1-Carboxy-3-phenylpropyl]-Ala-Pro (enalapril maleate, MK-421) was the prototype of the non-sulfhydryl, N-carboxymethyl-dipeptide series of inhibitors [47]. Bioactivation (de-esterification) in liver and blood results in the release of the active "carboxylate compound", MK-422 (K_i about 40 pM at pH 6.0). The inhibitor "351A", a p-hydroxybenzamidine derivative of enalapril, has been used in an inhibitor binding assay for PDP A [57]. The benzolactams [58] represent yet another class of carboxylate inhibitors.

Molecular properties: Generally found to be a large glycoprotein (M_r

140,000-150,000) comprised of 8-30% carbohydrate, a single polypeptide
chain, a single zinc atom, and having a pI (4.0-5.2) that tends to rise
during purification as a result of a loss of sialic acid [59]. $A_{280, 1\%} - 16$
for the human kidney enzyme [17].

Rabbit lung PDP A (M_r 129,000) contains 26% carbohydrate that is
comprised of 57 galactose, 53 N-acetylglucosamine, 43 mannose, 19
N-acetylneuraminic acid (sialic acid), and 4 fucose residues per molecule
[19]. Reports of unusually high M_r values (obtained by molecular
exclusion chromatography) are probably attributable to the glycoprotein
character of PDP A or to the various treatments used to solubilize this
membrane-bound enzyme. Whereas the rabbit lung [19] and hog kidney
enzymes [23] lack subunits, the hog lung enzyme may contain M_r 70,000
subunits [23], but this has been questioned [59a].

PDP A from human lung and kidney (the richest source in man [29a]) is
a M_r 150,000 glycoprotein [14] having a single polypeptide chain of
known amino acid composition and N-terminal sequence [17]. PDP A from
human lung, but not kidney, contains a significant amount of sialic acid
(20 residues per molecule) [17]. The highly-sialylated circulating form
of PDP A may originate in the lung. The sedimentation constant for the
human kidney enzyme is 7.9S, and its frictional ratio is 1.2 [17].
Consistent with its hydrophobic character, the enzyme from both sources
contains about 42% nonpolar residues [17,60]. Its high molar absorptivity
at 280 nm (2.4×10^5) is attributable to its high tryptophan content
[17]. A catalytically-active "light" form of PDP A (M_r 90,000) has been
generated from the native M_r 140,000 human plasma enzyme by proteolytic
or alkaline hydrolysis [61].

The catalytic mechanism of PDP A has many features in common with
that of tissue (pancreatic) carboxypeptidase A (tCPA) [62], but in
addition to functional Tyr, Arg and carboxylic amino acid residues, PDP
A of rabbit and human origins also contains one essential Lys residue
not found in tCPA [17]. Co^{2+} activates both enzymes [25].

Chloride ion appears to serve as an allosteric modifier of PDP A
[63]. K_m values are markedly lowered by Cl^-, and the pH optimum tends to
shift upward to pH 8.2.

Comment: There have been numerous reports of an activity (in mammalian
brain) that resembles that of PDP A in its inactivation of the endogenous
opiate peptides, Leu- and Met-enkephalin, by releasing the C-terminal
dipeptide [64]. The responsible enzyme is generally referred to as
"enkephalinase" or "enkephalinase A". (See Entry 13.03 for evidence that
"enkephalinase B" is dipeptidyl peptidase III.) Enkephalinase is known
to be distinct from PDP A [65], and has generally been regarded as
another peptidyl dipeptidase or "dipeptidyl carboxypeptidase". Like PDP
A, enkephalinase is also an integral membrane glycoprotein that is
sensitive to thiol compounds and metal chelating reagents. On the other
hand, it is not activated by Cl^-, and is relatively insensitive to
captopril. Recent studies with purified enzymes show that enkephalinase

is most probably identical to both the neutral metallo-endopeptidase found in beef brain and pituitary [66], and that (EC 3.4.24.11) known as "endopeptidase-24.11" present in kidney brush border and other membranes [67] (see Volume 1, Entry 4.06). This M_r 90,000 enzyme is neither a peptidyl dipeptidase, nor specific for the hydrolysis of enkephalins. Its specificity actually resembles that of thermolysin, a bacterial zinc metallo-endopeptidase that hydrolyzes peptide bonds involving the amino group of hydrophobic residues. Accordingly, a free carboxyl group is not required, and many neuropeptides are hydrolyzed, with predictable specificity [68-70].

In addition to PDP A, and "enkephalinase" which can mimic a dipeptidyl peptidase, distinct peptidyl dipeptidases may exist in brain [71,72], but further confirmation of this is needed. At present, cathepsin B is the only other enzyme in mammalian tissues known to possess true peptidyl dipeptidase activity, and accordingly the enzyme is described in the present volume under the name peptidyl dipeptidase B (PDP B, Entry 16.02).

Although the potential role of PDP A in blood pressure regulation has long been appreciated, recognition of its clinical impo:tance came with the discovery of specific inhibitors that were found useful for treating certain types of hypertension in man.

Serum levels of PDP A are significantly increased in 75% of patients with active pulmonary sarcoidosis [73]. This observation has resulted in a useful diagnostic test, especially since blood levels of PDP A are not elevated in patients afflicted with other pulmonary diseases such as lung cancer, tuberculosis, coccidioides, and other types of granulomatous disease. PDP A levels are almost always elevated in granulomatous lymph node tissue of patients with sarcoidosis. Blood levels fall during corticosteroid treatment or spontaneous remission. Macrophages present in the high turnover sarcoid granulomas are believed to be the source of the PDP A.

Some patients (24%) with diabetes mellitus show elevated blood levels of PDP A that correlate strongly with the presence of severe retinopathy. Elevated levels also occur in Gaucher's disease and leprosy too, so these diseases must be considered when confirming a diagnosis of sarcoidosis [74].

Cathepsin G (Volume 1, Entry 1.36) a major neutrophil granule proteinase, is capable of generating angiotensin II from angiotensin I or angiotensinogen at pH 7.0-7.5, and thus may be mistaken for PDP A. It does not, however, act on bradykinin or the synthetic substrate Bz-Gly-His-Leu, nor is it inhibited by captopril [75]. Human skin mast cell chymase also shows angiotensin-converting activity [76].

"Tonin" (Volume 1, Entry 5.04), a M_r 28,700 serine proteinase of the rat submaxillary gland that is clearly homologous with the γ-subunit of nerve growth factor [77] also resembles PDP A in its ability to generate angiotensin I from the synthetic tetradecapetide renin substrate and

angiotensin II. However, like cathepsin G, it lacks activity on
bradykinin and Bz-Gly-His-Leu.

References

[1] Skeggs *et al.* *J. Exp. Med.* **103**: 295-299, 1956.
[2] Caldwell *et al.* *Science* **191**: 1050-1051, 1976.
[3] Ward *et al.* *Biochem. J.* **157**: 642-650, 1976.
[4] Ward *et al.* *Biochem. Pharmacol.* **29**: 1525-1529, 1980.
[5] Ryan *et al.* *Tissue Cell* **8**: 125-145, 1976.
[6] Roth *et al.* *Experientia* **25**: 1247 only, 1969.
[7] Cushman & Cheung *Biochim. Biophys. Acta* **250**: 261-265, 1971.
[8] Cushman & Cheung In: *Hypertension* (Genest, J. ed.), pp. 532-541, Springer, New York, 1972.
[9] Yang & Neff *J. Neurochem.* **21**: 1035-1036, 1973.
[10] Chevillard & Saavedra *J. Neurochem.* **38**: 281-284, 1982.
[11] Kokubu *et al.* *Clin. Chim. Acta* **89**: 375-379, 1978.
[12] Erdös & Yang *Life Sci.* **6**: 569-574, 1967.
[13] Oshima *et al.* *Biochim. Biophys. Acta* **350**: 26-37, 1974.
[14] Stewart *et al.* *Peptides* **2**: 145-152, 1981.
[15] Elisseeva *et al.* *Clin. Chim. Acta* **31**: 413-419, 1971.
[16] Takada *et al.* *J. Biochem.* **90**: 1309-1319, 1981.
[17] Weare *et al.* *Anal. Biochem.* **123**: 310-319, 1982.
[18] Cushman & Cheung *Biochem. Pharmacol.* **20**: 1637-1648, 1971.
[19] Das & Soffer *J. Biol. Chem.* **250**: 6762-6768, 1975.
[20] Lanzillo & Fanberg *J. Biol. Chem.* **249**: 2312-2318, 1974.
[21] Lanzillo & Fanberg *Biochim. Biophys. Acta* **445**: 161-168, 1976.
[22] Dorer *et al.* *Circ. Res.* **31**: 356-366, 1972.
[23] Nakajima *et al.* *Biochim. Biophys. Acta* **315**: 430-438, 1973.
[24] Igic *et al.* *Circ. Res. (Suppl. II)* **31**: 51-61, 1972.
[25] Stevens *et al.* *Biochemistry* **11**: 2999-3007, 1972.
[26] Rohrbach *et al.* *J. Biol. Chem.* **256**: 225-230, 1981.
[26a] Harris & Wilson *Int. J. Pept. Protein Res.* **20**: 167-176, 1982.
[27] Fitz & Overturf *J. Biol. Chem.* **247**: 581-584, 1972.
[28] Overturf *et al.* *Life Sci.* **16**: 1669-1682, 1975.
[29] Gronhagen-Riska & Fyhrquist *Scand. J. Clin. Lab. Invest* **40**: 711-719, 1980.
[29a] Stewart *et al.* *Methods Enzymol.* **80**: 450-460, 1981.
[30] Yang *et al.* *Biochim. Biophys. Acta* **214**: 374-376, 1970.
[31] Fernley *Clin. Exp. Pharmacol. Physiol.* **4**: 267-281, 1977.
[32] Lanzillo *et al.* *Anal. Biochem.* **103**: 400-407, 1980.
[33] Yokasawa *et al.* *J. Neurochem.* **41**: 403-410, 1983.
[33a] Erdös *Handb. Exp. Pharmacol. (Suppl.)* **25**: 427-487, 1979.
[34] Piquilloud *et al.* *Biochim. Biophys. Acta* **206**: 136-142, 1970.
[35] Dorer *et al.* *Circ. Res.* **34**: 824-827, 1974.
[36] Beckner & Caprioli *Biochem. Biophys. Res. Commun.* **93**: 1290-1296, 1980.

[36a] Persson & Wilson *Anal. Biochem.* **83**: 296-303, 1977.
[37] Dorer *et al.* *Anal. Biochem.* **33**: 102-113, 1970.
[38] Neels *et al.* *Clin. Chem.* **29**: 1399-1403, 1983.
[39] Filipović *et al.* *Clin. Chem.* **128**: 177-180, 1983.
[40] Yang *et al.* *J. Pharmacol. Exp. Ther.* **177**: 291-300, 1971.
[41] Holmquist *et al.* *Anal. Biochem.* **95**: 540-548, 1979.
[42] Bunning *et al.* *Biochemistry* **22**: 103-110, 1983.
[43] Carmel *et al.* *Clin. Chim. Acta* **93**: 215-220, 1979.
[44] Persson & Wilson *Anal. Biochem.* **83**: 296-303, 1977.
[45] Cushman *et al.* *Biochemistry* **16**: 5484-5491, 1977.
[46] Funae *et al.* *Biochem. Pharmacol.* **29**: 1543-1547, 1980.
[46a] Harris *et al.* *Arch. Biochem. Biophys.* **206**: 105-112, 1981.
[47] Patchett *et al.* *Nature* **288**: 280-283, 1980.
[48] Galardy *Biochemistry* **21**: 5777-5781, 1982.
[49] Galardy *et al.* *Biochemistry* **22**: 1990-1995, 1983.
[50] Harris *et al.* *Biochem. Biophys. Res. Commun.* **116**: 394-399, 1983.
[51] Harris & Wilson *J. Biol. Chem.* **257**: 811-815, 1982.
[52] Ondetti *et al.* *Biochemistry* **10**: 4033-4039, 1971.
[53] McGeer & Singh *Neurosci. Lett.* **4**: 105-108, 1979.
[54] Kido *et al.* *J. Antibiot.* **36**: 1295-1299, 1983.
[55] Mikami & Suzuki *Agric. Biol. Chem.* **47**: 2693-2695, 1983.
[56] Singh & Johnson *J. Antibiot.* **37**: 336-343, 1984.
[57] Fyhrquist *et al.* *Clin. Chem.* **30**: 696-700, 1984.
[58] Parsons *et al.* *Biochem. Biophys. Res. Commun.* **117**: 108-113, 1983.
[59] Lanzillo *et al.* *Arch. Biochem. Biophys.* **227**: 434-439, 1983.
[59a] Lanzillo & Fanberg *Biochim. Biophys. Acta* **439**: 125-132, 1976.
[60] Soffer *Annu. Rev. Biochem.* **45**: 73-94, 1976.
[61] Yotsumoto *et al.* *Biochim. Biophys. Acta* **749**: 180-184, 1983.
[62] Bunning *et al.* *Biochem. Biophys. Res. Commun.* **83**: 1442-1449, 1978.
[63] Na & Lee *Arch. Biochem. Biophys.* **227**: 580-586, 1983.
[64] Malfroy *et al.* *Nature* **276**: 523-526, 1978.
[65] Swerts *et al.* *Eur. J. Pharmacol.* **57**: 279-281, 1979.
[66] Orlowski & Wilk *Biochemistry* **20**: 4942-4950, 1981.
[67] Kerr & Kenny *Biochem. J.* **137**: 477-488, 1974.
[68] Almenoff & Orlowski *Biochemistry* **22**: 590-599, 1983.
[69] Gafford *et al.* *Biochemistry* **22**: 3265-3271, 1983.
[70] Matsas *et al.* *Fed. Proc. Fed. Am. Soc. Exp. Biol.* **43**: 1546 only, 1984.
[71] Cushman *et al.* *Life Sci. (Suppl. I).* **33**: 25-28, 1983.
[72] Demmer & Brand *Biochem. Biophys. Res. Commun.* **114**: 804-812, 1983.
[73] Lieberman *et al.* *Am. Rev. Respir. Dis.* **120**: 329-335, 1979.

[74] Lieberman *Ann. Intern. Med.* **93**: 825-826, 1980.
[75] Klickstein *et al. J. Biol. Chem.* **257**: 15042-15046, 1982.
[76] Genest In: *Hypertension and the Angiotensin System* (Doyle, A. E. & Bearn, M. D. eds), pp. 93-107, Raven Press, New York, 1983.
[77] Lazure *et al. Nature* **292**: 383-384, 1981.

Bibliography

1956

Skeggs, L. T., Jr., Kahn, J. R. & Shumway, N. P. The preparation and function of the hypertensin converting enzyme. *J. Exp. Med.* **103**: 295-299.

1959

Shore, P. A., Burkhalter, A. & Cohn, V. H., Jr. A method for the fluorometric assay of histamine in tissues. *J. Pharmacol. Exp. Ther.* **127**: 182-186.

1967

Erdös, E. G. & Yang, H. Y. T. An enzyme in microsomal fraction of kidney that inactivates bradykinin. *Life Sci.* **6**: 569-574.
Yang, H. Y. T. & Erdös, E. G. Second kininase in human blood plasma. *Nature* **215**: 1402-1403.

1968

Bakhle, Y. S. Conversion of angiotensin I to angiotensin II by cell-free extracts of dog lung. *Nature* **220**: 919-921.
Huggins, C. G. & Thampi, N. S. A simple method for the determination of angiotensin I converting enzyme. *Life Sci.* **7**: 633-639.
Loyke, H. F. Converting enzyme in CCl$_4$-treated renal hypertension. *Am. J. Physiol.* **215**: 1334-1336.
Ng, K. K. F. & Vane, J. R. Fate of angiotensin I in the circulation. *Nature* **218**: 144-150.
Piquilloud, Y., Reinharz, A. & Roth, M. Action de l'enzyme de conversion ("converting enzyme") sur des substrats synthétiques. *Helv. Physiol. Pharmacol. Acta* **26**: CR231-CR232.

1969

Roth, M., Weitzman, A. F. & Piquilloud, Y. Converting enzyme content of different tissues of the rat. *Experientia* **25**: 1247-1248.

1970

Aiken, J. W. & Vane, J. R. The renin-angiotensin system: inhibition of converting enzyme in isolated tissues. *Nature* **228**: 30-34.
Dorer, F. E., Skeggs, L. T., Kahn, J. R., Lentz, K. E. & Levine, M.

Angiotensin converting enzyme: method of assay and partial purification. *Anal. Biochem.* **33**: 102-113.

Erdös, E. G. & Yang, H. Y. T. Kininases. In: *Handbook of Experimental Pharmacology* (Erdös, E. G. ed), vol. 25, pp. 289-323 (see pp. 294-299), Springer-Verlag, New York.

Ferreira, S. H., Greene, L. J., Alabaster, V. A., Bakhle, Y. S. & Vane, J. R. Activity of various fractions of bradykinin potentiating factor against angiotensin I converting enzyme. *Nature* **225**: 379-380.

Ng, K. K. F. & Vane, J. R. Some properties of angiotensin converting enzyme in the lung in vivo. *Nature* **225**: 1142-1144.

Piquilloud, Y., Reinharz, A. & Roth, M. Studies of the angiotensin converting enzyme with different substrates. *Biochim. Biophys. Acta* **206**: 136-142.

Yang, H. Y. T., Erdös, E. G. & Levin, Y. A dipeptidyl carboxypeptidase that converts angiotensin I and inactivates bradykinin. *Biochim. Biophys. Acta* **214**: 374-376.

1971

Cushman, D. W. & Cheung, H. S. Concentrations of angiotensin-converting enzyme in tissues of the rat. *Biochim. Biophys. Acta* **250**: 261-265.

Cushman, D. W. & Cheung, H. S. Spectrophotometric assay and properties of the angiotensin-converting enzyme of rabbit lung. *Biochem. Pharmacol.* **20**: 1637-1648.

Elisseeva, Y. E., Orekhovich, V. N., Pavlikhina, L. V. & Alexeenko, L. P. Carboxycathepsin - a key regulatory component of two physiological systems involved in regulation of blood pressure. *Clin. Chim. Acta* **31**: 413-419.

Lee, H.-J., Larue, J. N. & Wilson, I. B. Angiotensin-converting enzyme from guinea pig and hog lung. *Biochim. Biophys. Acta* **250**: 549-557.

Ondetti, M. A., Williams, N. J., Sabo, E. F., Pluscec, J., Weaver, E. R. & Kocy, O. Angiotensin-converting enzyme inhibitors from the venom of *Bothrops jararaca*. Isolation, elucidation of structure, and synthesis. *Biochemistry* **10**: 4033-4039.

Oparil, S., Tregear, G. W., Koerner, T., Barnes, B. A. & Haber, E. Mechanism of pulmonary conversion of angiotensin I to angiotensin II in the dog. *Circ. Res.* **29**: 682-690.

Ryan, J. W., Niemeyer, R. S., Goodwin, D. W., Smith, U. & Stewart, J. M. Metabolism of [8-L-[^{14}C]phenylalanine]-angiotensin I in the pulmonary circulation. *Biochem. J.* **125**: 921-923.

Sander, G. E., West, D. W. & Huggins, C. G. Peptide inhibitors of pulmonary angiotensin I converting enzyme. *Biochim. Biophys. Acta* **242**: 662-667.

Ueda, E., Akutsu, H., Kokubu, T. & Yamamura, Y. Partial purification and properties of angiotensin I converting enzyme from rabbit plasma. *Jpn. Circ. J.* **35**: 801-806.

Ueda, E., Kokubu, T., Akutsu, H. & Yamamura, Y. Inhibition of angiotensin

I converting enzyme and kininase in rabbit plasma by bradykinin potentiating peptide B (Pyr-Gly-Leu-Pro-Arg-Pro-Lys-Ile-Pro-Pro). *Experientia* **27**: 1020-1021.

Yang, H. Y. T., Erdös, E. G. & Levin, Y. Characterization of a dipeptide hydrolase (kininase II: angiotensin I converting enzyme). *J. Pharmacol. Exp. Ther.* **177**: 291-300.

1972

Aiken, J. W. & Vane, J. R. Inhibition of converting enzyme of the renin-angiotensin system in kidneys and hindlegs of dogs. *Circ. Res.* **30**: 263-273.

Angus, C. W., Lee, H.-J. & Wilson, I. B. Angiotensin-converting enzyme and a second dipeptidyl carboxypeptidase from hog plasma. *Biochim. Biophys. Acta* **276**: 228-233.

Cheung, H. S. & Cushman, D. W. Inhibition of homogeneous angiotensin-converting enzyme of rabbit lung by synthetic venom peptides of *Bothrops jararaca*. *Biochim. Biophys. Acta* **293**: 451-463.

Depierre, D. & Roth, M. Activity of a dipeptidyl carboxypeptidase (angiotensin converting enzyme) in lungs of different animal species. *Experientia* **28**: 154-155.

Dorer, F. E., Kahn, J. R., Lentz, K. E., Levine, M. & Skeggs, L. T. Purification and properties of angiotensin-converting enzyme of hog lung. *Circ. Res.* **31**: 356-366.

Fitz, A. & Overturf, M. Molecular weight of human angiotensin I. Lung converting enzyme. *J. Biol. Chem.* **247**: 581-584.

Igic, R., Erdös, E. G., Yeh, H. S. J., Sorrells, K. & Nakajima, T. Angiotensin I converting enzyme of the lung. *Circ. Res.* **31** Suppl. 2: 51-61.

Miller, E. D., Jr., Samuels, A. I., Haber, E. & Barger, A. C. Inhibition of angiotensin conversion in experimental renovascular hypertension. *Science* **177**: 1108-1109.

Osborn, E. C., Tildesley, G. & Pickens, P. T. Pressor response to angiotensin I and angiotensin II: the site of conversion of angiotensin I. *Clin. Sci.* **43**: 839-849.

Ryan, J. W., Smith, U. & Niemeyer, R. S. Angiotensin I: metabolism by plasma membrane of lung. *Science* **176**: 64-66.

Stevens, R. L., Micalizzi, E. R., Fessler, D. C. & Pals, D. T. Angiotensin I converting enzyme of calf lung. Method of assay and partial purification. *Biochemistry* **11**: 2999-3007.

Yang, H. Y. T. & Neff, N. H. Distribution and properties of angiotensin converting enzyme of rat brain. *J. Neurochem.* **19**: 2443-2450.

1973

Angus, C. W., Lee, H.-J. & Wilson, I. B. Substrate specificity of hog plasma angiotensin-converting enzyme. *Biochim. Biophys. Acta* **309**: 169-174.

Depierre, D., Roth, M. & Meylan, J. Some properties of the angiotensin-converting enzyme from human seminal plasma. *Experientia* **29**: 751 only.

Fanburg, B. L. & Glazier, J. B. Conversion of angiotensin 1 to angiotensin 2 in the isolated perfused dog lung. *J. Appl. Physiol.* **35**: 325-331.

Lehr, D., Goldman, H. W. & Casner, P. Renin-angiotensin role in thirst: paradoxical enhancement of drinking by angiotensin converting enzyme inhibitor. *Science* **182**: 1031-1034.

Nakajima, T., Oshima, G., Yeh, H. S. J., Igic, R. & Erdös, E. G. Purification of the angiotensin I-converting enzyme of the lung. *Biochim. Biophys. Acta* **315**: 430-438.

Oparil, S., Koerner, T., Tregear, G. W., Barnes, B. A. & Haber, E. Substrate requirements for angiotensin I conversion in vivo and in vitro. *Circ. Res.* **32**: 415-423.

Yang, H. Y. T. & Neff, N. H. Differential distribution of angiotensin converting enzyme in the anterior and posterior lobe of the rat pituitary. *J. Neurochem.* **21**: 1035-1036.

1974

Dorer, F. E., Kahn, J. R., Lentz, K. E., Levine, M. & Skeggs, L. T. Hydrolysis of bradykinin by angiotensin-converting enzyme. *Circ. Res.* **34**: 824-827.

Dorer, F. E., Ryan, J. W. & Stewart, J. M. Hydrolysis of bradykinin and its higher homologues by angiotensin-converting enzyme. *Biochem. J.* **141**: 915-917.

Grandino, A. & Paiva, A. C. M. Isolation of angiotensin-converting enzyme without kininase activity from hog and guinea pig plasma. *Biochim. Biophys. Acta* **364**: 113-119.

Lanzillo, J. J. & Fanburg, B. L. Membrane-bound angiotensin-converting enzyme from rat lung. *J. Biol. Chem.* **249**: 2312-2318.

Oparil, S., Koerner, T. & O'Donoghe, J. K. Structural requirements for substrates and inhibitors of angiotensin I-converting enzyme in vivo and in vitro. *Circ. Res.* **34**: 19-26.

Oshima, G. & Erdös. Inhibition of the angiotensin I converting enzyme of the lung by a peptide fragment of bradykinin. *Experientia* **30**: 733-734.

Oshima, G., Gecse, A. & Erdös, E. G. Angiotensin I-converting enzyme of the kidney cortex. *Biochim. Biophys. Acta* **350**: 26-37.

Soffer, R. L., Reza, R. & Caldwell, P. R. B. Angiotensin-converting enzyme from rabbit pulmonary particles. *Proc. Natl. Acad. Sci. U.S.A.* **71**: 1720-1724.

1975

Beckerhoff, R., Uhlschmid, G., Vetter, W., Armbruster, H., Nussberger, J., Reck, G., Schmied, U. & Siegenthaler, W. Effect of angiotensin II and of an angiotensin II analogue (Sar[1]-Ile[8]-angiotensin II) on blood pressure, plasma aldosterone and plasma renin activity in the dog. *Clin. Sci. Mol. Med.* **48**: 41-44.

Das, M. & Soffer, R. L. Pulmonary angiotensin-converting enzyme. Structural and catalytic properties. *J. Biol. Chem.* **250**: 6762-6768.

Depierre, D. & Roth, M. Fluorimetric determination of dipeptidyl carboxypeptidase (angiotensin-I-converting enzyme). *Enzyme* **19**: 65-70.

Erdös, E. G. Angiotensin I converting enzyme. *Circ. Res.* **36**: 247-255.

Gavras, H., Brunner, H. R., Laragh, J. H., Gavras, I. & Vukovich, R. A. The use of angiotensin-converting enzyme inhibitor in the diagnosis and treatment of hypertension. *Clin. Sci. Mol. Med.* **48**: 57-60.

Johnson, J. G., Black, W. D., Vukovich, R. A., Hatch, F. E., Jr., Friedman, B. I., Blackwell, C. F., Shenouda, A. N., Share, L., Shade, R. E., Acchiardo, S. R. & Muirhead, E. E. Treatment of patients with severe hypertension by inhibition of angiotensin-converting enzyme. *Clin. Sci. Mol. Med.* **48**: 53-56.

Overturf, M., Wyatt, S., Boaz, D. & Fitz, A. Angiotensin I [Phe^8-His^9] hydrolase and bradykininase from human lung. *Life Sci.* **16**: 1669-1682.

Ryan, J. W., Ryan, U. S., Schultz, D. R., Whitaker, C., Chung, A. & Dorer, F. E. Subcellular localization of pulmonary angiotensin-converting enzyme (kininase II). *Biochem. J.* **146**: 497-499.

1976

Caldwell, P. R. B., Seegal, B. C., Hsu, K. C., Das, M. & Soffer, R. L. Angiotensin-converting enzyme: vascular endothelial localization. *Science* **191**: 1050-1051.

Conroy, J. M., Hoffman, H., Kirk, E. S., Hirzel, H. O., Sonnenblick, E. H. & Soffer, R. L. Pulmonary angiotensin-converting enzyme. Interspecies homology and inhibition by heterologous antibody in vivo. *J. Biol. Chem.* **251**: 4828-4832.

Das, M. & Soffer, R. L. Pulmonary angiotensin-converting enzyme antibody. *Biochemistry* **15**: 5088-5094.

Dorer, F. E., Kahn, J. R., Lentz, K. E., Levine, M. & Skeggs, L. T. Kinetic properties of pulmonary angiotensin-converting enzyme. Hydrolysis of hippurylglycylglycine. *Biochim. Biophys. Acta* **429**: 220-228.

Eliseeva, Y. E., Orekhovich, V. N. & Pavlikhina, L. V. Isolation and properties of carboxycathepsin (peptidyl-dipeptidase) from bovine lung tissue. *Biochemistry U.S.S.R. (Engl. Transl.)* **41**: 417-422.

Erdös, E. G. Conversion of angiotensin I to angiotensin II. *Am. J. Med.* **60**: 749-759.

Lanzillo, J. J. & Fanberg, B. L. The estimation and comparison of molecular weight of angiotensin I converting enzyme by sodium dodecyl sulfate-polyacrylamide gel electrophoresis. *Biochim. Biophys. Acta* **439**: 125-132.

Lanzillo, J. J. & Fanburg, B. L. Angiotensin I-converting enzyme from guinea pig lung and serum. A comparison of some kinetic and inhibition properties. *Biochim. Biophys. Acta* **445**: 161-168.

Massey, T. H. & Fessler, D. C. Substrate binding properties of converting enzyme using a series of p-nitrophenylalanyl derivatives of angiotensin

I. *Biochemistry* **15**: 4906-4912.

Nishimura, K., Hiwada, K., Ueda, E. & Kokubu, T. Solubilization of angiotensin I-converting enzyme from rabbit lung using trypsin treatment. *Biochim. Biophys. Acta* **452**: 144-150.

Oshima, G., Nagasawa, K. & Kato, J. Renal angiotensin I-converting enzyme as a mixture of sialo- and asialo-enzyme, and a rapid purification method. *J. Biochem.* **80**: 477-483.

Ryan, J. W., Day, A. R., Schultz, D. R., Ryan, U. S., Chung, A., Marlborough, D. I. & Dorer, F. E. Localization of angiotensin converting enzyme (kininase II). 1. Preparation of antibody-heme-octapeptide conjugates. *Tissue Cell* **8**: 111-124.

Ryan, U. S., Ryan, J. W., Whitaker, C. & Chiu, A. Localization of angiotensin converting enzyme (kininase II). II. Immunocytochemistry and immunofluorescence. *Tissue Cell* **8**: 125-145.

Soffer, R. L. Angiotensin-converting enzyme and the regulation of vasoactive peptides. *Annu. Rev. Biochem.* **45**: 73-94.

1977

Cushman, D. W., Cheung, H. S., Sabo, E. F. & Ondetti, M. A. Design of potent competitive inhibitors of angiotensin-converting enzyme. Carboxyalkanoyl and mercaptoalkanoyl amino acids. *Biochemistry* **16**: 5484-5491.

Fernley, R. T. Equine angiotensin converting enzyme: a zinc metalloenzyme. *Clin. Exp. Pharmacol. Physiol.* **4**: 267-281.

Fitz, A., Wyatt, S., Boaz, D. & Fox, B. Peptide inhibitors of converting enzyme. *Life Sci.* **21**: 1179-1185.

Kokubu, T., Ueda, E., Nishimura, K. & Yoshida, N. Angiotensin I converting enzyme activity in pulmonary tissues of fetal and newborn rabbits. *Exprientia* **33**: 1137-1138.

Lanzillo, J. J. & Fanburg, B. L. Angiotensin I converting enzyme from human plasma. *Biochemistry* **16**: 5491-5495.

Lanzillo, J. J. & Fanburg, B. L. Low molecular weight angiotensin I converting enzyme from rat lung. *Biochim. Biophys. Acta* **491**: 339-344.

McDonald, J. K. & Schwabe, C. Intracellular exopeptidases. In: *Proteinases in Mammalian Cells and Tissues* (Barrett, A. J. ed.), pp. 311-391 (see pp. 377-381), North-Holland Publishing Co., Amsterdam.

Nishimura, K., Hiwada, K., Ueda, E. & Kokubu, T. Purification and properties of angiotensin I-converting enzyme from rabbit lung. *Jpn. Circ. J.* **41**: 864-866.

Nishimura, K., Yoshida, N., Hiwada, K., Ueda, E. & Kokubu, T. Purification of angiotensin I-converting enzyme from human lung. *Biochim. Biophys. Acta* **483**: 398-408.

Ondetti, M. A., Rubin, B. & Cushman, D. W. Design of specific inhibitors of angiotensin-converting enzyme: new class of orally active antihypertensive agents. *Science* **196**: 441-444.

Persson, A. & Wilson, I. B. A fluorogenic substrate for angiotensin-

converting enzyme. *Anal. Biochem.* **83**: 296-303.
Ryan, J. W., Chung, A., Ammons, C. & Carlton, M. L. A simple radioassay
for angiotensin-converting enzyme. *Biochem. J.* **167**: 501-504.

1978

Brecher, P., Tercyak, A., Gavras, H. & Chobanian, A. V. Peptidyl
dipeptidase in rabbit brain microvessels. *Biochim. Biophys. Acta* **526**:
537-546.
Bünning, P., Holmquist, B. & Riordan, J. F. Functional residues at the
active site of angiotensin converting enzyme. *Biochem. Biophys. Res.
Commun.* **83**: 1442-1449.
Carmel, A. & Yaron, A. An intramolecularly quenched fluorescent tripeptide
as a fluorogenic substrate of angiotensin-I-converting enzyme and of
bacterial dipeptidyl carboxypeptidase. *Eur. J. Biochem.* **87**: 265-273.
Depierre, D., Bargetzi, J.-P. & Roth, M. Dipeptidyl carboxypeptidase from
human seminal plasma. *Biochim. Biophys. Acta* **523**: 469-476.
Friedland, J. & Silverstein, E. Angiotensin converting enzyme (ACE):
apparent identity of rabbit macrophage and lung enzymes and proteolytic
processing of predominant large cellular to small extracellular form.
Fed. Proc. Fed. Am. Soc. Exp. **37**: 1332 only.
Hayes, L. W., Goguen, C. A., Ching, S.-F. & Slakey, L. L. Angiotensin-
converting enzyme: accumulation in medium from cultured endothelial
cells. *Biochem. Biophys. Res. Commun.* **82**: 1147-1153.
Kenny, A. J. & Booth, A. G. Microvilli: their ultrastructure, enzymology
and molecular organization. *Essays Biochem.* **14**: 1-44.
Kokubu, T., Kato, I., Nishimura, K., Hiwada, K. & Ueda, E. Angiotensin
I-converting enzyme in human urine. *Clin. Chim. Acta* **89**: 375-379.
Litorowicz, A. & Malofiejew, M. Kininases and converting enzyme in human
placenta. *Biochem. Pharmacol.* **27**: 2829-2834.
Markle, R. A., Sonnenblick, E. H., Conroy, J. M. & Soffer, R. L. Reversal
of renovascular hypertension by antibodies specific for angiotensin-
converting enzyme. *Proc. Natl. Acad. Sci. USA* **75**: 5702-5705.
Nishimura, K., Yoshida, N., Hiwada, K., Ueda, E. & Kokubu, T. Properties
of three different forms of angiotensin I-converting enzyme from human
lung. *Biochim. Biophys. Acta* **522**: 229-237.
Persson, A. V., Russo, S. F. & Wilson, I. B. A new chromogenic substrate
for angiotensin-converting enzyme. *Anal. Biochem.* **91**: 674-683.
Riordan, J. F. & Holmquist, B. Kinetic analysis of dipeptidyl
carboxypeptidase with chromophoric substrates. *Fed. Proc. Fed. Am. Soc.
Exp. Biol.* **37**: 1286 only.
Rohrbach, M. S. [Glycine-1-^{14}C]hippuryl-L-histidyl-L-leucine: a substrate
for the radiochemical assay of angiotensin converting enzyme. *Anal.
Biochem.* **84**: 272-276.
Russo, S. F., Persson, A. V. & Wilson, I. B. A fluorogenic substrate for
angiotensin-converting enzyme in plasma. *Clin. Chem.* **24**: 539-1542.
Ryan, J. W., Chung, A., Martin, L. C. & Ryan, U. S. New substrates for

the radioassay of angiotensin converting enzyme of endothelial cells in culture. *Tissue Cell* **10**: 555-562.

Spadaro, A. C. C., Martins, A. R., Berti, J. D. & Greene, L. J. Chromatographic determination of angiotensin-converting enzyme and angiotensinase activity. *Anal. Biochem.* **91**: 410-420.

1979

Benuck, M. & Marks, N. Co-identity of brain angiotensin converting enzyme with a membrane bound dipeptidyl carboxypeptidase inactivating Met-enkephalin. *Biochem. Biophys. Res. Commun.* **88**: 215-221.

Carmel, A., Ehrlich-Rogozinsky, S. & Yaron, A. A fluorimetric assay for angiotensin-I converting enzyme in human serum. *Clin. Chim. Acta* **93**: 215-220.

Chiknas, S. G. A liquid chromatography-assisted assay for angiotensin-converting enzyme (peptidyl dipeptidase) in serum. *Clin. Chem.* **25**: 1259-1262.

Cushman, D. W., Cheung, H. S., Sabo, E. F., Rubin, B. & Ondetti, M. A. Development of specific inhibitors of angiotensin I converting enzyme (kininase II). *Fed. Proc. Fed. Am. Soc. Exp. Biol.* **38**: 2778-2782.

Depierre, D., Bargetzi, J.-P. & Roth, M. Study of peptides inhibiting the angiotensin-converting enzyme of human seminal plasma. *Enzyme* **24**: 362-365.

Ercan, Z. S., Öner, G., Türker, R. K. & Bor, N. Zinc deficiency and lung converting enzyme activity in rats. *Experientia* **35**: 215-216.

Erdös, E. G. Kininases. *Handb. Exp. Pharmacol. (Suppl.)* **25**: 427-487.

Fisher, G. H. & Ryan, J. W. Superactive inhibitors of angiotensin converting enzyme. Analogs of BPP$_{9a}$ containing dehydroproline. *FEBS Lett.* **107**: 273-276.

Gimbrone, M. A. Jr., Majeau, G. R., Atkinson, W. J., Sadler, W. & Cruise, S. A. Angiotensin-converting enzyme activity in isolated brain microvessels. *Life Sci.* **25**: 1075-1083.

Gorenstein, C. & Snyder, S. H. Two distinct enkephalinases: solubilization, partial purification and separation from angiotensin converting enzyme. *Life Sci.* **25**: 2065-2070.

Hinman, L. M., Stevens, C., Matthay, R. A. & Gee, J. B. L. Angiotensin convertase activities in human alveolar macrophages: effects of cigarette smoking and sarcoidosis. *Science* **205**: 202-203.

Holmquist, B. & Vallee, B. L. Metal-coordinating substrate analogs as inhibitors of metalloenzymes. *Proc. Natl. Acad. Sci. USA* **76**: 6216-6220.

Holmquist, B., Bunning, P. & Riordan, J. F. A continuous spectrophotometric assay for angiotensin converting enzyme. *Anal. Biochem.* **95**: 540-548.

Klauser, R. J., Robinson, C. J. G., Marinkovic, D. V. & Erdös, E. G. Inhibition of human peptidyl dipeptidase (Angiotensin I converting enzyme: kininase II) by human serum albumin and its fragments. *Hypertension* **1**: 281-286.

Lieberman, J., Nosal, A., Schlessner, L. A. & Sastre-Foken, A. Serum angiotensin-converting enzyme for diagnosis and therapeutic evaluation of sarcoidosis. *Am. Rev. Respir. Dis.* **120**: 329-335.

McGeer, E. G. & Singh, E. A. Inhibition of angiotensin converting enzyme by substance P. *Neurosci. Lett.* **14**: 105-108.

Oshima, G. & Nagasawa, K. Stereospecificity of peptidyl dipeptide hydrolase (angiotensin I-converting enzyme). *J. Biochem.* **86**: 1719-1724.

Oshima, G., Shimabukuro, H. & Nagasawa, K. Peptide inhibitors of angiotensin I-converting enzyme in digests of gelatin by bacterial collagenase. *Biochim. Biophys. Acta* **566**: 128-137.

Polsky-Cynkin, R. & Fanburg, B. L. Immunochemical comparison of angiotensin I converting enzymes from different rat organs. *Int. J. Biochem.* **10**: 669-674.

Stalcup, S. A., Lipset, J. S., Woan, J.-M., Leuenberger, P. & Mellins, R. B. Inhibition of angiotensin converting enzyme activity in cultured endothelial cells by hypoxia. *J. Clin. Invest.* **63**: 966-976.

Ward, P. E., Stewart, T. A., Hammon, K. J., Reynolds, R. C. & Igic, R. P. Angiotensin I converting enzyme (kininase II) in isolated retinal microvessels. *Life Sci.* **24**: 1419-1424.

1980

Beckner, C. F. & Caprioli, R. M. Proteolytic activity of dipeptidyl carboxypeptidase from human lung. *Biochem. Biophys. Res. Commun.* **93**: 1290-1296.

Cheung, H.-S., Wang, F.-L., Ondetti, M. A., Sabo, E. F & Cushman, D. W. Binding of peptide substrates and inhibitors of angiotensin-converting enzyme. *J. Biol. Chem.* **255**: 401-407.

Del Vecchio, P. J., Ryan, J. W., Chung, A. & Ryan, U. S. Capillaries of the adrenal cortex possess aminopeptidase A and angiotensin-converting-enzyme activities. *Biochem. J.* **186**: 605-608.

Funae, Y., Komori, T., Sasaki, D. & Yamamoto, K. Inhibitor of angiotensin I converting enzyme: (4R)-3-[(2S)-3-mercapto-2-methylpropanoyl]-4-thiazolidinecarboxylic acid (YS-980). *Biochem. Pharmacol.* **29**: 1543-1547.

Galardy, R. E. Inhibition of angiotensin converting enzyme with N^{α}-phosphoryl-L-alanyl-L-proline and N^{α}-phosphoryl-L-valyl-L-tryptophan. *Biochem. Biophys. Res. Commun.* **97**: 94-99.

Grönhagen-Riska, C. & Fyhrquist, F. Purification of human lung angiotensin-converting enzyme. *Scand. J. Clin. Lab. Invest.* **40**: 711-719.

Igic, R. & Kojovic, V. Angiotensin I converting enzyme (kininase II) in ocular tissues. *Exp. Eye Res.* **30**: 299-303.

Krutzsch, H. C. Determination of polypeptide amino acid sequences from the carboxyl terminus using angiotensin I converting enzyme. *Biochemistry* **19**: 5290-5296.

Lanzillo, J. J., Polsky-Cynkin, R. & Fanburg, B. L. Large-scale purification of angiotensin I-converting enzyme from human plasma utilizing an immunoadsorbent affinity gel. *Anal. Biochem.* **103**: 400-407.

Lieberman, J. & Sastre, A. Serum angiotensin-converting enzyme: elevations in diabetes mellitus. *Ann. Intern. Med.* **93**: 825-826.

Patchett, A. A., Harris, E., Tristram, E. W., Wyvratt, M.J., Wu, M. T., Taub, D., Peterson, E. R., Ikeler, T. J., ten Broeke, J., Payne, L. G., Ondeyka, D. L., Thorsett, E. D., Greenlee, W. J., Lohr, N. S., Hoffsommer, R. D., Joshua, H., Ruyle, W. V., Rothrock, J. W., Aster, S. D., Maycock, A. L., Robinson, F. M., Hirschmann, R., Sweet, C. S., Ulm, E. H., Gross, D. M., Vassil, T. C. & Stone, C. A. A new class of angiotensin-converting enzyme inhibitors. *Nature* **288**: 280-283.

Stone, C. A., Sander, G. E., Lorenz, P. E. & Verma, P. S. Inhibition of the partially purified canine lung angiotensin I converting enzyme by opioid peptides. *Biochem. Pharmacol.* **29**: 3115-3118.

Ward, P. E., Sheridan, M. A., Hammon, K. J. & Erdös, E. G. Angiotensin I converting enzyme (kininase II) of the brush border of human and swine intestine. *Biochem. Pharmacol.* **29**: 1525-1529.

Yokoyoma, M., Hiwada, K., Kokubu, T., Takaha, M. & Takeuchi, M. Angiotensin-converting enzyme in human prostate. *Clin. Chim. Acta* **100**: 253-258.

1981

Del Rio, C. G., Smellie, W. S. A. & Morton, J. J. Des-Asp-angiotensin I: its identification in rat blood and confirmation as a substrate for converting enzyme. *Endocrinology* **108**: 406-412.

Elisseeva, Y. E., Pavlikhina, L. V., Orekhovich, V. N., Giacomello, A., Salerno, C. & Fasella, P. Evidence for the presence of dipeptidyl carboxypeptidase and its inhibitors in inflammatory synovial fluids. *Biochim. Biophys. Acta* **658**: 165-168.

Fleminger, G., Goldenberg, D. & Yaron, A. Use of an intramolecularly quenched fluorogenic substrate for study of a thiol-dependent acidic dipeptidyl carboxypeptidase in cellular extracts and in living cells. *FEBS Lett.* **135**: 131-134.

Hara, A., Fukuyama, K. & Epstein, W. L. Angiotensin-converting enzyme measured in mouse tissue by inhibition of histidyl-leucine peptidase. *Biochem. Med.* **26**: 199-210.

Hara, A., Fukuyama, K. & Epstein, W. L. Studies of heterogeneity of angiotensin-converting enzyme and acid phosphatase in granulomatous lesions of skin. *Clin. Chim. Acta* **117**: 269-277.

Harris, R. B., Ohlsson, J. T. & Wilson, I. B. Inhibition and affinity chromatography of human serum angiotensin converting enzyme with cysteinyl-proline derivatives. *Arch. Biochem. Biophys.* **206**: 105-112.

Harris, R. B., Ohlsson, J. T. & Wilson, I. B. Purification of human serum angiotensin I-converting enzyme by affinity chromatography. *Anal. Biochem.* **111**: 227-234.

Hurst, P. L. & Lovell-Smith, C. J. Optimized assay for serum angiotensin-converting enzyme activity. *Clin. Chem.* **27**: 2048-2052.

Inokuchi, J.-I. & Nagamatsu, A. Tripeptidyl carboxypeptidase activity of kininase II (angiotensin-converting enzyme). *Biochim. Biophys. Acta* **662**: 300-307.

Kasahara, Y. & Ashihara, Y. Colorimetry of angiotensin-I converting enzyme activity in serum. *Clin. Chem.* **27**: 1922-1925.

Mendelsohn, F. A. O. & Kachel, C. Production of angiotensin converting enzyme by cultured bovine endothelial cells. *Clin. Exp. Pharmacol. Physiol.* **8**: 477-481.

Mendelsohn, F. A. O., Csicsmann, J. & Hutchinson, J. S. Complex competitive and non-competitive inhibition of rat lung angiotensin-converting enzyme by inhibitors containing thiol groups: captopril and SA 446. *Clin. Sci.* **61**: 277s-280s.

Okamura, T., Clemens, D. L. & Inagami, T. Renin, angiotensins, and angiotensin-converting enzyme in neuroblastoma cells: evidence for intracellular formation of angiotensins. *Proc. Natl. Acad. Sci. USA* **78**: 6940-6943.

Rohrbach, M. S., Williams, E. B. Jr. & Rolstad, R. A. Purification and substrate specificity of bovine angiotensin-converting enzyme. *J. Biol. Chem.* **256**: 225-230.

Ryder, K. W., Jay, S. J., Jackson, S. A. & Hoke, S. R. Characterization of a spectrophotometric assay for angiotensin converting enzyme. *Clin. Chem.* **27**: 530-534.

Silverstein, E., Fierst, S. M., Simon, M. R., Weinstock, J. V. & Friedland, J. Angiotensin-converting enzyme in Crohn's disease and ulcerative colitis. *Am. J. Clin. Pathol.* **75**: 175-178.

Stewart, T. A., Weare, J. A. & Erdös, E. G. Human peptidyl dipeptidase (converting enzyme, kininase II). *Methods Enzymol.* **80**: 450-460.

Stewart, T. A., Weare, J. A. & Erdös, E. G. Purification and characterization of human converting enzyme (kininase II). *Peptides* **2**: 145-152.

Takada, Y., Hiwada, K. & Kokubu, T. Isolation and characterization of angiotensin converting enzyme from human kidney. *J. Biochem.* **90**: 1309-1319.

1982

Antonaccio, M. J. Angiotensin converting enzyme (ACE) inhibitors. *Annu. Rev. Pharmacol. Toxicol.* **22**: 57-87.

Chevillard, C. & Saavedra, J. M. Distribution of angiotensin-converting enzyme activity in specific areas of the rat brain stem. *J. Neurochem.* **38**: 281-284.

Defendini, R., Zimmerman, E. A., Weare, J. A., Alhenc-Gelas, F. & Erdös, E. G. Hydrolysis of enkephalins by human converting enzyme and localization of the enzyme in neuronal components of the brain. In: *Regulatory Peptides: From Molecular Biology to Function* (Costa, E. & Trabucchi, M. eds), pp. 271-280, Raven Press, New York.

Fournié-Zaluski, M.-C., Soroca-Lucas, E., Waksman, G., Llorens, C., Schwartz, J.-C. & Roques, B. P. Differential recognition of "enkephalinase" and angiotensin-converting enzyme by new carboxyalkyl inhibitors. *Life Sci.* **31**: 2947-2954.

Galardy, R. E. Inhibition of angiotensin converting enzyme by phosphoramidates and polyphosphates. *Biochemistry* **21**: 5777-5781.

Harris, R. B. & Wilson, I. B. Irreversible inhibition of bovine lung angiotensin I-converting enzyme with *p*-[*N,N*-bis(chloroethyl)amino]phenylbutyric acid (Chlorambucil) and Chlorambucyl L-proline with evidence that an active site carboxyl group is labeled. *J. Biol. Chem.* **257**: 811-815.

Harris, R. B. & Wilson, I. B. Physicochemical characteristics of homogeneous bovine lung angiotensin I-converting enzyme. *Int. J. Peptide Protein Res.* **20**: 167-176.

Horiuchi, M., Fujimura, K.-I., Terashima, T. & Iso, T. Method for determination of angiotensin-converting enzyme activity in blood and tissue by high-perfomance liquid chromatography. *J. Chromatog.* **233**: 123-130.

Iwata, K., Lai, C.-Y., El-Dorry, H. A. & Soffer, R. L. The NH_2- and COOH-terminal sequences of the angiotensin converting enzyme isozymes from rabbit lung and testis. *Biochem. Biophys. Res. Commun.* **107**: 1097-1103.

Kamoun, P. P., Bardet, J. I., Di Giulio, S. & Grunfeld, J. P. Measurements of angiotensin converting enzyme in captopril-treated patients. *Clin. Chim. Acta* **118**: 333-336.

Klickstein, L. B., Kaempfer, C. E. & Wintroub, B. U. The granulocyte-angiotensin system. Angiotensin I-converting activity of cathepsin G. *J. Biol. Chem.* **257**: 15042-15046.

Larionova, N. I., Maslov, E. V., Eliseeva, Y. E. & Pavlikhina, L. V. Titration of active sites of dipeptidyl-carboxypeptidase by the reversible inhibitor SQ 20881 from snake (*Bothrops jararaca*) venom. *Biochemistry USSR (Engl. Transl.)* **47**: 1121-1126.

Rømer, F. K. Comparison of two methods for measurement of serum angiotensin-converting enzyme in sarcoidosis. *Scand. J. Clin. Lab. Invest.* **42**: 197-199.

Rømer, F. K. The level of angiotensin-converting enzyme as indicator of 2-year prognosis in untreated pulmonary sarcoidosis without erythema nodosum. *Acta Med. Scand.* **211**: 293-295.

Schweisfurth, H. Das Angiotensin-I-converting-Enzym. Physiologische Aspekte und klinische Bedeutung. *Dtsch. Med. Wochenschr.* **107**: 1815-1818.

Silverstein, E. & Friedland, J. Angiotensin converting enzyme in cultured fibroblasts in Gaucher and Niemann-Pick diseases. *Proc. Soc. Exp. Biol. Med.* **170**: 251-253.

Takada, Y., Hiwada, K., Unno, M. & Kokubu, T. Immunocytochemical localization of angiotensin converting enzyme at the ultrastructural level in the human lung and kidney. *Biomed. Res.* **3**: 169-174.

Taugner, R. & Ganten, D. The localization of converting enzyme in kidney vessels of the rat. *Histochemistry* **75**: 191-201.

Verma, P. S., Miller, R. L., Taylor, R. E., O'Donohue, T. L. & Adams, R. G. Inhibition of canine lung angiotensin converting enzyme by ACTH and structurally related peptides. *Biochem. Biophys. Res. Commun.* **104**: 1484-1488.

Ward, P. E. & Sheridan, M. A. Angiotensin I converting enzyme of rat intestinal and vascular surface membrane. *Biochim. Biophys. Acta* **716**: 208-216.

Weare, J. A. Activation/inactivation of human angiotensin I converting enzyme following chemical modifications of amino groups near the active site. *Biochem. Biophys. Res. Commun.* **104**: 1319-1326.

Weare, J. A., Gafford, J. T., Lu, H. S. & Erdös, E. G. Purification of human angiotensin I converting enzyme using reverse-immunoadsorption chromatography. *Anal. Biochem.* **123**: 310-319.

1983

Bünning, P. & Riordan, J. F. Activation of angiotensin converting enzyme by monovalent anions. *Biochemistry* **22**: 110-116.

Bünning, P., Holmquist, B. & Riordan, J. F. Substrate specificity and kinetic characteristics of angiotensin converting enzyme. *Biochemistry* **22**: 103-110.

Checler, F., Vincent, J.-P. & Kitabgi, P. Degradation of neurotensin by rat brain synaptic membranes: involvement of a thermolysin like metalloendopeptidase (enkephalinase), angiotensin-converting enzyme, and other unidentified peptidases. *J. Neurochem.* **41**: 375-384.

Cohen, M. L. & Kurz, K. Captopril and MK-421: stability on storage, distribution to the central nervous system, and onset of activity. *Fed. Proc. Fed. Am. Soc. Exp. Biol.* **42**: 171-175.

Corrdes, E. H., Bull, H. G. & Thornberry, N. A. The interaction of enalaprilic acid with angiotensin converting enzyme. In: *Hypertension and the Angiotensin System: Therapeutic Approaches* (Doyle, A. E. & Bearn, A. G. eds), pp. 167-178, Raven Press, New York.

Cushman, D. W., Gordon, E. M., Wang, F. L., Cheung, H. S., Tung, R. & Delaney, N. G. Purification and characterization of enkephalinase, angiotensin converting enzyme, and a third peptidyldipeptidase from rat brain. *Life Sci.* **33**, Suppl. 1: 25-28.

Filipović, N., Borčić, N. & Igić, R. A simple colormetric method for estimating plasma angiotensin I converting enzyme activity. *Clin. Chim. Acta* **128**: 177-180.

Galardy, R. E., Kontoyiannidou-Ostrem, V. & Kortylewicz, Z. P. Inhibition of angiotensin converting enzyme by phosphonic amides and phosphonic acids. *Biochemistry* **22**: 1990-1995.

Genest, J. Angiotensin-forming enzymes from extrarenal source. In: *Hypertension and the Angiotensin System: Therapeutic Approaches* (Doyle, A. E. & Bearn, A. G. eds), pp. 93-107, Raven Press, New York.

Gordon, E. M., Cushman, D. W., Tung, R., Cheung, H. S., Wang, F. L. & Delaney, N. G. Rat brain enkephalinase: characterization of the active site using mercaptopropanoyl amino acid inhibitors, and comparison with angiotensin-converting enzyme. *Life Sci.* **33**: 113-116.

Harris, R. B., Strong, P. D. M. & Wilson, I. B. Dipeptide-hydroxamates are good inhibitors of the angiotensin I-converting enzyme. *Biochem. Biophys. Res. Commun.* **116**: 394-399.

Kasai, Y., Abe, K., Yasujima, M., Tajima, J., Seino, M., Chiba, S., Sato, K., Goto, T., Omata, K., Tanno, M. & Yoshinaga, K. Acute effects of MK421, a new angiotensin converting enzyme inhibitor, in man. *Tohoku J. Exp. Med.* **141**: 417-422.

Kido, Y., Hamakado, T., Yoshida, T., Anno, M., Motoki, Y., Wakamiya, T. & Shiba, T. Isolation and characterization of ancovenin, a new inhibitor of angiotensin I converting enzyme, produced by actinomycetes. *J. Antibiot.* **36**: 1295-1299.

Lanzillo, J. J., Stevens, J., Tumas, J. & Fanburg, B. L. Spontaneous change of human plasma angiotensin I converting enzyme isoelectric point. *Arch. Biochem. Biophys.* **227**: 434-439.

Mikami, Y. & Suzuki, T. Novel microbial inhibitors of angiotensin-converting enzyme, aspergillomarasmines A and B. *Agric. Biol. Chem.* **47**: 2693-2695.

Na, K.-J. & Lee, H.-J. Role of chloride ion as an allosteric activator of angiotensin-converting enzyme. *Arch. Biochem. Biophys.* **227**: 580-586.

Neels, H. M., van Sande, M. E. & Scharpé, S. L. Sensitive colorimetric assay for angiotensin converting enzyme in serum. *Clin. Chem.* **29**: 1399-1403.

O'Brien, J. F., Forsman, R. W. & Rohrback, M. S. Spectrophotometric and radiometric assys of angiotensin-converting enzyme compared. *Clin. Chem.* **29**: 1990-1991.

Odya, C. E., Wilgis, F. P., Vavrek, R. J. & Stewart, J. M. Interactions of kinins with angiotensin I converting enzyme (kininase II). *Biochem. Pharmacol.* **32**: 3839-3847.

Parsons, W. H., Davidson, J. L., Taub, D., Aster, S. D., Thorsett, E. D. & Patchett, A. A. Benzolactams. A new class of converting enzyme inhibitors. *Biochem. Biophys. Res. Comm.* **117**: 108-113.

Ronca-Testoni, S. Direct spectrophotometric assay for angiotensin-converting enzyme in serum. *Clin. Chem.* **29**: 1093-1096.

Roques, B. P., Lucas-Soroca, E., Chaillet, P., Costentin, J. & Fournié-Zaluski, M.-C. Complete differentiation between enkephalinase and angiotensin-converting enzyme inhibition by *retro*-thiorphan. *Proc. Natl. Acad. Sci. USA* **80**: 3178-3182.

Roth, M. Biochemical and functional aspects of dipeptidyl carboxypeptidase (angiotensin converting enzyme). In: *Selected Topics in Clinical Enzymology* (Goldberg, D. M. & Werner, M. eds), pp. 125-130, Walter de Gruyter & Co., Berlin.

Schwab, A., Weinryb, I., Macerata, R., Rogers, W., Suh, J. & Khandwala, A. Inhibition of angiotensin-converting enzyme by derivatives of 3-mercapto-2-methylpropanoyl glycine. *Biochem. Pharmacol.* **32**: 1957-1960.

Strittmatter, S. M., Kapiloff, M. S. & Snyder, S. H. [^{3}H]Captopril binding to membrane associated angiotensin converting enzyme. *Biochem. Biophys. Res. Commun.* **112**: 1027-1033.

Studdy, P. R., Lapworth, R. & Bird, R. Angiotensin-converting enzyme and its clinical significance - a review. *J. Clin. Pathol.* **36**: 938-947.

Sweet, C. S. Pharmacological properties of the converting enzyme inhibitor, enalapril maleate (MK-421). *Fed. Proc. Fed. Am. Soc. Exp. Biol.* **42**: 167-170.

Thorsett, E. D., Harris, E. E., Aster, S., Peterson, E. R., Taub, D., Patchett, A. A., Ulm, E. H. & Vassil, T. C. Dipeptide mimics. Conformationally restricted inhibitors of angiotensin-converting enzyme. *Biochem. Biophys. Res. Comm.* **111**: 166-171.

Yokosawa, H., Endo, S., Ogura, Y. & Ishii, S.-I. A new feature of angiotensin-converting enzyme in the brain: hydrolysis of substance P. *Biochem. Biophys. Res. Commun.* **116**: 735-742.

Yokosawa, H., Ogura, Y. & Ishii, S.-I. Purification and inhibition by neuropeptides of angiotensin-converting enzyme from rat brain. *J. Neurochem.* **41**: 403-410.

Yotsumoto, H., Lanzillo, J. J. & Fanberg, B. L. Generation of a 90000 molecular weight fragment from human plasma angiotensin-I-converting enzyme by enzymatic or alkaline hydrolysis. *Biochim. Biophys. Acta* **749**: 180-184.

1984

Attwood, M. R., Francis, R. J., Hassall, C. H., Kröhn, A., Lawton, G., Natoff, I. L., Nixon, J. S., Redshaw, S. & Thomas, W. A. New potent inhibitors of angiotensin converting enzyme. *FEBS Lett.* **165**: 201-206.

Bush, K., Henry, P. R., Souser-Woehleke, M., Trejo, W. H. & Slusarchyk, D. S. Phenacein - an angiotensin-converting enzyme inhibitor produced by a streptomycete. I. Taxonomy, fermentation and biological properties. *J. Antibiot.* **37**: 1308-1312.

Cascieri, M. A., Bull, H. G., Mumford, R. A., Patchett, A. A., Thornberry, N. A. & Liang, T. Carboxyl-terminal tripeptidyl hydrolysis of substance P by purified rabbit lung angiotensin-converting enzyme and the potentiation of substance P activity *in vivo* by captopril and MK-422. *Mol. Pharmacol.* **25**: 287-293.

Fyhrquist, F., Tikkanen, I., Grönhagen-Riska, C., Hortling, L. & Hichens, M. Inhibitor binding assay for angiotensin-converting enzyme. *Clin. Chem.* **30**: 696-700.

Gordon, E. M., Natarajan, S., Pluscec, J., Weller, H. N., Godfrey, J. D., Rom, M. B., Sabo, E. F., Engebrecht, J. & Cushman, D. W. Ketomethyldipeptides II. Effect of modifications of the α-aminoketone portion on inhibition of angiotensin converting enzyme. *Biochem. Biophys. Res. Commun.* **124**: 148-155.

Greenlee, W. J., Thorsett, E. D., Springer, J. P. & Patchett, A. A. Azapeptides: a new class of angiotensin-converting enzyme inhibitors.

Biochem. Biophys. Res. Commun. **122**: 791-797.

Hurst, P. L. & Lovell-Smith, C. J. Comparison of two colorimetric assays for angiotensin-converting enzyme activity. *Clin. Chem.* **30**: 817 only.

Kapiloff, M. S., Strittmatter, S. M., Fricker, L. D. & Snyder, S. H. A fluorometric assay for angiotensin-converting enzyme activity. *Anal. Biochem.* **140**: 293-302.

Liu, W-C., Parker, W. L., Brandt, S. S., Atwal, K. S. & Ruby, E. P. Phenacein - an angiotensin-converting enzyme inhibitor produced by a streptomycete. II. Isolation, structure determination and synthesis. *J. Antibiot.* **37**: 1313-1319.

Natarajan, S., Gordon, E. M., Sabo, E. F., Godfrey, J. D., Weller, H. N., Pluščec, J., Rom, M. B. & Cushman, D. W. Ketomethyldipeptides I. A new class of angiotensin converting enzyme inhibitors. *Biochem. Biophys. Res. Commun.* **124**: 141-147.

Neels, H. M., Scharpé, S. L., Fonteyne, G. A., Yaron, A. & van Sande, M. E. Fluorometric assay for angiotensin converting enzyme in human serum by centrifugal analysis. *Clin. Chim. Acta* **141**: 281-286.

Ondetti, M. A. & Cushman, D. W. Angiotensin-converting enzyme inhibitors: biochemical properties and biological actions. *CRC Crit. Rev. Biochem.* **16**: 381-411.

Pandey, K. N., Misono, K. S. & Inagami, T. Evidence for intracellular formation of angiotensins: coexistence of renin and angiotensin-converting enzyme in Leydig cells of rat testis. *Biochem. Biophys. Res. Commun.* **122**: 1337-1343.

Pantoliano, M. W., Holmquist, B. & Riordan, J. F. Affinity chromatographic purification of angiotensin converting enzyme. *Biochemistry* **23**: 1037-1042.

Reynolds, C. H. Kinetics of inhibition of angiotensin converting enzyme by captopril and by enalapril diacid. *Biochem. Pharmacol.* **33**: 1273-1276.

Rohrbach, M. S. Metabolism and subcellular localization of angiotensin converting enzyme in cultured human monocytes. *Biochem. Biophys. Res. Commun.* **124**: 843-849.

Ryder, K. W., Thompson, H., Smith, D., Sample, M., Sample, R. B. & Oei, T. O. A radioassay for angiotensin converting enzyme. *Clin. Biochem.* **17**: 302-305.

Schweisfurth, H. & Schiöberg-Schiegnitz, S. Assay and biochemical characterization of angiotensin-I-converting enzyme in cerebrospinal fluid. *Enzyme* **32**: 12-19.

Shapiro, R. & Riordan, J. F. Inhibition of angiotensin converting enzyme: dependence on chloride. *Biochemistry* **23**: 5234-5240.

Shapiro, R. & Riordan, J. F. Inhibition of angiotensin converting enzyme: mechanism and substrate dependence. *Biochemistry* **23**: 5225-5233.

Singh, P. D. & Johnson, J. H. Muraceins - muramyl peptides produced by *Nocardia orientalis* as angiotensin-converting enzyme inhibitors. II. Isolation and structure determination. *J. Antibiot.* **37**: 336-343.

Strittmatter, S. M. & Snyder, S. H. Angiotensin-converting enzyme in the male rat reproductive system: autoradiographic visualization with [^{3}H]captopril. *Endocrinology* **115**: 2332-2341.

Ward, P. E. Immunoelectrophoretic analysis of vascular, membrane-bound angiotensin I converting enzyme, aminopeptidase M, and dipeptidyl(amino)peptidase IV. *Biochem. Pharmacol.* **33**: 3183-3193.

Weller, H. N., Gordon, E. M., Rom, M. B. & Pluščec, J. Design of conformationally constrained angiotensin-converting enzyme inhibitors. *Biochem. Biophys. Res. Commun.* **125**: 82-89.

1985

Skidgel, R. A. & Erdös, E. G. Novel activity of human angiotensin I converting enzyme: release of the NH_2- and COOH-terminal tripeptides from the luteinizing hormone-releasing hormone. *Proc. Natl. Acad. Sci. USA* **82**: 1025-1029.

Entry 16.02

Peptidyl Dipeptidase B

Summary

EC Number: This enzyme is cathepsin B, EC 3.4.22.1, which is a cysteine proteinase (Volume 1, Entry 2.01), but is included here in recognition of its peptidyldipeptidase activity.

Earlier names: Cathepsin II, cathepsin B1, cathepsin B_1, cathepsin B'.

Abbreviation: PDP B.

Distribution: Widely distributed in the lysosomes of most mammalian cells.

Source: Beef spleen, human liver [1,2], hog liver [3], and many other tissues.

Action: Releases dipeptides sequentially from the C-termini of glucagon [4], muscle aldolase [5] and other polypeptides [6] and proteins. Also acts as an endopeptidase on insulin B chain [7], collagen [8], proteoglycan [9] and many other proteins. Cleaves synthetic amides and esters especially of α-N-protected arginine.

Requirements: pH optima for protein substrates range from 3.5-6.0, but maximal activity is generally close to pH 6.0 with synthetic substrates. When assays are brief and at low temperatures, to minimize denaturation, a rather higher pH optimum is seen [10,11]. A thiol activator (e.g. 2 mM dithiothreitol) is required, and sometimes EDTA (1 mM). Full activation by 2 mM dithiothreitol takes about 7 min at pH 6.0 [12].

Substrate, usual: Bz-Arg-NPhNO$_2$ and Bz-Arg-NNap have seen much use, but are unsuitable, as they are also cleaved by cathepsin H (Entry 11.10). The best substrate is Z-Arg-Arg-NMec [13], and Z-Arg-Arg-NNap is also good [14,15]. Z-Phe-Arg-NMec is extremely sensitive to cathepsin B, but is also well hydrolyzed by cathepsin L (Volume 1, Entry 2.03). Active site titration is possible by use of E-64 (see below) in conjunction with a sensitive substrate [13,16].

Substrate, special: Z-Arg-NNapOMe for staining gels, with diazo-coupling [2,17], or Z-Ala-Arg-Arg-NNapOMe with a diazonium salt or nitrosalicylaldehyde for histochemical localization [18,19].

251

Inhibitors: Irreversible inhibition is produced slowly by non-specific thiol-alkylating reagents; iodoacetate is one of the faster-reacting of these, whereas iodoacetamide and N-ethylmaleimide are very slow. Reagents that first bind to specificity subsites of the active site react much more rapidly and selectively; these include Z-Phe-Ala-CHN$_2$, Z-Phe-Phe-CHN$_2$ [20,21] and Pro-Phe-Arg-CH$_2$Cl [16], as well as E-64 (3-carboxy-2,3-L-*trans*-epoxypropionyl-leucylamido[4-guanidino]butane) and related compounds [16].

Reversible inhibitors are leupeptin (K$_i$ about 6 μM) [12,15], and other peptide aldehydes [22]. Reversible covalent inhibition is produced by Hg^{2+} ions, but much more selectively by specific disulfides [3].

Protein inhibitors are the cystatins (K$_i$ 0.25-73 nM) [23,24], plasma kininogens (α-cysteine proteinase inhibitor) (K$_i$ 350 nM) [25] and α_2-macroglobulin [26].

Molecular properties: M_r about 27,000 [1,27], pI 4.5-5.5 (human: multiple forms [1]). A$_{280,1\%}$ 20 [1]. Amino acid sequence very clearly homologous with papain [28-32]. Contains little hexosamine, and has little affinity for concanavalin A-Sepharose [33], unlike other lysosomal endopeptidases. Rapidly and irreversibly denatured above pH 7 [1]. Lysosomal cathepsin B consists of a light chain of M_r about 5000 that contains the active site cysteine, and a heavy chain of M_r about 24,000; these separate in SDS/gel electrophoresis with reduction [28]. The cDNA sequence of human cathepsin B [31] confirms that the heavy and light chains arise from N- and C-terminal parts, respectively, of the single polypeptide chain by post-translational modification, and shows residues at both ends of the chain and between the heavy and light chains that were not found in the mature protein.

Comment: Antibodies raised against cathepsin B normally react only with the alkali-denatured form [1]. Cathepsin B is probably synthesized as a latent precursor of higher M_r; there are indications that this can go through several stages of limited proteolysis and conformational change in the transition to the form normally found in the lysosomes [34].

References
[1]   Barrett *Biochem. J.* **131**: 809-822, 1973.
[2]   Schwartz & Barrett *Biochem. J.* **191**: 487-497, 1980.
[3]   Evans & Shaw *J. Biol. Chem.* **258**: 10227-10232, 1983.
[4]   Aronson & Barrett *Biochem. J.* **171**: 759-765, 1978.
[5]   Bond & Barrett *Biochem. J.* **189**: 17-25, 1980.
[6]   Towatari & Katunuma *J. Biochem.* **93**: 1119-1128, 1983.
[7]   McKay *et al. Biochem. J.* **213**: 467-471, 1983.
[8]   Burleigh *et al. Biochem. J.* **137**: 387-398, 1974.
[9]   Roughley & Barrett *Biochem. J.* **167**: 629-637.
[10] A. J. Barrett & Y. Hojima *Unpublished results.*
[11] Willenbrock & Brocklehurst *Biochem. J.* **222**: 805-814, 1984.
[12] Baici & Gyger-Marazzi *Eur. J. Biochem.* **129**: 33-41, 1982.

[13] Barrett & Kirschke *Methods Enzymol.* **80**: 535-561, 1981.
[14] McDonald & Ellis *Life Sci.* **17**: 1269-1276, 1975.
[15] Knight *Biochem. J.* **189**: 447-453, 1980.
[16] Barrett *et al.* *Biochem. J.* **201**: 189-198, 1982.
[17] Mort & Leduc *Anal. Biochem.* **119**: 148-152, 1982.
[18] Dolbeare & Smith *Clin. Chem.* **23**: 1485-1491, 1977.
[19] Graf *et al.* *Histochemistry* **64**: 319-322, 1979.
[20] Shaw & Green *Methods Enzymol.* **80**: 820-826, 1981.
[21] Shaw *et al.* *Arch Biochem. Biophys.* **222**: 424-429, 1983.
[22] Kirschke *et al.* *Ciba Found. Symp.* **75**: 15-35, 1980.
[23] Green *et al.* *Biochem. J.* **218**: 939-946, 1984.
[24] Barrett In: *Intracellular Protein Catabolism* (Khairallah, E. A.,
      Bond, J. S. & Bird, J. W. C., eds), pp. 105-116, A. R. Liss, New
      York, 1985.
[25] Gounaris & Barrett *Biochem. J.* **221**: 445-452, 1984.
[26] Starkey & Barrett *Biochem. J.* **131**: 823-831, 1973.
[27] Kirschke & Barrett In: *Lysosomes: their Role in Protein Breakdown*
      (Glaumann, H. & Ballard, F. J. eds), Academic Press, New York, in the
      press.
[28] Takio *et al.* *Proc. Natl. Acad. Sci. USA* **80**: 3666-3670, 1983.
[29] Takahashi *et al.* *J. Biol. Chem.* **259**: 6059-6062, 1984.
[30] Ritonja *et al.* *FEBS Lett.* **181**: 169-171, 1985.
[31] San Segundo *et al.* *Proc. Natl. Acad. Sci. USA* **82**: 2320-2324,
      1985.
[32] Pohl *et al.* *FEBS Lett.* **142**: 23-26, 1982.
[33] Barrett In: *Proteinases in Mammalian Cells and Tissues* (Barrett
      A. J. ed.), pp. 181-207, North-Holland Publishing Co., Amstedam,
      1977.
[34] Recklies *et al.* *Biochem. J.* **207**: 633-636, 1982.

Bibliography

1970

Keilová, H. & Turková, J. Analogy between active sites of cathepsin B1
and papain. *FEBS Lett.* **11**: 287-288.
McDonald, J. K., Zeitman, B. B. & Ellis, S. Leucine naphthylamide: an
inappropriate substrate for the histochemical detection of cathepsin B
and B'. *Nature* **225**: 1048-1049.

1971

Keilová, H. On the specificity and inhibition of cathepsins D and B. In:
Tissue Proteinases (Barrett, A. J. and Dingle, J. T. eds), pp. 45-68,
North-Holland Publishing Co., Amsterdam.
Otto, K. Cathepsins B1 and B2. In: *Tissue Proteinases* (Barrett, A. J. &
Dingle, J. T. eds), pp. 1-28, North-Holland Publishing Co., Amsterdam.

1972

Barrett, A. J. A new assay for cathepsin B1 and other thiol proteinases. *Anal. Biochem.* **47**: 280-293.

1973

Barrett, A. J. Human cathepsin B1. Purification and some properties of the enzyme. *Biochem. J.* **131**: 809-822.
Keilová, H. & Tomášek, V. On the isozymes of cathepsin B1. *FEBS Lett.* **29**: 335-338.

1974

Burleigh, M. C., Barrett, A. J. & Lazarus, G. S. Cathepsin B1. A lysosomal enzyme that degrades native collagen. *Biochem. J.* **137**: 387-398.
Etherington, D. J. The purification of bovine cathepsin B1 and its mode of action on bovine collagens. *Biochem. J.* **137**: 547-557.
Keilová, H. & Tomásek, V. Effect of papain inhibitor from chicken egg white on cathepsin B1. *Biochim. Biophys. Acta* **334**: 179-186.

1975

McDonald, J. K. & Ellis, S. On the substrate specificity of cathepsin B1 and B2 including a new fluorogenic substrate for cathepsin B1. *Life Sciences* **17**: 1269-1276.

1976

Barrett, A. J. An improved color reagent for use in Barrett's assay of cathepsin B. *Anal. Biochem.* **76**: 374-376.
Etherington, D. J. Bovine spleen cathepsin B1 and collagenolytic cathepsin. A comparative study of the properties of the two enzymes in the degradation of native collagen. *Biochem. J.* **153**: 199-209.

1977

Barrett, A. J. Cathepsin B and other thiol proteinases. In: *Proteinases in Mammalian Cells and Tissues* (Barrett, A. J. ed.), pp. 181-207. North-Holland Publishing Co., Amsterdam.
Leary, R. & Shaw, E. Inactivation of cathepsin B_1 by diazomethyl ketones. *Biochem. Biophys. Res. Commun.* **79**: 926-931.

1978

Aronson, N. N., Jr. & Barrett, A. J. The specificity of cathepsin B. Hydrolysis of glucagon at the C-terminus by a peptidyldipeptidase mechanism. *Biochem. J.* **171**: 759-765.
Nakai, N., Wada, K., Kobashi, K. & Hase, J. The limited proteolysis of rabbit muscle aldolase by cathepsin B1. *Biochem. Biophys. Res. Commun.* **83**: 881-885.

1979

MacGregor, R. R., Hamilton, J. W., Kent, G. N., Shofstall, R. E. & Cohn, D. V. The degradation of proparathormone and parathormone by parathyroid and liver cathepsin B. *J. Biol. Chem.* **254**: 4428-4433.

MacGregor, R. R., Hamilton, J. W., Shofstall, R. E. & Cohn, D. V. Isolation and characterization of porcine parathyroid cathepsin B. *J. Biol. Chem.* **254**: 4423-4427.

Towatari, T., Kawabata, Y. & Katunuma, N. Crystallization and properties of cathepsin B from rat liver. *Eur. J. Biochem.* **102**: 279-289.

Watanabe, H., Green, G. D. J. & Shaw, E. A comparison of the behaviour of chymotrypsin and cathepsin B towards peptidyl diazomethyl ketones. *Biochem. Biophys. Res. Commun.* **89**: 1354-1360.

1980

Barrett, A. J. Fluorimetric assays for cathepsin B and cathepsin H with methylcoumarylamide substrates. *Biochem. J.* **187**: 909-912.

Bond, J. S. & Barrett, A. J. Degradation of fructose-1,6-bisphosphate aldolase by cathepsin B. *Biochem. J.* **189**: 17-25.

Knight, C. G. Human cathepsin B. Application of the substrate N-benzyloxycarbonyl-L-arginyl-L-arginine 2-naphthylamide to a study of the inhibition by leupeptin. *Biochem. J.* **189**: 447-453.

Takio, K., Towatari, T., Katunuma, N. & Titani, K. Primary structure study of rat liver cathepsin B. A striking resemblance to papain. *Biochem. Biophys. Res. Commun.* **97**: 340-346.

1981

Barrett, A. J. & Kirschke, H. Cathepsin B, cathepsin H and cathepsin L. *Methods Enzymol.* **80**: 535-561.

Graf, M., Baici, A. & Sträuli, P. Histochemical localization of cathepsin B at the invasion front of the rabbit V2 carcinoma. *Lab. Invest.* **45**: 587-596.

Kirschke, H. & Shaw, E. Rapid inactivation of cathepsin L by Z-Phe-Phe-CHN$_2$ and Z-Phe-Ala-CHN$_2$. *Biochem. Biophys. Res. Commun.* **101**: 454-458.

1982

Baici, A. & Gyger-Marazzi, M. The slow, tight-binding inhibition of cathepsin B by leupeptin. A hysteretic effect. *Eur. J. Biochem.* **129**: 33-41.

Barrett, A. J., Kembhavi, A. A., Brown, M. A., Kirschke, H., Knight, C. G., Tamai, M. & Hanada, K. L-*Trans*-epoxysuccinyl-leucylamido(4-guanidino)butane (E-64) and its analogues as inhibitors of cysteine proteinases including cathepsins B, H and L. *Biochem. J.* **201**: 189-198.

Mort, J. S. & Leduc, M. A simple, economical method for staining gels for cathepsin B-like activity. *Anal. Biochem.* **119**: 148-152.

Pohl, J., Baudyš, M., Tomášek, V. & Kostka, V. Identification of the active site cysteine and of the disulfide bonds in the N-terminal part of the molecule of bovine spleen cathepsin B. *FEBS Lett.* **142**: 23-26.

Recklies, A. D., Poole, A. R. & Mort, J. S. A cysteine proteinase secreted from human breast tumors is immunologically related to cathepsin B. *Biochem. J.* **207**: 633-636.

1983

Bajkowski, A. S. & Frankfater, A. Steady state kinetic evidence for an acyl-enzyme intermediate in reactions catalyzed by bovine spleen cathepsin B. *J. Biol. Chem.* **258**: 1645-1649.

Bajkowski, A. S. & Frankfater, A. The pH dependency of bovine spleen cathepsin B-catalyzed transfer of N^{α}-benzyloxycarbonyl-L-lysine from *p*-nitrophenol to water and dipeptide nucleophiles. Comparisons with papain. *J. Biol. Chem.* **258**: 1650-1655.

Evans, B. & Shaw, E. Inactivation of cathepsin B by active site-directed disulfide exchange. *J. Biol. Chem.* **258**: 10227-10232.

McKay, M. J., Offermann, M. K., Barrett, A. J. & Bond, J. S. Action of human liver cathepsin B on the oxidized insulin B chain. *Biochem. J.* **213**: 467-471.

Shaw, E., Wikstrom, P. & Ruscica, J. An exploration of the primary specificity site of cathepsin B. *Arch. Biochem. Biophys.* **222**: 424-429.

1984

Graf, F. M., Haemmerli, G. & Sträuli, P. Cathepsin B containing cells in the rabbit mesentery during invasion of V2 carcinoma cells. *Histochemistry* **80**: 509-515.

Hirao, T., Hara, K. & Takahashi, K. Purification and characterization of cathepsin B from monkey skeletal muscle. *J. Biochem. (Tokyo)* **95**: 871-879.

Takahashi, T., Dehdarani, A. H., Schmidt, P. G. & Tang, J. Cathepsins B and H from porcine spleen. Purification, polypeptide chain arrangements, and carbohydrate content. *J. Biol. Chem.* **259**: 9874-9882.

Takahashi, T., Schmidt, P. G. & Tang, J. Novel carbohydrate structures of cathepsin B from porcine spleen. *J. Biol. Chem.* **259**: 6059-6062.

Willenbrock, F. & Brocklehurst, K. Natural structural variation in enzymes as a tool in the study of mechanism exemplified by a comparison of the catalytic-site structure and characteristics of cathepsin B and papain. *Biochem. J.* **222**: 805-814.

Yanagisawa, K., Sato, S., Miyatake, T. & Kominami, E. Degradation of myelin proteins by cathepsin B and inhibition by E 64 analog. *Neurochem. Res.* **9**: 691-694.

1985

Ritonja, A., Popovic, T., Turk, V., Wiedenmann, K. & Machleidt, W. Amino acid sequence of human liver cathepsin B. *FEBS Lett.* **181**: 169-171.

San Segundo, B., Chan, S. J. & Steiner, D. F. Identification of cDNA clones encoding a precursor of rat liver cathepsin B. *Proc. Natl. Acad.*

Sci. USA **82**: 2320-2324.

Section 17
DIPEPTIDASES

Proline Dipeptidase

Summary

EC Number: 3.4.13.9

Earlier names: Prolidase, imidodipeptidase, γ peptidase, and peptidase D (PEPD).

Abbreviation: None.

Distribution: Found in all animal tissues thus far examined, including red and white blood cells. It is generally freely-soluble in the cytoplasm. In the rat, kidney is the richest source, followed, in decreasing order, by intestinal mucosa, spleen, liver, lung, heart, cervical spinal cord, sciatic nerve, pituitary, and plasma. Levels in pituitary and brain are 17% and 20%, respectively, of that in kidney [1]. A somewhat different distribution in the rat has also been reported [1a].

Source: Hog kidney [2-4]; hog [5,6], beef [7] and guinea pig [8] intestinal mucosa; guinea pig brain [9]; rabbit muscle [10], and horse erythrocytes [2,11].

Action: Catalyzes the hydrolysis of dipeptides only, and specifically those containing C-terminal proline or hydroxyproline. The relative rates at which such proline dipeptides are hydrolyzed (at pH 7.4) by the hog intestinal enzyme are reported [6] to be as follows: Ala-Pro 100, Ser-Pro 90, Met-Pro 40, Gly-Pro 30, Glu-Pro 30, and Phe-Pro 5. No action occurs on Pro-Pro or His-Pro, nor on Z-Gly-Pro, Z-Gly-Hyp, or Gly-Pro-NH$_2$. Tripeptides such as Gly-Pro-Gly, Gly-Pro-Ala and Gly-Gly-Pro are not acted upon. Dipeptides commonly used as substrates for other dipeptidases, e.g. Gly-Gly, Gly-Leu, Ala-Leu, Ala-Gly and Pro-Gly are not hydrolyzed.

For the beef intestinal enzyme, V_{max} values are equally high on Ala-Pro 100, Leu-Pro 100 and His-Pro 100, with lower rates seen on Met-Pro 50, Phe-Pro 50, Val-Pro 33, Gly-Pro 17, and Pro-Pro 17 [7].

Relative rates of hydrolysis of proline dipeptides by the guinea pig brain enzyme are as follows: Met-Pro 104, Leu-Pro 100, Ile-Pro 50, Phe-Pro 46, and His-Pro 23. Lesser rates occur on Ala-Pro, Gly-Pro, Val-Pro, Glu-Pro and Ser-Pro. Although this study was conducted with an enzyme preparation that was not homogeneous, it would appear that low rates may also occur on prolyl dipeptides such as Pro-Met and Pro-Leu, and even on some peptides that do not contain proline [9].

Requirements: pH 7.2-7.8 at 37°C, with preliminary activation for 1 h in 5 mM Mn^{2+}.

Substrate, usual: Ala-Pro in a colorimetric [12] procedure or in a continuous spectrophotometric assay that follows decrease in absorbance at 220 nm [13]. Gly-Pro has been used in conjunction with colorimetric procedures that selectively determine free proline [13a,13b].

Substrate, special: Leu-Pro in a discontinuous fluorometric assay [9,14].

Inhibitors: Sulfhydryl-blocking agents such as $IAcNH_2$ [2] and p-chloromercuribenzoate [7,9] show strong inhibition, reversible with the mercurial reagent. Pyrophosphate [1 mM) produces strong inhibition [2], but weak and variable degrees of inhibition are produced by other metal-binding agents such as EDTA, 1,10-phenanthroline, citrate, and fluoride [2,7]. Dip-F and Pms-F have no effect, nor does bestatin, puromycin, E-64 or Ep-475 [7]. Cu^{2+}, Zn^{2+}, Cd^{2+} and Hg^{2+} are inhibitory, whereas Ca^{2+} and Mg^{2+} have no effect [2,7].

Molecular properties: The enzyme from hog intestinal mucosa is a Mn^{2+}-dependent enzyme (M_r 108,000) consisting of 2 identical noncovalently-bound subunits; pI 4.4 [9,15]. The native molecule has a carbohydrate content of about 0.5% and contains about 6 thiol-groups and 15 disulfide bonds. This number of disulfide linkages should confer a rigid conformation on the molecule, and thereby contribute to its stability. Stoke's radius is about 4.2 nm, and $A_{280,1\%}$ is 10.4 [15]. Freeze-dried preparations of the purified enzyme are stable for many months at 4°C, and solutions can withstand heating to 50°C if the pH is approximately 7.0.

The enzyme from guinea pig intestinal mucosa is also comprised of two identical (M_r 57,000) subunits, and is similarly stimulated by Mn^{2+} [8], but the guinea pig brain enzyme reportedly contains two dissimilar subunits and is not stimulated by Mn^{2+} [9].

The M_r 116,000 enzyme from beef intestinal mucosa, which is also a dimer, possesses a high content of aspartic and glutamic acid residues, and appears to belong to the cysteine catalytic class. The enzyme is stable at pH 5.0-9.0 [7].

Comment: Some investigators have reported that preparations of proline dipeptidase from hog kidney [16] and guinea pig intestinal mucosa [8] hydrolyze tripeptides as well as assorted dipeptides lacking C-terminal proline residues. In view of the more restricted specificities observed with highly-purified preparations, it would appear that these discrepancies may be attributable to the use of impure preparations.

Whereas the urinary excretion of Gly-Pro, as well as other imidodipeptides, is normally negligible, a number of patients have been described who excreted massive amounts of these dipeptides, in particular Gly-Pro. Such cases of hyperimidodipeptiduria can result in a daily bound-proline excretion rate of over three grams, with an unusually high proline/hydroxyproline ratio, typically in the range 2.4 to 25.2 [17].

Patients exhibiting hyperimidodipeptiduria have been shown to lack proline dipeptidase in their red and white cells and skin fibroblasts [18,18a], and presumably in other cells and tissues as well. The syndrome resembles that of lathyrism and suggests a defect in collagen metabolism [17,18]. The skin collagen of such patients fails to follow the normal maturation process, and the composition of urinary collagen metabolites is unusual [19]. Skin manifestations such as fragility, scarring, soft and thin abdominal skin, and chronic recurring ulcerations of the legs and feet are the most typical clinical aspects of the disease [20], which is inherited in an autosomal recessive manner [17,18]. Proline dipeptidase deficiency is a rarely reported inborn error of metabolism that should be suspected as a possible cause of chronic skin ulcers, especially when accompanied by premature graying of the hair, recurring infections, splenomegaly, or mental retardation [21]. The gene responsible for proline dipeptidase synthesis has been assigned to chromosome 19 [22].

References
[1] Hui & Lajtha *J. Neurochem.* **30**: 321-327, 1978.
[1a] Imai *et al. Mol. Cell. Biochem.* **42**: 31-36, 1982.
[2] Smith *Methods Enzymol.* **2**: 100-105, 1955.
[3] Davis & Smith *J. Biol. Chem.* **224**: 261-275, 1957.
[4] Manao *et al. Physiol. Chem. Phys.* **4**: 75-87, 1972.
[5] Smith & Bergmann *J. Biol. Chem.* **153**: 627-651: 1944.
[6] Sjöström *et al. Biochim. Biophys. Acta* **327**: 457-470, 1973.
[7] Yoshimoto *et al. J. Biochem.* **94**: 1889-1896, 1983.
[8] O'Cuinn & Fottrell *Biochim. Biophys. Acta* **391**: 388-395, 1975.
[9] Brown & O'Cuinn *J. Biol. Chem.* **258**: 6147-6154, 1983.
[10] Smith *J. Biol. Chem.* **173**: 553-569, 1948.
[11] Adams & Smith *J. Biol. Chem.* **198**: 671-682, 1952.
[12] McDonald *et al. Biochem. Biophys. Res. Commun.* **46**: 62-70, 1972.
[13] Josefsson & Lindberg *Biochim. Biophys. Acta* **105**: 149-161, 1965.
[13a]Myara *et al. Clin. Chim. Acta* **125**: 193-205, 1982.
[13b]Priestman & Butterworth *Clin. Chim. Acta* **142**: 263-271, 1984.
[14] Nicholson & Peters *Anal. Biochem.* **87**: 418-424, 1978.
[15] Sjöström & Norén *Biochim. Biophys. Acta* **359**: 177-185, 1974.
[16] Hill & Schmidt *J. Biol. Chem.* **237**: 389-396, 1962.
[17] Isemura *et al. Clin. Chim. Acta* **93**: 401-407, 1979.
[18] Powell *et al. Metabolism* **23**: 505-513, 1974.
[18a]Butterworth & Priestman *J. Inherited Metab. Dis.* **7**: 32-34, 1984.

[19] Isemura *et al.* *Tohoku J. Exp. Med.* **134**: 21-28, 1981.
[20] Arata *et al.* *Arch. Dermatol.* **115**: 62-67, 1979.
[21] Der Kaloustian *et al.* *Dermatologica* **164**: 293-304, 1982.
[22] McAlpine *et al.* *Cytogenet. Cell Genet.* **16**: 204-205, 1976.

Bibliography

1937

Bergmann, M. & Fruton, J. S. On proteolytic enzymes. XII. Regarding the specificity of aminopeptidase and carboxypeptidase. A new type of enzyme in the intestinal tract. *J. Biol. Chem.* **117**: 189-202.

1944

Smith, E. L. & Bergmann, M. The peptidases of intestinal mucosa. *J. Biol. Chem.* **153**: 627-651.

1948

Hanson, H. T. & Smith, E. L. The application of peptides containing β-alanine to the study of the specificity of various peptidases. *J. Biol. Chem.* **175**: 833-848.
Smith, E. L. The peptidases of skeletal, heart, and uterine muscle. *J. Biol. Chem.* **173**: 553-569.

1951

Adams, E. & Smith, E. L. Proteolytic activity of pituitary extracts. *J. Biol. Chem.* **191**: 651-664.
Smith, E. L. The specificity of certain peptidases. *Adv. Enzymol.* **12**: 191-257 (see 219-221).

1952

Adams, E. & Smith, E. L. Peptidases of erthythrocytes. II. Isolation and properties of prolidase. *J. Biol. Chem.* **198**: 671-682.

1954

Adams, E., Davis, N. C. & Smith, E. L. Specificity of prolidase: Effect of alterations in the pyrrolidine ring of glycly-L-proline. *J. Biol. Chem.* **208**: 573-578.

1955

Smith, E. L. Dipeptidases. C. Prolidase (imidodipeptidase). *Methods Enzymol.* **2**: 100-105.

1957

Davis, N. C. & Smith, E. L. Purification and some properties of prolidase of swine kidney. *J. Biol. Chem.* **224**: 261-275.

1962

Hill, R. L. & Schmidt, W. R. The complete enzymic hydrolysis of proteins. *J. Biol. Chem.* **237**: 389-396.

1965

Josefsson, L. & Lindberg, T. Intestinal dipeptidases. I. Spectrophotometric determination and characterization of dipeptidase activity in pig intestinal mucosa. *Biochim. Biophys. Acta* **105**: 149-161.

1966

Hanson, H. Hydrolasen: Peptidasen (Exopeptidasen). In: *Hoppe-Seyler/Thierfelder-Handbuch der physiologisch- und pathologisch-chemischen Analyse* (Lang, K., Lehnartz, E., Hoffmann-Ostenhof, O. & Siebert, G. eds), 10th edn., vol. 6, part C, pp. 1-229 (see pp. 71-79), Springer-Verlag, Berlin.

1968

Goodman, S. I., Solomons, C. C., Muschenheim, F., McIntyre, C. A., Miles, B. & O'Brien, D. A syndrome resembling lathyrism associated with iminodipeptiduria. *Am. J. Med.* **45**: 152-159.

1969

Lewis, W. H. P. & Harris, H. Peptidase D (prolidase) variants in man. *Ann. Hum. Genet.* **32**: 317-322.

1971

Sjöström, H., Norén, O. & Josefsson, L. Isolation and purification of a prolidase from pig intestinal mucosa. *Acta Chem. Scand.* **25**: 1911-1913.

1972

Buist, N. R., Strandholm, J. J., Bellinger, J. F. & Kennaway, N. G. Further studies on a patient with iminodipeptiduria: A probable case of prolidase deficiency. *Metabolism* **21**: 1113-1123.

Manao, G., Nassi, P., Cappugi, G., Camici, G. & Ramponi, G. Swine kidney prolidase: assay, isolation procedure, and molecular properties. *Physiol. Chem. Phys.* **4**: 75-87.

1973

Sjöström, H., Norén, O. & Josefsson, L. Purification and specificity of pig intestinal prolidase. *Biochim. Biophys. Acta* **327**: 457-470.

1974

Baksi, K. & Radhakrishnan, A. N. Purification and properties of prolidase (imidodipeptidase) from monkey small intestine. *Indian J. Biochem. Biophys.* **11**: 7-11.

Johnstone, R. A. W., Povall, T. J., Baty, J. D., Pousset, J. L., Charpentier, C. & Lemonnier, A. Determination of dipeptides in urine. *Clin. Chim. Acta* **52**: 137-142.

Powell, G. F., Rasco, M. A. & Maniscalco, R. M. A prolidase deficiency in man with iminopeptiduria. *Metabolism* **23**: 505-513.

Sjöström, H. & Norén, O. Structural properties of pig intestinal proline dipeptidase. *Biochim. Biophys. Acta* **359**: 177-185.

1975

Jackson, S. H., Dennis, A. W. & Greenberg, M. Iminodipeptiduria: a genetic defect in recycling collagen: a method for determining prolidase in erythrocytes. *Can. Med. Assoc. J.* **113**: 759-763.

O'Cuinn, G. & Fottrell, P. F. Purification and characterization of an aminoacyl proline hydrolase from guinea-pig intestinal mucosa. *Biochim. Biophys. Acta* **391**: 388-395.

1976

Kodama, H., Umemura, S., Shimomura, M., Mizuhara, S., Arata, J., Yamamoto, Y., Yasutake, A. & Izumiya, N. Studies on a patient with iminopeptiduria. I. Identification of urinary iminopeptides. *Physiol. Chem. Phys.* **8**: 463-473.

McAlpine, P. J., Mohandas, T., Ray, M., Wang, H. & Hamerton, J. L. Assignment of the *peptidase D* gene locus (*PEPD*) to chromosome 19 in man. *Cytogenet. Cell Genet.* **16**: 204-205.

1977

Hui, K.-S., Weiss, B., Hui, M. & Lajtha, A. Covalent coupling of calf brain prolidase. *J. Neurosci. Res.* **3**: 231-239.

McDonald, J. K. & Schwabe, C. Intracellular exopeptidases. In: *Proteinases in Mammalian Cells and Tissues* (Barrett, A. J. ed.), pp. 311-391 (see pp. 365-366), North-Holland Publishing Co., Amsterdam.

Norén, O., Dabelsteen, E., Sjöström, H. & Josefsson, L. Histological localization of two dipeptidases in the pig small intestine and liver, using immunofluorescence. *Gastroenterology* **72**: 87-92.

Sheffield, L. J., Schlesinger, P., Faull, K.., Halpern, B. J., Schier, G. M., Cotton, R. G. H., Hammond, J. & Danks, D. M. Iminopeptiduria, skin ulcerations, and edema in a boy with prolidase deficiency. *J. Pediatr.* **91**: 578-583.

1978

Brown, S., Lalley, P. A. & Minna, J. D. Assignment of the gene for peptidase S (*PEPS*) to chromosome 4 in man and confirmation of peptidase D (*PEPD*) assignment to chromosome 19. *Cytogenet. Cell Genet.* **22**: 167-171.

Hui, K. S. & Lajtha, A. Prolidase activity in brain: comparison with other organs. *J. Neurochem.* **30**: 321-327.

Sjöström, H. & Norén, O. Size and shape of two intestinal dipeptidases. *Int. J. Pept. Protein Res.* **11**: 159-165.

Umemura, S. Studies on a patient with iminodipeptiduria. II. Lack of prolidase activity in blood cells. *Physiol. Chem. Phys.* **10**: 279-283.

1979

Arata, J., Umemura, S., Yamamoto, Y., Hagiyama, M. & Nohara, N. Prolidase deficiency. Its dermatological manifestations and some additional biochemical studies. *Arch. Dermatol.* **15**: 62-67.

Isemura, M., Hanyu, T., Gejyo, F., Nakazawa, R., Igarashi, R., Matsuo, S., Ikeda, K. & Sato, Y. Prolidase deficiency with imidodipeptiduria. A familial case with and without clinical symptoms. *Clin. Chim. Acta* **93**: 401-407.

1980

Hui, K.-S. & Lajtha, A. Activation and inhibition of cerebral prolidase. *J. Neurochem.* **35**: 489-494.

1981

Charpentier, C., Dagbovie, K., Larreque, M., Johnstone, R. A. W. & Lemonnier, A. Prolidase deficiency with iminodipeptiduria: biochemical investigations and first results of therapeutic assays. *J. Inherited Metab. Dis.* **4**: 77-78.

Endo, F. & Matsuda, I. Screening method for prolidase deficiency. *Hum. Genet.* **56**: 349-351.

Isemura, M., Hanyu, T., Ono, T., Igarashi, R., Sato, Y., Gejyo, F., Nakazawa, R., Miyakawa, T., Takagi, T., Kuboki, Y. & Sasaki, S. Studies on prolidase deficiency with a possible defect in collagen metabolism. *Tohoko J. Exp. Med.* **134**: 21-28.

Ogata, A., Tanaka, S., Tomoda, T., Murayama, E., Endo, F. & Kikuchi, I. Autosomal recessive prolidase deficiency. Arch. Dermatol. **117**: 689-694.

1982

Der Kaloustian, V. M., Freij, B. J. & Kurban, A. K. Prolidase deficiency: an inborn error of metabolism with major dermatological manifestations. *Dermatologica* **164**: 293-304.

Endo, F., Matsuda, I., Ogata, A. & Tanaka, S. Human erythrocyte prolidase and prolidase deficiency. Pediatr. Res. **16**: 227-231.

Imai, K., Nagatsu, T., Yajima, T., Maeda, N., Kumegawa, M. & Kato, T. Developmental changes in the activities of prolinase and prolidase in rat salivary glands, and the effect of thyroxine administration. *Mol. Cell. Biochem.* **42**: 31-36.

Myara, I., Charpentier, C. & Lemonnier, A. Optimal conditions for prolidase assay by proline colorimetric determination: application to iminodipeptiduria. *Clin. Chim. Acta* **125**: 193-205.

1983

Browne, P. & O'Cuinn, G. The purification and characterization of a proline dipeptidase from guinea pig brain. *J. Biol. Chem.* **258**: 6147-6154.

Myara, I., Charpentier, C., Wolfrom, C., Gautier, M., Lemonnier, A., Larregue, M., Chamson, A. & Frey, J. *In-vitro* responses to ascorbate and manganese in fibroblasts from a patient with prolidase deficiency and iminodipeptiduria: cell growth, prolidase activity and collagen metabolism. *J. Inherited Metab. Dis.* **6**: 27-31.

Yoshimoto, T. & Tsuru, D. Substrate specificity of aminopeptidase M: evidence that the commercial preparation is contaminated by dipeptidyl aminopeptidase IV and prolidase. *J. Biochem.* **94**: 619-622.

Yoshimoto, T., Matsubara, F., Kawano, E. & Tsuru, D. Prolidase from bovine intestine: purification and characterization. *J. Biochem.* **94**: 1889-1896.

1984

Butterworth, J. & Priestman, D. Substrate specificity of manganese-activated prolidase in control and prolidase-deficient cultured skin fibroblasts. *J. Inherited Metab. Dis.* **7**: 32-34.

Kanda, S., Maekawa, M., Kohno, H., Sudo, K., Hishiki, S., Nakamura, S. & Kanno, T. Examination of the subcellular distribution of tripeptide aminopeptidase and evaluation of its clinical usefulness in human serum. *Clin. Biochem.* **17**: 253-257.

Mikasa, H., Arata, J. & Kodama, H. Measurement of prolidase activity in erythrocytes using isotachophoresis. *J. Chromatog.* **310**: 401-406.

Myara, I., Charpentier, C. & Lemonnier, A. Prolidase and prolidase deficiency. *Life Sci.* **34**: 1985-1998.

Myara, I., Myara, A., Mangeot, M., Fabre, M., Charpentier, C. & Lemonnier, A. Plasma prolidase activity: a possible index of collagen catabolism in chronic liver disease. *Clin. Chem.* **30**: 211-215.

Myara, I., Wolfrom, C., Charpentier, C., Gautier, M. & Lemonnier, A. Prolidase and prolinase activity in Ehlers Danlos fibroblasts. *IRCS Med. Sci. Biochem.* **12**: 636 only.

Priestman, D. A. & Butterworth, J. Prolidase deficiency: characteristics of human skin fibroblast prolidase using colorimetric and fluorimetric assays. *Clin. Chim. Acta* **142**: 263-271.

Prolyl Dipeptidase

Summary

EC Number: 3.4.13.8.

Earlier names: Prolinase, iminodipeptidase and prolylglycine dipeptidase.

Abbreviation: None.

Distribution: In the rat, the richest source is kidney, followed by
jejunum, testis, ileum, spleen, duodenum, lung, liver, thymus,
epididymis, stomach, pancreas, heart, and skeletal muscle [6]. In the
kidney, the activity is located primarily in the soluble fraction of the
kidney cortex.

Source: Hog [1,2] and beef [3] kidney.

Action: Catalyzes specifically the hydrolysis of dipeptides with proline
or hydroxyproline at the N-terminus. The relative rates at which various
dipeptides are hydrolyzed (at pH 7.75) by the hog kidney enzyme are
reported [2] to be as follows: Pro-Gly 100, Pro-Ala 100, Pro-Met 100,
Pro-Lys 129, Pro-Glu 58, Hyp-Gly 30, Leu-Gly 19, Gly-Gly 18, Pro-Phe 7,
Pro-Leu 4, Pro-Hyp 1.4, and Pro-Pro 0.7. No action occurs on Z-Pro-Gly
or Pro-Gly-NH$_2$. Substantial rates of hydrolysis of Pro-Leu-NH$_2$ can now
be attributed to contamination by prolyl aminopeptidase. No action
occurs on tripeptides such as Gly-Gly-Gly, Pro-Gly-Gly or Gly-Pro-Hyp,
nor on Pro-NNap.

Beef prolyl dipeptidase hydrolyzes various dipeptides at the following
rates [3]: Pro-Gly 100, Pro-Ala 100, Pro-Met 100, Hyp-Gly 23, Leu-Gly
7.5, Gly-Ala 14.5, and Gly-Leu 8.0. No action occurs on Pro-Gly-Gly,
Pro-Val-Gly, polyproline, or Pro-NNap, but Pro-Leu-NH$_2$ was reportedly
hydrolyzed at 38% of the rate of Pro-Gly.

Requirements: Crude extracts show optimal activity at pH 8.0, purified
preparations at pH 8.7. The enzyme is extremely unstable above pH 9.0,
and shows negligible activity at pH 6.0. Activation and stabilization
are obtained by including 1 mM MnCl$_2$ in the assay buffer.

Substrate, usual: Pro-Gly in conjunction with a colorimetric procedure for
determining free proline [2], or a more sensitive fluorometric procedure

269

that detects the liberated C-terminal amino acid only [4]. Pro-NNap is not hydolyzed.

Substrate, special: Hyp-Gly (See Comment).

Inhibitors: Agents that bind Mn^{2+}: pyrophosphate, citrate, phosphate and fluoride. Unlike its counterpart, proline dipeptidase, sulfhydryl-blocking agents have no effect [2].

Molecular properties: A Mn^{2+}-activated enzyme showing some evidence of subunit structure [2] that is stabilized by Mn^{2+}; pI 4.25 [2,3]. Activation is also accomplished by Cd^{2+}, which is in contrast to its more characteristic inhibitory effect on aminopeptidases. Activity is completely destroyed by freezing and thawing, and about half the activity is lost when solutions are exposed to 40°C for 60 min [2].

Whereas the molecular mass of the hog enzyme (M_r 300,000) is reportedly much larger than that for the beef enzyme (M_r 100,000), this difference could simply reflect states of aggregation.

Comment: Prolyl dipeptidase is a very unstable enzyme that has not yet been purified to homogeneity. It is quite possible therefore that the low level of hydolysis of such dipeptides as Leu-Gly and Gly-Gly is attributable to contaminating dipeptidases or aminopeptidases. Indeed, the hydrolysis of Pro-Leu-NH2 by such preparations of prolyl dipeptidase constituted the initial discovery of prolyl aminopeptidase.

Although prolyl dipeptidase hydrolyzes Pro-Gly very effectively, so does prolyl aminopeptidase. Hyp-Gly, however, is only hydolyzed by the dipeptidase, and consequently should serve as a more reliable substrate [5].

Prolyl dipeptidase activity in the salivary glands of the rat steadily increases during the first 20-25 days after birth. In animals that receive thyroxine during that time, levels are precociously induced in the parotid gland, but not in the submandibular and sublingual glands [6].

Although deficiency states in humans have been described for proline dipeptidase, none has been reported for prolyl dipeptidase.

References
[1]   Smith *Methods Enzymol.* **2**: 97-100, 1955.
[2]   Mayer & Nordwig *Hoppe-Seyler's Z. Physiol. Chem.* **354**: 371-379, 1973.
[3]   Akrawi & Bailey *Biochim. Biophys. Acta* **422**: 170-178, 1976.
[4]   Butterworth & Priestman *Clin. Chim. Acta* **122**: 51-60, 1982.
[5]   Sarid *et al. J. Biol. Chem.* **237**: 2207-2212, 1962.
[6]   Imai *et al. Mol. Cell. Biochem.* **42**: 31-36, 1982.

Bibliography

1929

Grassmann, W., Dyckerhoff, H. & Schoenebeck, O. v. [Susceptibility of proline peptides to enzyme hydrolysis.] *Chem. Ber.* **62B**: 1307-1310.

1932

Grassmann, W., Schoenebeck, O.v. & Auerbach, G. [Enzymic cleavage of proline peptides. II.] *Hoppe-Seyler's Z. Physiol. Chem.* **210**: 1-14.

1951

Neuman, R. E. & Smith, E. L. Synthesis of proline and hydroxyproline peptides; their cleavage by prolinase. *J. Biol. Chem.* **193**: 97-111.

1953

Davis, N. C. & Smith, E. L. Partial purification and specificity of iminodipeptidase. *J. Biol. Chem.* **200**: 373-384.

1955

Davis, N. C. & Adams, E. Specificity of iminodipeptidase: effect of alterations in the pyrrolidine ring of L-prolylglycine. *Arch. Biochem. Biophys.* **57**: 301-305.
Smith, E. L. Dipeptidases. B. Iminodipeptidase (prolinase). *Methods Enzymol.* **2**: 97-100.

1962

Sarid, S., Berger, A. & Katchalski, E. Proline iminopeptidase. II. Purification and comparison with iminodipeptidase (prolinase). *J. Biol. Chem.* **237**: 2207-2212.

1966

Hanson, H. Hydrolasen: Peptidasen (Exopeptidasen). In: *Hoppe-Seyler/Thierfelder-Handbuch der physiologisch- und pathologisch-chemischen Analyse* (Lang, K., Lehnartz, E., Hoffmann-Ostenhof, O. & Siebert, G. eds), 10th edn., vol. 6, part C, pp. 1-229 (see pp. 67-71), Springer-Verlag, Berlin.

1973

Mayer, H. & Nordwig, A. The cleavage of prolyl peptides by kidney peptidases. Purification of iminodipeptidase (prolinase). *Hoppe-Seyler's Z. Physiol. Chem.* **354**: 371-379.

1976

Akrawi, A. F. & Bailey, G. S. Purification and specificity of prolyl dipeptidase from bovine kidney. *Biochim. Biophys. Acta* **422**: 170-178.

1977

Akrawi, A. F. & Bailey, G. S.   The separation of prolyl dipeptidase from other dipeptidases of bovine kidney. *Biochem. Soc. Trans.* **5**: 272-274.

McDonald, J. K. & Schwabe, C.   Intracellular exopeptidases.   In: *Proteinases in Mammalian Cells and Tissues* (Barrett, A. J. ed.), pp. 311-391 (see p. 365), North-Holland Publishing Co., Amsterdam.

1982

Butterworth, J. & Priestman, D.   Fluorometric assay for prolinase and partial characterisation in cultured skin fibroblasts. *Clin. Chim. Acta* **122**: 51-60.

Butterworth, J. & Priestman, D.   Fluorometric assay for prolinase. *J. Inherited Metab. Dis.* **5**, Suppl. 1: 9-10.

Imai, K., Nagatsu, T., Yajima, T., Maeda, N., Kumegawa, M. & Kato, T. Developmental changes in the activities of prolinase and prolidase in rat salivary glands, and the effect of thyroxine administration. *Mol. Cell. Biochem.* **42**: 31-36.

1983

King, G. F., York, M. J., Chapman, B. E. & Kuchel, P. W.   Proton NMR spectroscopic studies of dipeptidase in human erythrocytes. *Biochem. Biophys. Res. Commun.* **110**: 305-312.

1984

Myara, I., Wolfrom, C., Charpentier, C., Gautier, M. & Lemonnier, A. Prolidase and prolinase activity in Ehlers Danlos fibroblasts. *IRCS Med. Sci. Biochem.* **12**: 636 only.

Glycylglycine Dipeptidase

Summary

EC Number: 3.4.13.11.

Earlier names: None.

Abbreviation: GGD.

Distribution: Widely distributed in animal tissues, i.e. liver, kidney, and intestinal mucosa of rabbit, guinea pig and mouse [1].

Source: Skeletal muscle of the rat [2], human uterus [2], and hog erythrocytes [3].

Action: Catalyzes the hydrolysis of only Gly-Gly and Sar-Gly. The rate on the latter is about 8% of the former, and very early studies showed no action on Gly-Sar [4]. The K_m value was determined to be 2.7 mM for the hydrolysis of Gly-Gly by rat brain GGD at pH 7.9 and 37°C [5]. Bz-Gly-Gly, Gly-Gly-NH$_2$, Gly-Gly-Gly and N-dimethylglycylglycine are not hydrolyzed, hence the classification as a dipeptidase.

Requirements: pH 7.6 and 1 mM Co^{2+} for activation and stabilization. Crude preparations of GGD show a 10- to 50-fold activation upon addition of 1 mM CoCl$_2$. Mn^{2+} also shows activation, but to a lesser degree. Mg^{2+} has no effect.

Substrate, usual: Gly-Gly in conjunction with a method for determining the liberated amino acids colorimetrically with ninhydrin [6] or trinitrobenzenesulfonic acid [7], spectrophotometrically [8], fluorometrically [9], or by amino acid analysis [10].

Inhibitors: Compounds which form Co^{2+} chelates are inhibitory. Such agents include leucine, histidine, imidazole, tryptophan, and indole. Sulfhydryl-blocking reagents are usually inhibitory, as are metals such as Zn^{2+}, Cd^{2+}, Hg^{2+} and Ag^{2+}.

Molecular properties: GGD is an exceptionally labile enzyme. The rat muscle enzyme loses about 50% of its activity during overnight storage at 4°C and pH 7.5. Inactivation is especially rapid above pH 8.0 and below pH 6.0. Human uterine GGD, in contrast to the enzyme found in most tissues, is stable for many weeks at 4°C.

Comment: The identification of glycylglycine dipeptidase is facilitated by its very restricted substrate specificity and by its Co^{2+}-dependence.

It has been proposed that the function of Co^{2+} is to act as a bridge by forming chelate complexes wherein Co^{2+} coordinates with the enzyme and with the polar groups of the substrate.

Early reports of an activation of Gly-Gly hydrolysis by dialysates or boiled extracts are most probably attributable to the presence of activating metal ions, i.e. Co^{2+} or Mn^{2+}. Whereas GGD found in most tissues is extremely labile, the dipeptidase of human uterus is relatively stable and may be prepared as an acetone powder.

Glycylglycolic acid is not hydrolyzed, which is consistent with the extremely narrow substrate specificity of GGD, and shows a lack of esterase activity on a substrate with the highest structural similarity to Gly-Gly.

References

[1] Maschmann *Biochem. Z.* **309**: 28-41, 1941.
[2] Smith *J. Biol. Chem.* **173**: 571-583, 1948.
[3] Bidlingmaier & Schneider *Hoppe-Seyler's Z. Physiol. Chem.* **349**: 1529-1536, 1968.
[4] Levene & Simms *J. Biol. Chem.* **41**: 711-724, 1925.
[5] Stern & Marks *Brain Res. Bull.* **4**: 49-55, 1979.
[6] Matheson & Tattrie *Can. J. Biochem.* **42**: 95-103, 1964.
[7] McDonald *et al.* *J. Biol. Chem.* **243**: 4143-4150, 1968.
[8] Josefsson & Lindberg *Biochim. Biophys. Acta* **105**: 149-161, 1965.
[9] Palekar *Anal. Biochem.* **104**: 200-204, 1980.
[10] Spackman *et al.* *Anal. Chem.* **30**: 116-126, 1958.

Bibliography

1925

Levene, P. A. & Simms, H. S. The relation of chemical structure to the rate of hydrolysis of peptides. II. Hydrolysis with enzyme (erepsin). *J. Biol. Chem.* **41**: 711-724.

1926

Euler, H. v. & Josephson, K. [Enzyme splitting of dipeptides I.] *Chem. Ber.* **59B**: 226-233.
Euler, H. v. & Josephson, K. [Enzymic cleavage of dipeptides. II.] *Hoppe-Seyler's Z. Physiol. Chem.* **157**: 122-139.
Josephson, K. & Euler, H. v. [Enzymic cleavage of dipeptides. IV. The mode of action of intestinal erepsin (peptidase).] *Hoppe-Seyler's Z. Physiol. Chem.* **162**: 85-94.

1941

Gailey, F. B. & Johnson, M. J. The dipeptidases of intestinal mucosa. *J. Biol. Chem.* **141**: 921-929.

Maschmann, E. [Animal peptidases. V.] *Biochem. Z.* **310**: 28-41.

1948

Smith, E. L. Studies on dipeptidases. III. Hydrolysis of methylated peptides; the role of cobalt in the action of glycylglycine dipeptidase. *J. Biol. Chem.* **176**: 21-32.

Smith, E. L. The glycylglycine dipeptidases of skeletal muscle and human uterus. *J. Biol. Chem.* **173**: 571-584.

1951

Smith, E. L. The specificity of certain peptidases. *Adv. Enzymol.* **12**: 211-216.

1955

Smith, E. L. Dipeptidases. E. Glycylglycine dipeptidase. *Methods Enzymol.* **2**: 107-109.

1965

Josefsson, L. & Lindberg, T. Intestinal dipeptidases. I. Spectrophotometric determination and characterization of dipeptidase activity in pig intestinal mucosa. *Biochim. Biophys. Acta* **105**: 149-161.

1968

Bidlingmaier, F. & Schneider, F. [Studies on the glycylglycine dipeptidase (EC 3.4.3.1) from porcine erythrocytes.] *Hoppe-Seyler's Z. Physiol. Chem.* **349**: 1529-1536.

1969

Schwabe, C. Peptide hydrolases in mammalian connective tissue. I. Survey of activities and preliminary characterization of certain peptidases. *Biochemistry* **8**: 771-782.

1979

Stern, F. & Marks, N. Glycyl-glycine hydrolase in rat brain: distribution and role in cleavage of glycine-rich oligopeptides. *Brain Res. Bull.* **4**: 49-55.

1980

Palekar, A. G. A fluorometric method for measurement of dipeptidase activity. *Anal. Biochem.* **104**: 200-204.

Glycylleucine Dipeptidase

Summary

EC Number: 3.4.13.11 (formerly 3.4.13.2, see Comment)

Earlier names: None.

Abbreviation: GLD.

Distribution: An activity that hydrolyzes glycylleucine and shows strong activation by Zn^{2+} or Mn^{2+} is widely distributed in mammalian tissues. The activity is predominantly located in the soluble fraction of tissue homogenates, as shown for the intestinal mucosa of monkey, guinea pig, rabbit and rat. The soluble enzyme is most abundant in the proximal and middle regions of the small intestine, whereas the minor fraction of particulate activity is most prominent in the distal region of the intestine.

Source: Hog intestinal mucosa and human uterus [1-3].

Action: Catalyzes the hydrolysis of many neutral dipeptides, excluding Gly-Gly. Relative rates at which some dipeptides are hydrolyzed at pH 7.8 by highly purified monkey intestinal enzyme are as follows: Gly-Leu 100, Gly-Val 76, Gly-Ile 70, Ala-Leu 25, and Leu-Gly 18 [4]. Many other neutral dipeptides are hydrolyzed, e.g. Gly-Phe, Leu-Leu, Leu-Gly and Ala-Gly [5]. The N-methylated peptide sarcosylleucine is slowly hydrolyzed. Most arginine- and a few lysine-containing dipeptides are resistant to attack [5]. There is no action on Gly-D-Leu, Z-Gly-Leu, Z-Gly-Leu-NH$_2$, Gly-Leu-NH$_2$, Gly-Leu-Gly, Gly-Gly-Leu, Gly-Pro or Pro-Gly [1,4].

Requirements: pH 7.8 and preincubation at 37°C in 1 mM Zn^{2+} (for activation of the human uterine and rat muscle enzymes) or in 50 mM Mn^{2+} (for activation of the hog intestinal enzyme), with 1 mM and 10 mM concentrations of the activators, respectively, in the assay [2]. Crude preparations give higher rates in phosphate buffer than in Tris buffer, but the opposite is true for purified preparations [5].

Substrate, usual: Gly-Leu in conjunction with a method for determining the liberated amino acids colorimetrically [6,7], spectrophoto-metrically

[8], fluorometrically [9], or by amino acid analysis [10]. Leucine dehydrogenase can be employed to determine free leucine selectively [12].

Inhibitors: Activity is completely inhibited by 10 mM EDTA or 1 mM 1,10-phenanthroline. Co^{2+} is inhibitory, in contrast to its action on Gly-Gly dipeptidase. Gly-Leu dipeptidase is also inhibited by Ca^{2+}, and the observation that crude preparations of the enzyme usually give the highest rates in phosphate buffer may be a consequence of the precipitation of Ca^{2+}.

Molecular properties: The native enzyme (M_r 107,000) consists of two equal subunits [5]. A somewhat higher value (M_r 115,000) has been determined by a sedimentation equilibrium technique [13]]. The K_m for Gly-Leu is 10.6 mM at pH 7.8 and 37°C [4]. Stability is far greater than that of Gly-Gly dipeptidase. Highly active preparations of GLD have been obtained in aqueous extracts of acetone-dried powders of human uterus [1]. Stable preparations of the hog intestinal enzyme have been obtained by precipitation with ammonium sulfate and dialysis. Highly purified preparations of GLD from monkey small intestine are extremely unstable, and are stabilized to only a limited extent in 12.5% (v/v) glycerol [4].

Comment: Gly-Gly dipeptidase and Gly-Leu dipeptidase have been shown to be truly distinct dipeptidases. Accordingly, they were originally given separate EC numbers, i.e. 3.4.13.1 (Gly-Gly dipeptidase) and 3.4.13.2 (Gly-Leu dipeptidase). However, for reasons that are unclear, both enzymes were given the same number (EC 3.4.13.11) in the 1978 recommendations [11].

The hydrolysis of Leu-Gly, which was once attributed to yet another dipeptidase, has been shown to be catalyzed by an aminopeptidase, namely leucyl aminopeptidase.

References
[1] Smith *J. Biol. Chem.* **176**: 9-19, 1948.
[2] Smith *Methods Enzymol.* **2**: 105-107, 1955.
[3] Norén *et al.* *Acta Chem. Scand.* **25**: 1913-1915, 1971.
[4] Das & Radhakrishnan *Biochem. J.* **128**: 463-465, 1972.
[5] Das & Radhakrishnan *Biochem. J.* **135**: 609-615, 1973.
[6] Matheson & Tattrie *Can. J. Biochem.* **42**: 95-103, 1964.
[7] McDonald *et al.*. *J. Biol. Chem.* **243**: 4143-4150, 1968.
[8] Josefsson & Lindberg *Biochim. Biophys. Acta* **105**: 149-161, 1965.
[9] Palekar *Anal. Biochem.* **104**: 200-204, 1980.
[10] Spackman *et al.* *Anal. Chem.* **30**: 116-126, 1958.
[11] IUB IUPAC Nomenclature Committee. *Enzyme Nomenclature 1978.* Academic Press, New York, 1979.

Bibliography

1948

Smith, E. L. Studies on dipeptidases. II. Some properties of the glycyl-L-leucine dipeptidases of animal tissues. *J. Biol. Chem.* **176**: 9-19.

1951

Smith, E. L. The specificity of certain peptidases. *Adv. Enzymol.* **12**: 191-196.

1955

Smith, E. L. Dipeptidases. D. Glycyl-L-leucine dipeptidase. *Methods Enzymol.* **2**: 105-107.

1971

Norén, O., Sjöström, H. & Josefsson, L. Preparation of highly purified glycyl-L-leucine dipeptidase from pig intestinal mucosa. *Acta Chem. Scand.* **25**: 1913-1915.

1972

Das, M. & Radhakrishnan, A. N. Substrate specificity of a highly active dipeptidase purified from monkey small intestine. *Biochem. J.* **128**: 463-465.

1973

Das, M. & Radhakrishnan, A. N. Glycyl-L-leucine hydrolase, a versatile aster' dipeptidase from monkey small intestine. *Biochem. J.* **135**: 609-615.

1974

Das, M. & Radhakrishnan, A. N. A comparative study of the distribution of soluble and particulate glycyl-L-leucine hydrolase in the small intestine. *Clin. Sci. Mol. Med.* **46**: 501-510.

1978

Sjöström, H. & Norén, O. Size and shape of two intestinal dipeptidases. *Int. J. Pept. Protein Res.* **11**: 159-165.

1983

Takamiya, S., Ohshima, T., Tanizawa, K. & Soda, K. A spectrophotometric method for the determination of aminopeptidase activity with leucine dehydrogenase. *Anal. Biochem.* **130**: 266-270.

Cysteinylglycine Dipeptidase

Summary

EC Number: 3.4.13.6

Earlier names: Cysteinylglycinase.

Abbreviation: CGD.

Distribution: Cys-Gly dipeptidase activity has been detected in kidney, muscle and liver of the rat [1], and is said to be present in all tissues other than nervous tissue [2].

Source: Hog kidney [3].

Action: Catalyzes the hydrolysis of Cys-Gly. Inactive against Gly-Gly and Leu-Gly at pH 7.3.

Requirements: pH 7.3 at 37°C with 1 mM Mn^{2+} for activation [3,4].

Substrate, usual: Cysteinylglycine or a partial hydrolysate of glutathione [4]. Activity has been measured as the rate of increase in absorbance at 500 nm relative to that at 580 nm, i.e. the A_{500}/A_{580} ratio [1].

Inhibitors: None reported, but inhibition by metal chelators should be anticipated.

Molecular properties: Physicochemical characteristics have not been reported. CGD is extremely labile below pH 5.5, but tolerates heating to 70°C for 10 min at pH 9.0. Activity is lost rapidly above pH 10.5. Tolerates extraction with chloroform or octanol and fractionation with ethanol [2]. It is inactive in 6 M urea, but the effect is reversible.

Comment: The breakdown of glutathione (γ-Glu-Cys-Gly) in the tissues, in particular kidney, is a two-step process involving the removal of the γ-glutamyl group by γ-glutamyl transpeptidase (γ-GTP, Entry 18.03) and the subsequent hydrolysis of Cys-Gly by the dipeptidase.

It should be noted that microsomal alanine aminopeptidase (mAAP, Entry 11.06) is also capable of hydrolyzing Cys-Gly. The high activity of this enzyme in the renal brush border, together with its ability to hydrolyze S-Bz-Cys-Gly and S-Bz-Cys-NPhNO$_2$, raises the possibility that it, too, may participate along with γ-GPT in the renal degradation of glutathione and its S-conjugates [5].

References
[1] Binkley & Nakamura *J. Biol. Chem.* **173**: 411-421, 1948.
[2] Binkley *Nature* **167**: 888-889, 1951.
[3] Semenza *Biochim. Biophys. Acta* **24**: 401-413, 1957.
[4] Olson & Binkley *J. Biol. Chem.* **186**: 731-735, 1950.
[5] Rankin *et al. Biochem. Biophys. Res. Commun.* **96**: 991-996, 1980.

Bibliography

1948

Binkley, F. & Nakamura, K. Metabolism of glutathione. I. Hydrolysis by tissues of the rat. *J. Biol. Chem.* **173**: 411-421.

1950

Olson, C. K. & Binkley, F. Metabolism of glutathione. III. Enzymatic hydrolysis of cysteinylglycine. *J. Biol. Chem.* **186**: 731-735.

1951

Binkley, F. Metabolism of glutathione. *Nature* **167**: 888-88.

1952

Binkley, F. Evidence for the polynucleotide nature of cysteinylglycinase. *Exp. Cell Res. Suppl.* **2**: 145-160.

1957

Semenza, G. Chromatographic purification of cysteinyl-glycinase. *Biochim. Biophys. Acta* **24**: 401-413.

1980

Rankin, B. B., McIntyre, T. M. & Curthoys, N. P. Brush border membrane hydrolysis of S-benzyl-cysteine-p-nitroanilide, an activity of aminopeptidase M. *Biochem. Biophys. Res. Commun.* **96**: 991-996.

Arginine Dipeptidase

Summary

EC Number: 3.4.13.4

Earlier names: γ-Aminobutyryl-lysine dipeptidase, aminoacyl-lysine
dipeptidase, lysine dipeptidase.

Abbreviation: None.

Distribution: Widely distributed in the tissues of the rat and rabbit. An
unusual aspect of its distribution in the rat is its abundance in muscle
tissue. Its tissue distribution relative to skeletal muscle (100) is as
follows: heart muscle 57, kidney 30, testis 8, small intestine 7.3,
pancreas 4.7, lung 4, spleen 3.5, liver 3, brain stem 3, and stomach
2.5. In the rabbit, kidney is the richest source, followed by skeletal
muscle, heart muscle, testis and liver.

Source: Hog kidney cortex [1].

Action: Catalyzes the hydrolysis of dipeptides, specifically those
containing C-terminal basic amino acids, and with a preference for
arginine. A wide range of residues is accepted at the N-terminus, but a
strikingly unusual preference is exhibited for dipeptides having a γ-
aminobutyric acid (γ-Abu) moiety located at the N-terminus.

The relative rates at which a range of dipeptides are hydrolyzed (at
pH 7.0) by the hog kidney enzyme are as follows: γ-Abu-Arg 100, γ-Abu-Lys
81, Ala-Arg 48, Gly-Arg 40, Phe-Lys 39, β-Ala-Lys 39, Leu-Lys 30, β-
Ala-Arg 29, β-Ala-Orn 22, γ-Abu-Orn 14, Ile-Arg 8.1, Ala-Lys 7.2, Ser-Arg
6.4 and Ile-Lys 2.4. The K_m value for γ-Abu-Lys is 9.8 mM at pH 7.0 and
37°C.

β-Ala-His (carnosine) and γ-Abu-His (homocarnosine), which are
substrates for related dipeptidases found in muscle, are not hydrolyzed
by arginine dipeptidase.

Consistent with its dipeptidase character, arginine dipeptidase does
not hydrolyze Z-Leu-Arg, γ-Abu-Lys-NH₂, N-Ac-γ-Abu-Lys, D-Phe-Pro-Val-Orn
or Bz-Arg-NH₂. Neither is there action on dipeptides wherein the peptide
linkage is through the ε-amino of lysine, e.g. ε-N-(γ-aminobutyryl)lysine.

Requirements: Tris-maleate buffer, pH 6.5. The activity of the purified enzyme is enhanced by Mn^{2+} or Mg^{2+} at 1 mM.

Substrate, usual: α-N-(γ-aminobutyryl)lysine in conjunction with a method for detecting the liberated amino acid colorimetrically with ninhydrin [2] or trinitrobenzenesulfonic acid [3] or by amino acid analysis [4]. A method has been reported for synthesizing the substrate [5].

Substrate, special: γ-Aminobutyryl-arginine.

Inhibitors: 1,10-Phenanthroline inhibits strongly, but EDTA and citrate have no effect. Cd^{2+}, Cu^{2+}, Zn^{2+} and Hg^{2+} are all inhibitory, as is phosphate. p-Chloromercuribenzoate (0.4 mM) shows 50% inhibition, but IAcOH (1 mM) and Dip-F (10 mM) have no effect. Product inhibition should be expected since arginine, lysine and ornithine are inhibitory.

Molecular properties: Very little information is available. Solutions of the enzyme purified 62-fold retained 89% of the original activity during one week of storage at 4°C [1].

Comment: γ-Aminobutyryl-lysine ("GABA lysine") occurs only in the mammalian brain, where it may serve as an important physiological substrate for arginine dipeptidase. This particular focus has resulted in the enzyme being named aminobutyryl-lysine dipeptidase and lysine dipeptidase. However, the enzyme is more abundant in almost all other tissues than in brain. Because of this, and its relatively broad substrate specificity, it seems reasonable to assume that the enzyme acts on a variety of dipeptides possessing C-terminal basic residues. From this standpoint, and because arginine is released at the highest rate, we propose that the enzyme be named arginine dipeptidase, rather than aminoacyl-lysine dipeptidase as has been suggested [6].

As is explained in the Introduction, the name **arginine** dipeptidase signifies that the enzyme releases C-terminal arginine from dipeptides. If the specificity were for dipeptides possessing N-terminal arginyl residues, resulting in the scission of arginyl linkages, the enzyme would have been named **arginyl** dipeptidase.

It is thought that a major reason for the restriction of activity of dipeptidases to hydrolysis of dipeptides is a requirement for the presence in the substrate of the correctly spaced terminal (amino and carboxyl) charged groups. Interestingly, though, the binding site on arginine dipeptidase lacks such a requirement, as is shown by its ability to hydrolyze γ-Abu-Arg, β-Ala-Arg and Ala-Arg.

References
[1] Kumon *et al. Biochim. Biophys. Acta* **200**: 455-474, 1970.
[2] Matheson & Tattrie *Can. J. Biochem.* **42**: 95-103, 1964.
[3] McDonald *et al. J. Biol. Chem.* **243**: 4143-4150, 1968.
[4] Spackman *et al. Anal. Chem.* **30**: 116-126, 1958.
[5] Nakajima *et al. J. Neurochem.* **16**: 417-422, 1969.
[6] IUB-IUPAC Nomenclature Committee. *Enzyme Nomenclature 1978.*

Academic Press, New York, 1979.

Bibliography

1970

Kumon, A., Matsuoka, Y., Kakimoto, Y., Nakajima, T. & Sano, I.   A peptidase that hydrolyzes α-N-(γ-aminobutyryl)lysine. *Biochim. Biophys. Acta* **200**: 466-474.

β-Alanylhistidine Dipeptidase

Summary

EC Number: 3.4.13.3

Earlier names: Carnosinase and aminoacyl-histidine dipeptidase.

Abbreviation: β-AHD.

Distribution: The relative activities of various rat tissues are reported
to be as follows: uterus 100, kidney 84, liver 66, and lung 56 [1].
Other tissues, including spleen, heart, brain, skeletal muscle, and
small intestine fall in the range 14-34 on this scale. Activity is not
detectable in the plasma or red cells of the rat, nor in the plasma of
the hog. However, in man, substantial amounts of β-AHD are present in
plasma [2]. An earlier report [3] describing two electrophoretic forms
could not be confirmed [2]. The enzyme is generally cytosolic, and is
especially plentiful in the proximal tubular epithelium of the kidney.
It also occurs in the glandular cells of several tissues such as the
uterine epithelium and the secretory epithelium of the nasal cavity
[12].

Source: Hog [1,4,5] and mouse kidney [2].

Action: Catalyzes the hydrolysis of dipeptides containing C-terminal
histidine, but with a very restricted specificity. It hydrolyzes β-Ala-
His (carnosine), β-Ala-1-methyl-His (anserine), and Gly-His, the relative
rates being 100, 30 and 28, respectively. (Both carnosine and anserine
are present in high concentrations in vertebrate skeletal muscle.)
γ-Aminobutyryl-His (homocarnosine) and Ala-His are not hydrolyzed,
nor are α-aminobutyryl-His, His-His and Ac-His [1].

Requirements: pH 8.0-8.5; preincubation with 5 mM Mn^{2+} for activation and
stabilization. Cd^{2+} and Zn^{2+} also provide activation, but not
stabilization. In the absence of added metal, the pH optimum is closer
to pH 7.5.

Substrate, usual: β-Ala-His (carnosine) in conjunction with a fluorometric
method of determining free histidine [3,6]

284

Inhibitors: EDTA, 1,10-phenanthroline, and 8-hydroxyquinoline-5-sulfonic acid show strong inhibition [12]. The Mn^{2+}-activated enzyme is inhibited by citrate and phosphate, but the unactivated enzyme is not affected.

Molecular properties: The hog kidney enzyme (M_r 84,000) has a pI of about 5.8 [1]. A M_r of 90,000 has been reported for β-AHD of human kidney [2,3] whereas a value of 160,000 has been reported for the plasma form of the enzyme [2]. Mouse kidney β-AHD (M_r 112,00) is comprised of M_r 58,000 subunits [12]. The enzyme present in human plasma shows only one electrophoretic form. Solutions of the enzyme are most stable when stored with Mn^{2+} at pH 7-8; about 30% of the activity is lost during freeze drying [5]. Mn^{2+} shifts the K_m for β-Ala-His from 60 μM to 2 mM and increases V_{max} about 50% [12].

Comment: It was once thought that β-alanylhistidine dipeptidase had a relatively broad specificity [4,7], but this impression was gained from work with enzyme preparations contaminated with dipeptidases. It has now been shown that the enzyme has a very restricted specificity with a marked preference for β-Ala-His [1], hence its specific name.

Recent studies [2] have revealed that β-AHD of human blood plasma is not identical to the more familiar tissue enzyme. Unlike the latter, it hydrolyzes γ-Abu-His (homocarnosine), has a different pH optimum, a pI of about 4.7, and a M_r of about 160,000. Additionally, Cd^{2+} is a more effective activator than Mn^{2+}. The plasma form of the enzyme becomes detectable at about 10 months of age and does not reach maximal levels until about 13 to 15 years [2,8]. The low levels of the plasma enzyme in early life may explain why carnosine is detectable in the blood and urine of young children. Carnosine is normally found in adult blood plasma following the eating of meat. It is believed that the high levels of β-AHD in the kidney may be involved in the hydrolysis of dietary carnosine that enters the blood, thus providing a means of retaining dietary histidine.

The physiological role of tissue β-AHD remains unclear because its distribution differs from that of carnosine (β-AH). However, its prevalence in glandular cells of several tissues suggests a role in their secretory functions.

A second β-AHD activity has been described, but its Mn^{2+} dependence, membrane association, and bestatin sensitivity [12,13], suggest that it may be attributable to an aminopeptidase.

In contrast to normal adults, who typically show relatively high levels of plasma β-AHD, patients afflicted with carnosinemia associated with mental retardation and neurological disturbance show pronounced deficiencies of the plasma enzyme [9]. The status of the tissue form of the enzyme in these patients remains to be elucidated.

Carnosinuria, also associated with neurological disfunction and certain types of myopathy, has been observed in patients lacking the plasma enzyme and also deficient in urea cycle enzymes [10].

An earlier report [11] of two "carnosinases" in hog kidney is now believed to be attributable to the partial separation of β-AHD and γ-homoalanylhistidine dipeptidase (γ-HHD).

References

[1] Lenney *Biochim. Biophys. Acta* **429**: 214-219, 1976.
[2] Lenney *et al. Clin. Chim. Acta* **123**: 221-231, 1982.
[3] Murphy *et al. Clin. Chim. Acta* **42**: 309-314, 1972.
[4] Hanson & Smith *J. Biol. Chem.* **179**: 789-801, 1949.
[5] Rosenberg *Arch. Biochem. Biophys.* **88**: 83-93, 1960.
[6] Ambrose *et al. Clim. Chem.* **15**: 361-366, 1969.
[7] Smith *Methods Enzymol.* **2**: 93-96, 1955.
[8] Van Munster *et al. Clin. Chim. Acta* **29**: 243-248, 1970.
[9] Perry *New Eng. J. Med.* **277**: 1219-1227, 1967.
[10] Burgess *et al. Clin. Chim. Acta* **61**: 215-218, 1975.
[11] Wolos *et al. Int. J. Biochem.* **9**: 57-62, 1978.
[12] Margolis *et al. Biochim. Biophys. Acta* **744**: 237-248, 1983.
[13] Hirsch *et al. Brain Res.* **158**: 407-422, 1978.

Bibliography

1937

Garkavi, P. G. The fate of carnosine in the animal organism. III. The splitting of carnosine under the influence of the peptidases of organs and tissues of the animal organism. *Bull. Biol. Med. Exp. URSS* **4**: 57-61.

Meshkova, N. P. & Zolotarevskaya, A. I. The fate of carnosine in the animal organism. I. The action of carnosine upon autolytic processes in muscle tissue. *Bull. Biol. Med. Exp. URSS* **4**: 50-52.

Severin, S. E. & Georgievskaya, E. F. The fate of carnosine in the animal organism. II. The splitting of carnosine by kidney tissue enzymes. *Bull. Biol. Med. Exp. URSS* **4**: 53-56.

du Vigneaud, V., Sifferd, R. H. & Irving, G. W., Jr. The utilization of *l*-carnosine by animals on a histidine-deficient diet. *J. Biol. Chem.* **117**: 589-597.

1939

du Vigneaud, V. & Behrens, O. K. Carnosine and anserine. *Engeb. Physiol. Biol. Chem. Exp. Pharmakol.* **41**: 917-973.

1949

Hanson, H. T. & Smith, E. L. Carnosinase: an enzyme of swine kidney. *J. Biol. Chem.* **179**: 789-801.

1951

Smith, E. L. The specificity of certain peptidases. VIII. Carnosinase. *Adv. Enzymol.* **12**: 221-225.

1955

Smith, E. L. Dipeptidases. A. Carnosinase. *Methods Enzymol.* **2**: 93-96.

1956

Davis, N. C. Action of proteolytic enzymes on some peptides and derivatives containing histidine. *J. Biol. Chem.* **223**: 935-947.

1957

Wood, T. Carnosine and carnosinase in rat tissue. *Nature* **180**: 39-40.

1960

Rosenberg, A. Purification and some properties of carnosinase of swine kidney, *Arch. Biochem. Biophys.* **88**: 83-93.

Rosenberg, A. Studies on carnosinase. III. Metal-ion stabilization of the enzyme. *Ark. Kemi* **17**: 25-40.

Rosenberg, A. Studies on carnosinase. IV. The interaction of the activating metal ions with the substrate. *Ark. Kemi* **17**: 41-50.

Rosenberg, A. The activation of carnosinase by divalent metal ions. *Biochim. Biophys. Acta* **45**: 297-316.

1964

Smith, L. C. & Koostra, W. L. Carnosinase activity in tissues of normal and dystrophic rabbits. *Proc. Soc. Exp. Biol. Med.* **116**: 102-103.

1967

Perry, T. L., Hansen, S., Tischler, B., Bunting, R. & Berry, K. Carnosinemia. A new metabolic disorder associated with neurologic disease and mental defect. *New Eng. J. Med.* **277**: 1219-1227.

1968

Perry, T. L., Hansen, S. & Love, D. L. Serum-carnosinase deficiency in carnosinaemia. *Lancet* **1**: 1229-1230.

1969

Ambrose, J. A., Crimm, A., Burton, J., Paullin, K. & Ross, C. Fluorometric determination of histidine. *Clin. Chem.* **15**: 361-366.

Heeswijk, P. J. van, Trijbels, J. M. F., Schretlen, E. D. A. M., Munster, P. J. J. van & Monnens, L. A. H. A patient with a deficiency of serum carnosinase activity. *Acta Paediatr. Scand.* **58**: 584-592.

1971

Zoch, E. & Müller, H.   Nachweiss und Bestimmung von Carnosinase-Aktivität in der menschlichen Placenta. *Enzymologia* **40**: 199-208.

1972

Murphey, W. H., Patchen, L. & Lindmark, D. G.   Carnosinase: a fluorometric assay and demonstration of two electrophoretic forms in human tissue extracts. *Clin. Chim. Acta* **42**: 309-314.

1973

Murphey, W. H., Lindmark, D. G., Patchen, L. I., Housler, M. E., Harrod, E. K. & Mosovich, L.   Serum carnosinase deficiency concomitant with mental retardation. *Pediatr. Res.* **7**: 601-606.

1975

Burgess, E. A., Oberholzer, V. G., Palmer, T. & Levin, B.   Plasma carnosinase deficiency in patients with urea cycle defects. *Clin. Chim. Acta* **61**: 215-218.

1976

Lenney, J. F.   Specificity and distribution of mammalian carnosinase. *Biochim. Biophys. Acta* **429**: 214-219.

1977

Lenney, J. F., Kan, S.-C., Siu, K. & Sugiyama, G. H.   Homocarnosinase: a hog kidney dipeptidase with a broader specificity than carnosinase. *Arch. Biochem. Biophys.* **184**: 257-266.

1978

Wolos, A., Piekarska, K., Glogowski, J. & Konieczka, I.   Two molecular forms of swine kidney carnosinase. *Int. J. Biochem.* **9**: 57-62.

1979

Harding, J. W. & O'Fallon, J. V.   The subcellular distribution of carnosine, carnosine synthetase, and carnosinase in mouse olfactory tissues. *Brain Res.* **173**: 99-109.

Margolis, F. L., Grillo, M., Brown, C. E., Williams, T. H., Pitcher, R. G. & Elgar, G. J.   Enzymatic and immunological evidence for two forms of carnosinase in the mouse. *Biochim. Biophys. Acta* **570**: 311-323.

1980

Fleisher, L. D., Rassin, D. K., Wisniewski, K. & Salwen, H. R.   Carnosinase deficiency: a new variant with high residual activity. *Pediatr. Res.* **14**: 269-271.

1982

Hartlage, P. L., Roesel, R. A., Eller, A. G. & Hommes, F. A. Serum carnosinase deficiency: decreased affinity of the enzyme for the substrate. *J. Inherited Metab. Dis.* **5**, Suppl. 1: 13-14.

Lenney, J. F., George, R. P., Weiss, A. M., Kucera, C. M., Chan, P. W. H. & Rinzler, G. S. Human serum carnosinase: characterization, distinction from cellular carnosinase, and activation by cadmium. *Clin. Chim. Acta.* **123**: 221-231.

1983

Margolis, F. L., Grillo, M., Grannot-Reisfeld, N. & Farbman, A. I. Purification, characterization and immunocytochemical localization of mouse kidney carnosinase. *Biochim. Biophys. Acta* **744**: 237-248.

1984

Bando, K., Shimotsuji, T., Toyoshima, H., Hayashi, C. & Miyai, K. Fluorometric assay of human serum carnosinase activity in normal children, adults and patients with myopathy. *Ann. Clin. Biochem.* **21**: 510-514.

1985

Jablonowska, C., Piechocki, D. & Wolos, A. Swine uterus carnosinase activity in oestrous cycle and early pregnancy. *Comp. Biochem. Physiol.* **80B**: 381-383.

Lenney, J. F., Peppers, S. C., Kucera-Orallo, C. M. & George, R. P. Characterization of human tissue carnosinase. *Biochem. J.* **228**: 653-660.

γ-Homoalanylhistidine Dipeptidase

Summary

EC Number: 3.4.13.13

Earlier names: Homocarnosinase, γ-aminobutyrylhistidine dipeptidase.

Abbreviation: γ-HHD.

Distribution: Activity is measurable in only four tissues of the rat,
their relative activities being: kidney 100, uterus 55, lung 50 and
liver 33 [1]. The activity appears to be cytosolic, and is not detectable
in the blood cells or plasma of the rat. Although γ-HHD activity could
not be detected in rat brain, it is readily detectable in human brain
[2], and its presence in brain (and possibly cerebrospinal fluid) is of
interest because brain is the only tissue known to contain homocarnosine
[3]. The level of homocarnosine in human brain white matter is at least
100-fold greater than that in brain of the hog, dog or cat [3].

Source: Hog kidney [1].

Action: γ-HHD has a much broader specificity than β-alanylhistidine
dipeptidase (β-AHD). Both enzymes hydrolyze β-Ala-His, but only γ-HHD
is
active on γ-Hal-His (homocarnosine, also known as γ-aminobutyryl-His,
γ-Abu-His and GABA-His). The ability of γ-HHD to hydrolyze homocarnosine
is unique, and it is this characteristic that serves as the basis for
naming the enzyme. Relative to its rate of hydrolysis of γ-Hal-His
(100), rates on other substrates are reported to be as follows: β-Ala-His
213, β-Ala-methyl-His 208, Gly-His 164, Gly-Leu 164, β-Ala-Gly 108, β-
Ala-Ala 104, Ala-His 59 and β-Ala-Trp 15 [1]. No action is detectable on
His-Ala, Ala-Gly, Tyr-Gly, or Gly-γ-Hal.

Requirements: pH 7.2 and 1 mM Co^{2+} for activation and stabilization. Mn^{2+}
is a less effective activator and does not provide stabilization.
N-Ethylmorpholine is the preferred buffer. Activity is suppressed by
phosphate and Tris-maleate buffers [1].

Substrate, usual: γ-Hal-His (homocarnosine) in conjunction with a
fluorometric method for determining free histidine [4].

Inhibitors: Chelating agents such as EDTA and 1,10-phenanthroline. Cd^{2+} also is inhibitory.

Molecular properties: The hog kidney enzyme (M_r 57,000) has a pI of about 5.6 [1]. It is a single polypeptide chain which seems not to occur in multiple molecular forms; it does not contain cysteine or serine residues essential for activity. The enzyme is more unstable than β-AHD.

Comment: Although γ-HHD hydrolyzes β-Ala-His (carnosine) as well as γ-Hal-His (homocarnosine), the latter substrate should be used when it is necessary to avoid interference from β-AHD. Assays with either substrate require a method for detecting liberated histidine. Assays on crude tissue extracts may be complicated by the presence of histidase. It is possible, however, to inhibit selectively histidase by incorporating NH_2OH (0.2 mM) into the assay system, without affecting γ-HHD or β-AHD [1].

γ-HHD is a much less stable enzyme than β-AHD, and is largely inactivated during the heat treatment step that is usually used in the purification of β-AHD.

Unlike most dipeptidases, which hydrolyze only dipeptides consisting of two α-amino acids, γ-HHD prefers dipeptides in which the free amino group is in the β or γ position. The broad specificity of hog kidney γ-HHD shows a marked resemblance to that of anserinase of fish brain [5].

A progressive neurological disorder has been described in which homocarnosine is elevated 20- to 40-fold in brain and cerebrospinal fluid. This disorder, homocarnosinosis, has been linked to a deficiency in brain of γ-HHD [2,6], apparently inherited as an autosomal recessive trait. However, more recent studies [7] have shown that the γ-HHD activity attributed to human brain tissue may actually have been due to plasma β-AHD contained in the extracts. Thus, γ-HHD may not be present in any human tissue. It is believed that γ-HHD activity was found to be lacking in a brain biopsy sample [2] from a homocarnosinosis patient because the individual was lacking plasma β-AHD.

References
[1]   Lenney *et al. Arch. Biochem,. Biophys.* **184**: 257-266, 1977.
[2]   Perry *et al. J. Neurochem.* **32**: 1637-1640, 1979.
[3]   Abraham *et al. Arch. Biochem. Biophys.* **99**: 210-213, 1962.
[4]   Murphey *et al. Clin. Chim. Acta* **42**: 309-314, 1972.
[5]   Lenney *et al. Comp. Biochem. Physiol.* **61B**: 253-258, 1978.
[6]   Sjaastad *et al. Acta Neurol. Scand.* **55**: 158-162, 1977.
[7]   Lenney *et al. Clin. Chim. Acta* **132**: 157-165, 1983.

Bibliography

1962

Abraham, D., Pisano, J. J. & Udenfriend, S. The distribution of homocarnosine in mammals. *Arch. Biochem. Biophys.* **99**: 210-213.

1977

Lenney, J. F., Kan, S.-C., Siu, K. & Sugiyama, G. H. Homocarnosinase: a hog kidney dipeptidase with a broader specificity than carnosinase. *Arch. Biochem. Biophys.* **184**: 257-266.

Sjaastad, O., Gjessing, L., Berstad, J. R. & Gjesdahl, P. Homocarnosinosis. 3. Spinal fluid amino acids in familial spastic paraplegia. *Acta Neurol. Scand.* **55**: 158-162.

1979

Kish, S. J., Perry, T. L. & Hansen, S. Regional distribution of homocarnosine, homocarnosine-carnosine synthetase and homocarnosinase in human brain. *J. Neurochem.* **32**: 1629-1636.

Perry, T. L., Kish, S. J., Sjaastad, O., Gjessing, L. R., Nesbakken, R., Schrader, H. & Loken, A. C. Homocarnosinosis: increased content of homocarnosine and deficiency of homocarnosinase in brain. *J. Neurochem.* **32**: 1637-1640.

1983

Lenney, J. F., Peppers, S. C., Kucera, C. M. & Sjaastad, O. Homocarnosinosis: lack of serum carnosinase is the defect probably responsible for elevated brain and CSF homocarnosine. *Clin. Chim. Acta* **132**: 157-165.

Microsomal Dipeptidase

Summary

EC Number: 3.4.13.11

Earlier names: Dehydropeptidase I (DPH I), renal dipeptidase.

Abbreviation: MDP.

Distribution: Crude extracts of kidney, pancreas, spleen and liver of the rat catalyze the hydrolysis of glycyldehydrophenylalanine. Brain and muscle lack this activity [1]. In kidney, the most plentiful source, it is located exclusively in the cortex. It is operationally found in the microsomal fraction [2], and morphologically within the microvillus of the proximal tubular cells [9]. Its distribution differs from microsomal alanyl aminopeptidase (Entry 11.06) in that microsomal dipeptidase is not on the luminal surface [10].

Source: Hog kidney cortex [3].

Action: Catalyzes the hydrolysis of a range of dipeptides including Gly-Gly, Gly-Leu and Leu-Gly. Relative rates of hydrolysis at pH 8.0 established with a ninhydrin assay procedure [5] for a variety of dipeptides are as follows: Gly-Gly 100, Ala-Gly 51, Leu-Gly 46, Gly-Phe 45, Gly-Leu 41, Gly-Ala 35, Gly-Ser 34, Ala-Leu 34, Leu-Met 34, Gly-D-Leu 27, Gly-His 22, Gly-Tyr 22, Leu-Ala 21, Gly-Lys 16, Leu-Ser 16, Ser-Gly 9, Phe-Gly 8, Leu-Leu 7 and Gly-Asp 5 [4]. Subsequent rate studies with a titrimetric assay procedure [6] revealed that Ala-Gly was actually hydrolyzed more rapidly than any of the others; the rate was estimated to be about 230 μmoles hydrolyzed per min per mg enzyme [3]. Although activity is restricted to dipeptides possessing an unsubstituted N-terminal residue having the L-configuration, it includes dipeptides in which the C-terminal residue has the D-configuration. The enzyme has no esterase activity, and no action on leucinamide, tripeptides or proteins such as casein and hemoglobin.

Requirements: pH 8.0 and 37°C, usually in a Tris-HCl buffer. Activation is produced by Zn^{2+}, Mg^{2+} and Co^{2+}.

Substrate, usual: Ala-Gly, like other saturated dipeptides, is used in conjunction with a ninhydrin [5] or titrimetric [6] assay procedure.

Substrate, special: Glycyldehydrophenylalanine, the rate of hydrolysis being measured spectrophotometrically at pH 7.6, as the fall in absorbance at 275 nm [3].

Inhibitors: Metal chelators such as 1,10-phenanthroline and EDTA. Inorganic phosphate and nucleotides show reversible inhibition. Monovalent anions inhibit competitively, the most effective being CN^-, followed in decreasing order by SCN^-, N_3^-, I^-, NO_3^-, HCO_3^-, Br^-, AcO^-, Cl^- and F^-. Cilastatin, a substituted amino-propenoate, is a potent and specific competitive inhibitor of renal MDP [11].

Molecular properties: MDP from hog kidney is a Zn^{2+}-metalloprotein, M_r about 94,000 and pI 4.89. Each mole of enzyme contains two g-atoms of tightly-bound Zn^{2+}. The apoenzyme regains 24% of its original activity upon the addition of the first g-atom of Zn^{2+}, and full activity with the incorporation of a second. An artificial form of the enzyme containing two atoms of Co^{2+} per molecule has 42% of the activity of the Zn-enzyme [7].

Human renal MDP is a M_r 220,000 enzyme consisting of four M_r 59,000 subunits, each containing a Zn^{2+} atom. In contrast to the hog renal enzyme, human MDP is not a glycoprotein [7a].

Comment: More recently, the isolation of a dipeptidase from rat kidney microvillus membrane, which also contain γ-glutamyl transpeptidase (γ-GTP) and microsomal alanyl aminopeptidase (mAAP), was reported [8]. This enzyme hydrolyzes S-conjugates of Cys-Gly, including S-methyl-Cys-Gly, Cys-*bis*-Gly, and leukotriene D_4. The latter, a spasmogenic and vasoactive S-conjugate of Cys-Gly, is hydrolyzed to leukotriene E_4, a less potent metabolite. Cys-Gly itself is a potent inhibitor. The activity may be attributable to microsomal dipeptidase: like the latter, the microvillus enzyme is a Zn^{2+}-metalloprotein composed of two M_r 50,000 subunits, and the purified dipeptidase shows a broad specificity, with Ala-Gly being the preferred substrate.

Human renal MDP possesses β-lactamase activity [7a, 11] and is thus able to inactivate naturally occuring β-lactam antibiotics such as imipenem (N-formimidoyl thienamycin). This class of antibiotics exhibits structural homology to dehydropeptides. It appears that MDP located in the brush-border microvilli of the proximal renal tubule is responsible for the observed inactivation of β-lactam antibiotics by the kidney [12]. When cilastatin, Z-S-[6-carboxy-6-([2,2-dimethyl-(S)-cyclopropyl)carboxyl]-amino)-5-hexenyl]-L-cysteine, a specific inhibitor of the β-lactamase activity of MDP, is administered with these antibiotics, the urinary recovery of the intact antibiotics is greatly enhanced [12].

Two cytosolic dipeptidases that have been isolated from human kidney [13] are believed to be solubilized components of the kidney microvilli [7a].

References

[1] Greenstein *Adv. Enzymol.* **8**: 117-169, 1948.
[2] Harper *et al.* *Biochim. Biophys. Acta* **242**: 446-458, 1971.
[3] Campbell *Methods Enzymol.* **19**: 722-729, 1970.
[4] Campbell *et al.* *Biochim. Biophys. Acta* **118**: 371-386, 1966.
[5] Matheson & Tattrie *Can J. Biochem. Physiol.* **42**: 95-103, 1964.
[6] Bryce & Rabin *Biochem. J.* **90**: 509-512, 1964.
[7] Armstrong *et al.* *Biochemistry* **13**: 1745-1750, 1974.
[7a] Campbell *et al.* *J. Biol. Chem.* **259**: 14586-14590, 1984.
[8] Kozak & Tate *J. Biol. Chem.* **257**: 6322-6327, 1982.
[9] Welch & Campbell *J. Membr. Biol.* **54**: 39-50, 1980.
[10] Kim & Campbell *J. Membr. Biol.* **75**: 115-122, 1983.
[11] Kim & Campbell *Biochem. Biophys. Res. Commun.* **108**: 1638-1642, 1982.
[12] Kropp *et al.* *Antimicrob. Agents Chemother.* **22**: 62-70, 1982.
[13] Sugiura *et al.* *Biochim. Biophys. Acta* **522**: 541-550, 1978.

Bibliography

1930

Bergmann, M., Schmitt, V. & Miekeley, A. [Peptide-like substances. XXX. Polypeptides of dehydrogenated amino acids, their behaviour toward pancreatic enzymes and their application to peptide synthesis.] *Z. Physiol. Chem.* **187**: 264-276.

1932

Bergmann, M. & Schleich, H. [Dehydrodipeptidase. Enzymic cleavage of compounds of pyruvic acid and amino acids.] *Z. Physiol. Chem.* **207**: 235-240.
Bergmann, M. & Schleich, H. [Enzymic cleavage of dehydrogenated peptides. Discovery of a dehydrodipeptidase.] *Z. Physiol. Chem.* **205**: 65-75.

1947

Yudkin, W. H. & Fruton, J. S. On the proteolytic enzymes of animal tissues. VI. Dehydropeptidase. *J. Biol. Chem.* **169**: 521-529.
Yudkin, W. H. & Fruton, J. S. The activation of dehydropeptidase by zinc. *J. Biol. Chem.* **170**: 421-422.

1948

Greenstein, J. P. Dehydropeptidases. *Adv. Enzymol.* **8**: 117-169.
Greenstein, J. P., Price, V. E. & Leuthardt, F. M. Studies on the possible multiple nature of dehydropeptidase. I. *J. Biol. Chem.* **175**: 953-962.
Price, V. E. & Greenstein, J. P. Enzymatic hydrolysis of analogous saturated and unsaturated peptides. *J. Biol. Chem.* **175**: 969-974.

1953

Robinson, D. S., Birnbaum, S. M. & Greenstein, J. P. Purification and properties of an aminopeptidase from kidney cellular particulates. *J. Biol. Chem.* **202**: 1-26.

1963

Campbell, B. J., Lin, Y.-C. & Bird, M. E. Renal aminopeptidase- and copper-activated peptide hydrolysis. *J. Biol. Chem.* **238**: 3632-3639.

1964

Bryce, G. F. & Rabin, B. R. The function of the metal ion in leucine aminopeptidase and the mechanism of action of the enzyme. *Biochem. J.* **90**: 513-518.

Matheson, A. T. & Tattrie, B. L. A modified Yemm and Cocking ninhydrin reagent for peptidase assay. *Can. J. Biochem.* **42**: 95-103.

1966

Campbell, B. J., Lin, Y.-C., Davis, R. V. & Ballew, E. The purification and properties of a particulate renal dipeptidase. *Biochim. Biophys. Acta* **118**: 371-386.

1969

René, A. M. & Campbell, B. J. Amino acid composition and effect of pH on the kinetic parameters of renal dipeptidase. *J. Biol. Chem.* **244**: 1445-1450.

1970

Campbell, B. J. Renal dipeptidase. *Methods Enzymol.* **19**: 722-729.

1971

Harper, C., René, A. & Campbell, B. J. Renal dipeptidase: localization and inhibition. *Biochim. Biophys. Acta* **242**: 446-458.

1974

Armstrong, D. J., Mukhopadhyay, S. K. & Campbell, B. J. Physicochemical characterization of renal dipeptidase. *Biochemistry* **13**: 1745-1750.

1975

Rene, A. M. Renal dipeptidase of swine and man. *J. Am. Osteopath. Assoc.* **74**: 672-673.

1977

McDonald, J. K. & Schwabe, C. Intracellular exopeptidases. In: *Proteinases in Mammalian Cells and Tissues* (Barrett, A. J. ed.), pp. 311-391, (see pp. 381-384), North-Holland Publishing Co., Amsterdam.

1978

Sugiura, M., Ito, Y., Hirano, K. & Sawaki, S. Purification and properties of human kidney dipeptidases. *Biochim. Biophys. Acta* **522**: 541-550.

1980

Mullins, J. M. & Campbell, B. J. Urinary renal dipeptidase: partial purification and characterization. *Fed. Proc. Fed. Am. Soc. Exp. Biol.* **39**: 1686 only.

Welch, C. L. & Campbell, B. J. Uptake of glycine from L-alanylglycine into renal brush border vesicles. *J. Membr. Biol.* **54**: 39-50.

1982

Kim, H. S. & Campbell, B. J. β-Lactamase activity of renal dipeptidase against N-formimidoyl-thienamycin. *Biochem. Biophys. Res. Commun.* **108**: 1638-1642.

Kozak, E. M. & Tate, S. S. Glutathione-degrading enzymes of microvillus membranes. *J. Biol. Chem.* **257**: 6322-6327.

Kropp, H., Sundelof, J. G., Hajdu, R. & Kahan, F. M. Metabolism of thienamycin and related carbapenem antibiotics by the renal dipeptidase, dehydropeptidase-I. *Antimicrob. Agents Chemother.* **22**: 62-70.

1983

Kahan, F. M., Kropp, H., Sundelof, J. G. & Birnbaum, J. Thienamycin: development of imipenem-cilastatin. *J. Antimicrob. Chemother.* **12**, Suppl. D: 1-35.

Kim, H. S. & Campbell, B. J. Association of renal dipeptidase with the Triton-insoluble fraction of kidney microvilli. *J. Membr. Biol.* **75**: 115-122.

1984

Campbell, B. J., Forrester, L. J., Zahler, W. L. & Burks, M. β-Lactamase activity of purified and partially characterized human renal dipeptidase. *J. Biol. Chem.* **259**: 14586-14590.

1985

Basker, M. J., Coulton, S. & Southgate, R. Mutual pro-drugs of the olivanic acids and renal dipeptidase inhibitors. *J. Antibiot.* **38**: 70-74.

Lysosomal Dipeptidase I

Summary

EC Number: 3.4.13.-

Earlier names: Ser-Met dipeptidase.

Abbreviation: LDP I.

Distribution: The lysosomes of liver and spleen and most probably other tissues as well.

Source: Beef spleen [1,2] and rat liver [1].

Action: Catalyzes the hydrolysis of a wide variety of dipeptides, Ser-Met being one of the most susceptible. No action is detectable on Ser-Met-NH$_2$, Z-Ser-Met, Ser-Met-Glu or Ser-NNap [1]. Other dipeptides showing significant rates of hydrolysis include Ser-Ala, Ser-Tyr, Ser-Phe, Ser-Leu, Ser-His, Ser-Ser, Ala-Met, Ala-Phe, Gly-Met, Met-Ser, Met-Phe, Thr-Phe, His-Phe, Arg-Phe and Asp-Phe. Little or no action has been detected on Gly-Gly, Gly-Leu, Ser-Pro, Pro-Phe, Phe-Ser and Lys-Lys.

Requirements: pH 5.5 and 37°C. Citrate buffer is strongly inhibitory; sodium cacodylate and pyridine acetate buffers are innocuous.

Substrate, usual: Ser-Met.

Inhibitors: EDTA, 1,10-phenanthroline, citrate and metals such as Hg^{2+}, Cu^{2+} and Fe^{3+}. Thiol-blocking reagents and Dip-F have no effect.

Molecular properties: Possibly a Zn^{2+}-metalloprotein. M_r about 180,000 and pI about 5.4 [1].

Comment: Lysosomal dipeptidase I was discovered as a contaminant in highly-purified preparations of dipeptidyl peptidase I [1]. Its presence in such preparations indicates that it is relatively resistent to the rigorous treatments used in the purification of DPP I. These include a 22 h autolysis at pH 3.5 and 37°C, and a 40 min heat treatment at 65°C.

   Although a specific procedure has not yet been developed for the purification of LDP I, it is possible to isolate it from DPP I preparations by isoelectric focusing in a pH 5-7 gradient [1]. DPP I preparations used for sequencing can be freed of dipeptidase activity by treatment with EDTA [2].

The provisional classification of LDP I as a Zn^{2+}-metalloprotein is based upon its sensitivity to chelating agents and its reactivation with zinc acetate. Such treatment sometimes unmasks a significant amount of latent dipeptidase activity. Mg^{2+} and Mn^{2+} are less effective than Zn^{2+}. The fact that the EDTA-treated enzyme can be reactivated by dialysis alone raises the possibility that its metal is either very tightly bound, or that the chelating agent acts directly as an inhibitor.

A metal-dependent dipeptidase with a pH 5.3 optimum has also been isolated from beef thyroid acid proteinase preparations [3]. Its activity was originally detected on Cys-Gly, hence its name *cysteinylglycinase*. Other dipeptides that are readily hydrolyzed include Ala-Ala, Ala-Gly, Glu-Tyr, Leu-Ala, Leu-Gly, Leu-Leu, Leu-Phe, Leu-Tyr, Leu-Trp, Met-Met and Tyr-Leu. Lower, but still appreciable rates occur on Gly-Leu, Gly-Phe, Val-Val and Gly-Asn. Little or no action was detected on Gly-Leu-Tyr, Gly-Phe-NH₂, Ac-Phe-Tyr, Leu-NH₂, Leu-NNapOMe, Gly-Gly, Lys-Gly or Gly-Pro. The thyroid dipeptidase resembles LDP I in that it is inhibited by EDTA, and is most effectively reactivated by Zn^{2+}. Its subcellular localization has not been reported, but its pH 5.3 optimum suggests that it is lysosomal in the beef thyroid gland.

Another dipeptidase activity that may be attributable to LDP I is that of "Leu-Gly dipeptidase". It occurs in rat liver lysosomes, and, like LDP I, it is acid stable, has no sulfhydryl requirement, and is EDTA-sensitive [4].

References

[1] McDonald *et al* *Biochem. Biophys. Res. Commun.* **46**: 62-70, 1972.

[2] McDonald *et al.* *Methods Enzymol.* **25B**: 272-281, 1972.

[3] Loughlin & Trikojus *Biochim. Biophys. Acta* **92**: 529-542, 1964.

[4] Bouma *et al.* *Biochim. Biophys. Acta* **444**: 853-862, 1976.

Bibliography

1964

Loughlin, R. E. & Trikojus, V. M. A metal-dependent peptidase from thyroid glands. *Biochim. Biophys. Acta* **92**: 529-542.

1972

McDonald, J. K., Callahan, P. X. & Ellis, S. Preparation and specificity of dipeptidyl aminopeptidase I. *Methods Enzymol.* **25**: 272-281.

McDonald, J. K., Zeitman, B. B. & Ellis, S. Detection of a lysosomal carboxypeptidase and a lysosomal dipeptidase in highly-purified dipeptidyl aminopeptidase I (cathepsin C) and the elimination of their activities from preparations used to sequence peptides. *Biochem.*

Biophys. Res. Commun. **46**: 62-70.

1976

Bouma, J. M. W., Scheper, A., Duursma, A. & Gruber, M. Localization and some properties of lysosomal dipeptidases in rat liver. *Biochim. Biophys. Acta* **444**: 853-862.

1977

McDonald, J. K. & Schwabe, C. Intracellular exopeptidases. In: *Proteinases in Mammalian Cells and Tissues* (Barrett, A. J. ed.), pp. 311-391 (see pp. 346-348), North-Holland Publishing Co., Amsterdam.

Lysosomal Dipeptidase II

Summary

EC Number: 3.4.13.-

Earlier names: Ile-Glu-dipeptidase.

Abbreviation: LDP II.

Distribution: Rat liver lysosomes and probably the lysosomes of other tissues.

Source: Rat liver lysosomal fraction.

Action: Appears to have a wide range of action on dipeptide substrates, but its specificity cannot be defined because it has not yet been separated from other lysosomal peptide hydrolases.

Requirements: pH 4.5 and 37°C, after a 30 min preincubation with dithiothreitol (1 mM).

Substrate, usual: Ile-Glu.

Inhibitors: Iodoacetate. Unaffected by pepstatin and leupeptin.

Molecular properties: An acid-stable, cysteine dipeptidase; M_r about 120,000.

Comment: LDP II has been characterized only with lysosomal fractions. It appears to have a broad range of action on dipeptide substrates, but has not yet been fully characterized as a true dipeptidase through the use of suitable derivatives of Ile-Glu. Its sensitivity to thiol-blocking reagents and its insensitivity to EDTA distinguish it from LDP I.

Bibliography

1976

Bouma, J. M. W., Scheper, A., Duursma, A. & Gruber, M. Localization and some properties of lysosomal dipeptidases in rat liver. *Biochim. Biophys. Acta* **444**: 853-862.

Section 18
OMEGA PEPTIDASES

Pyroglutamyl Peptidase I

Summary

EC Number: 3.4.19.3 (formerly 3.4.11.8)

Earlier names: Pyrrolidonecarboxylyl peptidase, pyrrolidone carboxylate (PCA) peptidase, pyroglutamate aminopeptidase, pyrrolidonyl peptidase.

Abbreviation: PGP I.

Distribution: Widely distributed as a cytosolic enzyme in the tissues of mammals, birds, fish, plants and bacteria. In mammals, relatively high levels of activity occur in liver, kidney and spleen [1], as well as the anterior pituitary gland [2].

Source: Rat [3] and calf [4] liver, beef [2] and rat [5] pituitary gland, and guinea pig brain [6]. Certain strains of *Pseudomonas* [7] and *Bacillus* [8] are rich sources of PGP I.

Action: Catalyzes the hydrolytic release of N-terminal pyroglutamyl (pGlu-) groups from a range of polypeptides and arylamide derivatives. The relative rates of hydrolysis of three pyroglutamyl amino acids by rat liver PGP I are [3]: pGlu-Ala 100, pGlu-Ile 35, pGlu-Phe 29. (The bacterial enzyme shows the same order of specificity.) No action occurs on Glu-Ala [2]. K_m values on pGlu-Ala are 1.5 mM for the rat liver enzyme [3] and 0.23 mM for the beef pituitary enzyme [2]. pGlu-Val is also hydrolysed [6], but not pGlu-Pro [2,6].

The pGlu-His linkage in pGlu-His-Pro-NH$_2$, also known as thyrotropin-releasing hormone (TRH) and thyroliberin, is hydrolyzed [2], as is the pGlu-Ser linkage at the N-terminus of the B chain of relaxin [9].

PGP I purified from guinea pig brain catalyzed the hydrolysis of the pGlu-His linkage in thyroliberin, luliberin (decapeptide) and the anorexogenic peptide (pGlu-His-Gly), the pGlu-Leu linkage in neurotensin (tridecapeptide), the pGlu-Gln linkage in bombesin (tetradecapeptide), and the pGlu-Gly linkage in bradykinin-potentiating peptide B (decapeptide). Free pyroglutamic acid and the respective des-pyroglutamyl peptides are the only products. The pGlu-Pro terminus in eledoisin (undecapeptide) is resistent to attack [6].

Analogs of TRH ("pseudo-hormones") such as pGlu-His-amphetamine and derivatives thereof that possess psychotropic effects, although lacking TSH-releasing activity, are also attacked by PGP I [10].

PGP I appears capable of releasing pyroglutamyl groups from the N-termini of high M_r proteins, including the β-chain of native bovine fibrinogen (M_r 340,000) [11].

Requirements: pH 7.5 and 37°C in phosphate buffer containing 2 mM dithiothreitol and 2 mM EDTA [6].

Substrate, usual: pGlu-Ala in conjunction with a ninhydrin procedure to detect free alanine [12], or pGlu-NNap in a continuous fluorometric assay [9].

Substrate, special: pGlu-NMec (K_m 0.15 mM) in a continuous fluorometric assay of high sensitivity [6].

Inhibitors: pGlu-CH$_2$Cl [13] and thiol-blocking reagents, especially IAcNH$_2$ [6]. Although EDTA has a stabilizing effect, 1,10-phenanthroline shows strong inhibition. Pms-F, puromycin, bestatin, and bacitracin have no effect [6]. 2-Pyrrolidone (0.1 M), a reversible noncompetitive inhibitor [3,12], also serves as an effective stabilizing agent [3,2].

Molecular properties: The mammalian (guinea pig brain) enzyme (M_r 24,000) appears to be monomeric, pI 4.6 [6], but the native bacterial (*Bacillus*) enzyme occurs as a M_r 72,000 trimer, pI 5.4 [8]. PGP I is somewhat unstable when stored in assay buffer at 4°C, but can be stabilized with (0.1 M) 2-pyrrolidone [3] (see Inhibitors).

Comment: The distribution and levels of pyroglutamyl peptidase activity in brain are currently receiving special attention as a result of the recent recognition of several neurological functions for thyroliberin (pGlu-His-Pro-NH$_2$) and its metabolites [14]. These functions suggest neurotransmitter or neuromodulator roles for thyroliberin in addition to its well known adenohypophyseal activity. Endogenous pyroglutamyl peptidase activity is responsible, at least in part, for the rapid inactivation of TRH by adenohypophyseal extracts [5].

Because of its highly selective action, PGP has proved to be a useful tool for the sequence analysis of peptides and proteins.

References
[1]   Szewczuk & Kwiatkowska *Eur. J. Biochem.* **15**: 92-96, 1970.
[2]   Mudge & Fellows *Endocrinology* **93**: 1428-1434, 1973.
[3]   Armentrout *Biochim. Biophys. Acta* **191**: 756-759, 1969.
[4]   Bauer *et al. Eur. J. Biochem.* **118**: 173-176, 1981.
[5]   Bauer & Kleinkauf *Eur. J. Biochem.* **106**: 107-117, 1980.
[6]   Browne & O'Cuinn *Eur. J. Biochem.* **137**: 75-87, 1983.
[7]   Doolittle *Methods Enzymol.* **19**: 555-569, 1970.
[8]   Tsuru *et al. J. Biochem.* **84**: 467-476, 1978.
[9]   Schwabe & McDonald *Biochem. Biophys. Res. Commun.* **74**: 1501-1504, 1977.
[10] Morier *et al. Int. J. Biochem.* **10**: 769-783, 1979.

[11] Armentrout & Doolittle *Arch. Biochem. Biophys.* **132**: 80-90, 1969.

[12] Doolittle & Armentrout *Biochemistry* **7**: 516-521, 1968.

[13] Fujiwara *et al. Biochim. Biophys. Acta* **655**: 10-16, 1981.

[14] Jackson & Reichlin. In: *Central Nervous System Effects of Hypothalamic Hormones and Other Peptides* (Collu *et al.* eds), pp. 3-54, Raven Press, New York, 1979.

Bibliography

1969

Armentrout, R. W. Pyrrolidonecarboxylyl peptidase from rat liver. *Biochim. Biophys. Acta* **191**: 756-759.

1970

Doolittle, R. F. Pyrrolidonecarboxylyl peptidase. *Methods Enzymol.* **19**: 555-569.

Szewczuk, A. & Kwiatkowska, J. Pyrrolidonyl peptidase in animal, plant and human tissues. Occurrence and some properties of the enzyme. *Eur. J. Biochem.* **15**: 92-96.

1971

De Lange, R. J. & Smith, E. L. Leucine aminopeptidase and other N-terminal exopeptidases. In: *The Enzymes* (Boyer, P. D. ed.), 3rd edn., vol. 3, pp. 81-118 (see pp. 113-114), Academic Press, New York.

1973

Mudge, A. W. & Fellows, R. E. Bovine pituitary pyrrolidonecarboxylyl peptidase. *Endocrinology* **93**: 1428-1434.

1976

Prasad, C. & Peterkofsky, A. Demonstration of pyroglutamylpeptidase and amidase activities toward thyrotropin-releasing hormone in hamster hypothalamus extracts. *J. Biol. Chem.* **251**: 3229-3234.

1978

Fujiwara, K. & Tsuru, D. New chromogenic and fluorogenic substrates for pyrrolidonyl peptidase. *J. Biochem.* **83**: 1145-1149.

1979

Bauer, K. & Nowak, P. Characterization of a thyroliberin-degrading serum enzyme catalyzing the hydrolysis of thyroliberin at the pyroglutamyl-histidine bond. *Eur. J. Biochem.* **99**: 239-246.

Morier, E., Moreau, O., Masson, M. A., Han, K.-K. & Rips, R. Evidence for the enzymic degradation of thyrotropin-releasing hormone (TRH) and pseudo-hormone (pyroglutamyl-histidyl-amphetamine) by calf liver pyroglutamine-amino-peptidase. *Int. J. Biochem.* **10**: 769-783.

1980

Bauer, K. & Kleinkauf, H. Catabolism of thyroliberin by rat adenohypophyseal tissue extract. *Eur. J. Biochem.* **106**: 107-117.

Fujiwara, K., Matsumoto, E., Kitagawa, T. & Tsuru, D. N-α-Carbobenzoxy pyroglutamyl diazomethyl ketone as active-site-directed inhibitor for pyroglutamyl peptidase. *Biochim. Biophys. Acta* **702**: 149-154.

1981

Fujiwara, K., Kitagawa, T. & Tsuru, D. Inactivation of pyroglutamyl aminopeptidase by L-pyroglutamyl chloromethyl ketone. *Biochim. Biophys. Acta* **655**: 10-16.

Fujiwara, K., Matsumoto, E., Kitagawa, T. & Tsuru, D. Inactivation of pyroglutamyl aminopeptidase by Nα-carbobenzoxy-L-pyroglutamyl chloromethyl ketone. *J. Biochem.* **90**: 433-437.

1983

Browne, P. & O'Cuinn, G. An evaluation of the role of a pyroglutamyl peptidase, a post-proline cleaving enzyme and a post-proline dipeptidyl amino peptidase, each purified from the soluble fraction of guinea-pig brain, in the degradation of thyroliberin *in vitro*. *Eur. J. Biochem.* **137**: 75-87.

Pyroglutamyl Peptidase II

Summary

EC Number: 3.4.19.-

Earlier names: Thyroliberin (TRH) pyroglutamyl peptidase, TRH-specific pyrrolidonecarboxylyl peptidase, TRH pyroglutamate aminopeptidase, thyroliberinase, thyroliberin-degrading enzyme.

Abbreviation: PGP II.

Distribution: Membrane bound in the hypothalamus, thalamus, cortex and cerebellum of the brain [1], where it is located predominantly in synaptosomal membranes [2,3,3a]. An enzyme with similar properties has been described in rat [4,5] and hog [6,7] serum.

Source: Rat [5] and hog [6] serum, and a synaptosomal-membrane preparation from guinea pig brain [3,3a,8].

Action: Has a high degree of specificity for the release of the pyroglutamyl (pGlu-) group contained in thyroliberin (pGlu-His-Pro-NH$_2$, TRH). Also acts on pGlu-His-Pro and pGlu-His-Gly, but has no action on closely related pyroglutamyl peptides such as luliberin, neurotensin, bombesin, and thyroliberin analogs such as pGlu-His-Pro-NHCH$_3$, pGlu-His-Pro-Gly-NH$_2$ and pGlu-Phe-Pro-NH$_2$ [3a]. These compounds are, however, effective inhibitors of thyroliberin degradation by PGP II from rat serum [7] and brain [1]. No action occurs on pGlu-Ala, pGlu-Val or pGlu-Pro [3a].

Requirements: Imidazole or Tris buffer at pH 7.5 and 37°C. In contrast to PGP I requirements, dithiothreitol and EDTA should not be included (see Inhibitors).

Substrate, usual: [^{14}C]Pyroglutamyl thyroliberin (K$_m$ 50 μM) in a radiochemical assay with chromotographic separation of the split products [5,6]. Thyroliberin (pGlu-His-Pro-NH$_2$) degradation rates have also been followed by direct radioimmunoassay [1]. pGlu-NNap and pGlu-NMec are not useful substrates [3a,6].

Inhibitors: PGP II, in contrast to PGP I, is inhibited by dithiothreitol and EDTA [6,8]. 1,10-Phenanthroline is also a good inhibitor [6], but thiol-blocking reagents such as IAcNH$_2$ and MalNEt have little effect [6], as do the serine proteinase ihibitors Dip-F and Pms-F [5,6]. Heavy metals such as Hg^{2+}, Cu^{2+} and Cd^{2+} show strong inhibition at the 1 mM level [5,6].

Molecular properties: Relatively high molecular masses have been reported for both the membrane bound (synaptosomal) enzyme (M_r 230,000) solubilized with papain [3a], and the circulating form (M_r 260,000) present in hog [6] and rat [5] sera. The synaptosomal enzyme is resistant to solubilzation by sonication, or extraction with Triton X-100 or deoxycholate [3a,8].

Comment: The high degree of specificity shown by PGP II for thyroliberin inactivation, together with its synaptosomal localization, is strongly suggestive of a physiological function. His-Pro diketopiperazine, a major metabolite of TRH formed by a rapid intramolecular cyclization of the His-Pro-NH$_2$ product of PGP II action, suppresses TRH-stimulated prolactin secretion in rats. These findings, as well as other properties shared with PIF (prolactin release inhibiting factor), suggest a possible physiological role for PGP II in the control of prolactin secretion [4].

References
[1] Griffiths *et al. Biochem. Soc. Trans.* **7**: 74-75, 1979.
[2] Hayes *et al. Biochem. Soc. Trans.* **7**: 59-62, 1979.
[3] Greaney *et al. Biochem. Soc. Trans.* **8**: 423 only, 1980.
[3a] O'Connor & O'Cuinn *Eur. J. Biochem.* **144**: 271-278, 1984.
[4] Bauer *et al. Nature* **274**: 174-175, 1978.
[5] Taylor & Dixon *J. Biol. Chem.* **253**: 6934-6940, 1978.
[6] Bauer & Nowak *Eur. J. Biochem.* **99**: 239-246, 1979.
[7] Bauer *et al. Eur. J. Biochem.* **118**: 173-176, 1981.
[8] Browne *et al. Ir. J. Med. Sci.* **150**: 348-349, 1981.

Bibliography

1978

Bauer, K., Graf, K. J., Faivre-Bauman, A., Beier, S., Tixier-Vidal, A. & Kleinkauf, H. Inhibition of prolactin secretion by histidyl-proline-diketopiperazine. *Nature* **274**: 174-175.

Taylor, W. L. & Dixon, J. E. Characterization of a pyroglutamate aminopeptidase from rat serum that degrades thyrotropin-releasing hormone. *J. Biol. Chem.* **253**: 6934-6940.

1979

Bauer, K. & Nowak, P. Characterization of a thyroliberin-degrading serum enzyme catalyzing the hydrolysis of thyroliberin at the pyroglutamyl-histidine bond. *Eur. J. Biochem.* **99**: 239-246.

Griffiths, E. C., Kelly, J. A., White, N. & Jeffcoate, S. L. Hypothalamic inactivation of thyroliberin (thyrotropin-releasing hormone). *Biochem. Soc. Trans.* **7**: 74-75.

Hayes, D. J., Phelan, J. J. & O'Cuinn, G. The metabolism of thyrotropin-releasing hormone by guinea-pig brain. *Biochem. Soc. Trans.* **7**: 59-62.

1980

Greaney, A., Phelan, J. & O'Cuinn, G. Localization of thyroliberin pyroglutamyl peptidase on synaptosomal-membrane preparations of guinea-pig brain tissue. *Biochem. Soc. Trans.* **8**: 423 only.

1981

Bauer, K., Nowak, P. & Kleinkauf, H. Specificity of a hog serum peptidase hydrolyzing thyroliberin at the pyroglutamyl-histidine bond. *Eur. J. Biochem.* **118**: 173-176.

Browne, P., Phelan, J. & O'Cuinn, G. TRH pyroglutamate aminopeptidase activity in two different subcellular fractions of guinea pig brain. *Ir. J. Med. Sci.* **150**: 348-349.

1984

O'Connor, B. & O'Cuinn, G. Localization of a narrow-specificity thyroliberin hydrolyzing pyroglutamate aminopeptidase in synaptosomal membranes of guinea-pig brain. *Eur. J. Biochem.* **144**: 271-278.

Entry 18.03

γ-Glutamyl Transpeptidase

Summary

EC Number: 2.3.2.2

Earlier names: γ-Glutamyl transferase, γ-glutamyl hydrolase, phosphate-independent glutaminase, maleate-stimulated glutaminase.

Abbreviation: γ-GTP.

Distribution: Widely distributed in animal tissues as an ectoenzyme associated with the plasma membranes of many epithelial cells located in organs and tissues that have absorptive or secretory functions, i.e., kidney, liver, pancreas, jejunum, bronchi, lactating mammary gland, uterus, oviduct, seminal vesicles, retina, and choroid plexus. Activity in human serum is believed to arise from the bile ductules of the liver. A mitogen-responsive γ-GTP has recently been detected on human lymphoid cells. Immunocytochemical studies have localized γ-GTP of the central nervous system to the capillaries and the luminal side of the ependymal cells, in addition to the choroid plexus, and to glial cells primarily, throughout both grey and white matter.

Source: Beef kidney [1], liver [1a] and parotid gland [1b], hog kidney [2], rat kidney [3] and human kidney [4] and liver [4a].

Action: Catalyzes the transfer of L- (and D-) γ-glutamyl groups from a broad range of peptide donors such as γ-glutamyl amino acids or peptides, e.g., glutathione, to a wide variety of amino acid and peptide acceptors [5]. Peptides comprised of D-amino acids are ineffective as acceptors. Catalyzes autotranspeptidation reactions in which the substrate itself serves as the acceptor of the γ-glutamyl group. With only water as an acceptor γ-GTP hydrolyzes D- and L-γ-glutamyl compounds, including glutamine and γ-glutamyl-4-nitroanilide (K_m 5 μM), to form glutamate. In the presence of hydroxylamine, γ-glutamylhydroxamate is formed [2,6]. Maleate is thought to dissociate the catalytic functions of γ-GTP, causing an inhibition of transpeptidation and a 5- to 10-fold enhancement of hydrolysis rates [6]. γ-GTP is believed to participate in the translocation of amino acids across cellular membranes as a consequence of its role in the γ-glutamyl cycle [7,9]. The light subunit possesses

312

latent proteinase activity [23].

As described under Comments, γ-GTP is capable of catalyzing several biologically significant leukotriene conversions.

Requirements: pH 8.0 at 37°C, 75 mM NaCl [10], and a γ-glutamyl acceptor, usually Gly-Gly [5]. The optimum pH, between pH 7.5 and 9.0, depends on the γ-glutamyl donor and acceptor [5].

Substrate, usual: L-γ-Glu-NPhNO$_2$ [11].

Substrate, special: α-Methyl-γ-glutamyl-α-aminobutyrate [12], which precludes autotranspeptidation; γ-Glu-NMec for fluorometry [13], and S-acetophenone-glutathione for spectrophotometry [3,5].

Inhibitors: Inhibition, competitive with respect to the γ-glutamyl donor, is shown by serine in the presence of borate [14,15], and for the γ-glutamyl hydrazones of α-keto acids, especially the derivative of α-ketoglutarate [5]. Sulfophthalein derivatives such as bromosulfophthalein are also inhibitory [16]. Maleate, which binds to the cysteinylglycine binding site [17], inhibits transpeptidation, but promotes the hydrolysis of γ-glutamyl donors [3,6], thereby causing γ-GTP to behave as a glutathionase or glutaminase. Pms-F is inhibitory, but only in the presence of maleate [18]. γ-Glutamyl analogs such as 6-diazo-5-oxo-L-norleucine and azaserine (O-diazoacetyl-L-serine) serve as affinity labels which react specifically and covalently with the γ-glutamyl site of the enzyme [19]. The D-enantiomer of the norleucine derivative, which also acts as an affinity label [20], is expected to serve as a highly-specific inhibitor in studies of the physiological role of γ-GTP since no other mammalian enzymes are known to react with D-glutamine. A fermentation product of *Streptomyces sviceus*, known as AT-125 (αS,5S-amino-3-chloro-4,5-dihydro-5-isoxazoleacetic acid), structurally resembling the above analogs of glutamine, is reported to be an inhibitor of γ-GTP [21] that exhibits antitumor activity.

Molecular properties: M_r 68,000-90,000 for the protease-solubilized enzyme (P-γ-GTP); M_r 169,000-200,000 for the detergent-solubilized enzyme (D-γ-GTP) [3,3a]. The former contains about 18.5% carbohydrate (10% hexose, 7% hexosamine and 1.5% sialic acid), whereas the latter contains about 35% carbohydrate. The specificities of the two forms are virtually identical. Each consists of two noncovalently associated glycoprotein subunits; the smaller subunit in each form is about the same size (M_r 22,000), and contains the glutamylation site, whereas the larger subunit ranges in M_r from about 46,000 in rat P-γ-GTP to at least 62,000 in human D-γ-GTP. Rat kidney D-γ-GTP contains a hydrophobic M_r 6,000 segment at the N-terminus (Met-Lys-Asn-Arg-Phe-Leu-) of the heavy subunit that serves to anchor the enzyme to the membrane [21a,21b]. P-γ-GTP usually exhibits a higher specific activity, and numerous isozymic forms differing in sialic acid content are separable by isoelectric focusing. pI values range from 4.4 (for highly-sialylated P-γ-GTP) to 8.3 (for poorly-sialylated). pI is 8.6 for the asialo-form of P-γ-GTP.

Whereas the isolated light subunit exhibits proteolytic activity, the heavy subunit appears to contain the binding domains for γ-glutamyl substrates. It has been proposed that intact γ-GTP shows only transpeptidase activity as a consequence of the imposed specificity of the heavy subunit [23].

Comment: D-γ-glutamyl-NPhNO$_2$ and D-glutamine serve as substrates (donors), but a lack of activity on asparagine and homoglutamine indicates that a 5-carbon side chain is required. Autotranspeptidation can be prevented by using D-γ-Glu-NPhNO$_2$ as substrate (K_m 30 μM) since, like other D-amino acids, it cannot serve as an acceptor. The activation of transpeptidation that has been observed with Mg^{2+} [2] and Na^+ [10] is reported to occur with model substrates such as γ-Glu-NPhNO$_2$, but not with glutathione [5]. A recent report [22] suggests that endogenously-formed benzoylglycine, which acts like maleate, may function *in vivo* to regulate γ-GTP activity.

Highly purified γ-GTP from rat kidney is reported to possess latent proteinase activity that is unmasked by the dissociation of the two subunits [23]. The isolated light subunit (M_r 22,000) exhibits proteolytic activity on bovine serum albumin at pH 7.5, and is unaffected by Tos-Lys-CH$_2$Cl, Tos-Phe-CH$_2$Cl, antipain, leupeptin, and 6-diazo-5-oxo-L-norleucine, a γ-glutamyl analog. The proteolytic activity manifest at pH 5.0 is 33% of that observed at pH 7.0. No activity is exhibited on Bz-Arg-NPhNO$_2$ or Glt-Phe-NPhNO$_2$. When native γ-GTP is incubated in 6 M urea, the dissociated light subunit catalyzes a rapid degradation of the heavy subunit. The isolated, renatured heavy subunit is reported [24] to possess an active site with transpeptidase properties that are indistinguishable from those of the native enzyme, wherein the light subunit is the catalytic component. In the native oligomer, however, the active site on the heavy subunit remains latent.

γ-GTP contributes to the production of leukotrienes, biologically active compounds derived from polyunsaturated fatty acids [25], by virtue of its ability to remove (transfer) the γ-glutamyl group from leukotriene C_4 (LTC_4, or SRS-GSH), a glutathione (GSH) containing derivative of arachidonic acid [26]. LTC_4 is the parent compound of a major class of leukotrienes that comprise the "slow reacting substance" (SRS) of anaphylaxis [27]. The precursor is converted (reversibly) by γ-GTP to leukotriene D_4 (LTD_4; SRS-Cys-Gly). The potent smooth muscle-contracting activity of this product (which is believed to be responsible for hypersensitivity symptoms such as bronchoconstriction in asthma) can in turn be converted to the less active peptidolipid leukotriene E_4 (LTE_4: SRS-Cys) by other exopeptidases, in particular microsomal alanyl aminopeptidase (Entry 11.06) and cysteinylglycine dipeptidase (Entry 17.05).

Since plasma amino acids capable of serving as γ-glutamyl acceptors are readily accessible to γ-GTP located on the external surface of cell membranes, it is believed that γ-GTP is capable of carrying out leukotriene conversion under physiological conditions.

Tumors of the liver, colon and skin often show elevated γ-GTP activaties that have been attributed to a general expression of onco-fetal genes. Accordingly, it has been proposed that γ-GTP be used as a pre-neoplastic marker.

References

[1] Szewczuk & Baranowski *Biochem. Z.* **338**: 317-329, 1963.

[1a] Furakawa *et al. J. Biochem.* **93**: 839-846, 1983.

[1b] Hata *et al. Int. J. Biochem.* **13**: 681-692, 1981.

[2] Orlowski & Meister *J. Biol. Chem.* **240**: 338-347, 1965.

[3] Tate & Meister *J. Biol. Chem.* **250**: 4619-4627, 1975.

[3a] Hughey & Curthoys *J. Biol. Chem.* **251**: 7863-7870, 1976.

[4] Miller *et al. J. Biol. Chem.* **251**: 2271-2278, 1976.

[4a] Huseby *Biochim. Biophys. Acta* **483**: 46-56, 1977.

[5] Tate & Meister *J. Biol. Chem.* **249**: 7593-7602, 1974.

[6] Tate & Meister *Proc. Natl. Acad. Sci. USA* **71**: 3329-3333. 1974.

[7] Orlowski & Meister *Proc. Natl. Acad. Sci. USA* **67**: 1248-1255, 1970.

[8] Meister *Science* **180**: 33-39, 1973.

[9] Meister & Tate *Annu. Rev. Biochem.* **45**: 559-604, 1976.

[10] Orlowski *et al. FEBS Lett.* **31**: 237-240, 1973.

[11] Orlowski & Meister *Biochim. Biophys. Acta* **73**: 679-681, 1963.

[12] Karkowsky *et al. J. Biol. Chem.* **251**: 4736-4743, 1976.

[13] Smith *et al. Anal. Biochem.* **100**: 136-139, 1979.

[14] Revel & Ball *J. Biol. Chem.* **234**: 577-582, 1959.

[15] Tate & Meister *Proc. Natl. Acad. Sci. USA* **75**: 4806-4809, 1978.

[16] Binkley *J. Biol. Chem.* **236**: 1075-1082, 1961.

[17] Thompson & Meister *J. Biol. Chem.* **254**: 2956-2960, 1979.

[18] Elce *Biochem. J.* **185**: 473-481, 1980.

[19] Tate & Meister *Proc. Natl. Acad. Sci. USA* **74**: 931-935, 1977.

[20] Inoue *et al. Eur. J. Biochem.* **99**: 169-177, 1979.

[21] Allen *et al. Chem. Biol. Interact.* **33**: 361-365, 1981.

[21a] Nash & Tate *J. Biol. Chem.* **257**: 585-588, 1982.

[21b] Matsuda *et al. J. Biochem.* **93**: 1427-1433, 1983.

[22] Thompson & Meister *J. Biol. Chem.* **255**: 2109-2113, 1980.

[23] Gardell & Tate *J. Biol. Chem.* **254**: 4942-4945, 1979.

[24] Horiuchi *et al. Eur. J. Biochem.* **105**: 93-102, 1980.

[25] Samuelsson *et al. Allergy* **35**: 375-381, 1980.

[26] Andersen *et al. Proc. Natl. Acad. Sci. USA* **79**: 1088-1091, 1982.

[27] Lewis & Austen *J. clin. Invest.* **73**: 889-897, 1984.

Bibliography

1948

Binkley, F. & Nakamura, K. Metabolism of glutathione. I. Hydrolysis by tissues of the rat. *J. Biol. Chem.* **173**: 411-421.

1950

Hanes, C. S., Hird, F. J. R. & Isherwood, F. S. Synthesis of peptides in enzymic reactions involving glutathione. *Nature* **166**: 288-292.

1951

Binkley, F. & Olson, C. K. Metabolism of glutathione. IV. Activators and inhibitors of the hydrolysis of glutathione. *J. Biol. Chem.* **188**: 451-457.

1952

Hanes, C. S., Hird, F. J. R. & Isherwood, F. A. Enzymic transpeptidation reactions involving γ-glutamyl peptides and α-amino-acyl peptides. *Biochem. J.* **51**: 25-35.

1954

Hird, F. J. R. & Springell, P. H. The enzymic hydrolysis of the γ-glutamyl bond in glutathione. *Biochim. Biophys. Acta* **15**: 31-37.

1959

Revel, J. P. & Ball, E. G. The reaction of glutathione with amino acids and related compounds as catalyzed by γ-glutamyl transpeptidase. *J. Biol. Chem.* **234**: 577-582.

1960

Goldbarg, J. A., Friedman, O. M., Pineda, E. P., Smith, E. E., Chatterji, R., Stein, E. H. & Rutenburg, A. M. The colorimetric determination of γ-glutamyl transpeptidase with a synthetic substrate. *Arch. Biochem. Biophys.* **91**: 61-70.

Szewczuk, A. & Orlowski, M. The use of α-(N-γ-DL-glutamyl)-aminonitriles for the colorimetric determination of a specific peptidase in blood serum. *Clin. Chim. Acta* **5**: 680-688.

1961

Binkley, F. Purification and properties of renal glutathionase. *J. Biol. Chem.* **236**: 1075-1082.

Glenner, G. G. & Folk, J. E. Glutamyl peptidases in rat and guinea pig kidney slices. *Nature* **192**: 338-340.

Orlowski, M. & Szewczuk, A. A note on the occurrence of γ-glutamyl transpeptidase in human serum. *Clin. Chim. Acta* **6**: 430-432.

Orlowski, M. & Szewczuk, A. Colorimetric determination of γ-glutamyl transpeptidase activity in human serum and tissue with synthetic substrates. *Acta Biochim. Pol.* **8**: 189-200.

Szczeklik, E., Orlowski, M. & Szewczuk, A. Serum γ-glutamyl transpeptidase activity in liver disease. *Gastroenterol.* **41**: 353-359.

1963

Orlowski, M. & Meister, A. γ-Glutamyl-p-nitroanilide: A new convenient substrate for determination and study of L- and D-γ-glutamyltranspeptidase activities. *Biochim. Biophys. Acta* **73**: 679-681.

Szewczuk, A. & Baranowski, T. Purification and properties of γ-glutamyl transpeptidase from beef kidney. *Biochem. Z.* **338**: 317-329.

1964

Szewszuk, A. & Connell, G. E. The effect of neuraminidase on the properties of γ-glutamyl transpeptidase. *Biochim. Biophys. Acta* **83**: 218-223.

1965

Orlowski, M. & Meister, A. Isolation of γ-Glutamyl transpeptidase from hog kidney. *J. Biol. Chem.* **240**: 338-347.

1966

Greenberg, E., Wollaeger, E. E., Fleisher, C. A. & Engstrom, G. W. Demonstration of γ-glutamyl transpeptidase activity in human jejunal mucosa. *Clin. Chim. Acta* **16**: 79-89.

Katunuma, N., Tomino, I. & Nishino, H. Glutaminase isozymes in rat kidney. *Biochem. Biophys. Res. Commun.* **22**: 321-328.

1967

Katunuma, N., Huzino, A. & Tomino, I. Organ specific control of glutamine metabolism. *Adv. Enzyme Regul.* **5**: 55-69.

1968

Leibach, F. H. & Binkley, F. γ-Glutamyl transferase of swine kidney. *Arch. Biochem. Biophys.* **127**: 292-301.

1969

Rutenburg, A. M., Kim, H., Fischbein, J. W., Hanker, J. S., Wasserkrug, H. L. & Seligman, A. M. Histochemical and ultrastructural demonstration of γ-glutamyl transpeptidase activity. *J. Histochem. Cytochem.* **17**: 517-526.

Naftalin, L., Child, V. J., Morley, D. A. & Smith, D. A. Observations on the site of origin of serum γ-glutamyl-transpeptidase. *Clin. Chim. Acta* **26**: 297-300.

Naftalin, L., Sexton, M., Whitaker, J. F. & Tracey, D. A routine procedure for estimating serum γ-glutamyltranspeptidase activity. *Clin. Chim.*

Acta **26**: 293-296.

Richter, R. Some properties of γ-glutamyl transpeptidase from human kidney. *Arch. Immunol. Ther. Exp.* **17**: 476-495.

Szasz, G. A kinetic photometric method for serum γ-glutamyl transpeptidase. *Clin. Chem.* **15**: 124-136.

1970

Connell, G. E. & Adamson, E. D. γ-glutamyl transpeptidase. *Methods Enzymol.* **19**: 782-789.

Orlowski, M. & Meister, A. The γ-glutamyl cycle: a possible transport system for amino acids. *Proc. Natl. Acad. Sci. U.S.A.* **67**: 1248-1255.

1971

Van der Werf, P., Orlowski, M. & Meister, A. Enzymatic conversion of 5-oxo-L-proline (L-pyrrolidone carboxylate) to L-glutamate coupled with cleavage of adenosine triphosphate to adenosine diphosphate, a reaction in the γ-glutamyl cycle. *Proc. Natl. Acad. Sci. U.S.A.* **68**: 2982-2985.

1972

Bodnaryk, R. P. Membrane-bound γ-glutamyl transpeptidase. Evidence that it is a component of the 'amino acid site' of certain neutral amino acid transport systems. *Can. J. Biochem.* **50**: 524-528.

Ceriotti, G. & De Nadai-Frank, A. γ-Glutamyl transpeptidase. A simple method for routine microdetermination. *Enzyme* **14**: 221-228.

Lum, G. & Gambino, S. R. Serum gamma-glutamyl transpeptidase activity as an indicator of disease of liver, pancreas, or bone. *Clin. Chem.* **18**: 358-362.

1973

George, S. G. & Kenny, A. J. Studies on the enzymology of purified preparations of brush border from rabbit kidney. *Biochem. J.* **134**: 43-57.

Keane, P. M., Garcia, L., Gupta, R. N. & Walker, W. H. C. Serum gamma glutamyl transpeptidase in liver disorders. *Clin. Biochem.* **6**: 41-45.

Meister, A. Glutathione; metabolism and function via the γ-glutamyl cycle. *Life Sci.* **15**: 177-190.

Meister, A. On the enzymology of amino acid transport. *Science* **180**: 33-39.

Orlowski, M., Okonkwo, P. O. & Green, J. P. Activation of γ-glutamyl transpeptidase by monovalent cations. *FEBS Lett.* **31**:237-240.

Ross, L. L., Barber, L., Tate, S. S. & Meister, A. Enzymes of the γ-glutamyl cycle in the ciliary body and lens. *Proc. Natl. Acad. Sci. U.S.A.* **70**: 2211-2214.

Tate, S. S., Ross, L. L. & Meister, A. The γ-glutamyl cycle in the choroid plexus: its possible function in amino acid transport. *Proc. Natl. Acad. Aci. U.S.A.* **70**: 1447-1449.

1974

Binkley, F. & Johnson, J. D.   γ-Glutamyl cycle.   *Science* **184**: 586-587.
Elce, J. S., Bryson, J. & McGirr, L. G.   γ-Glutamyl transpeptidase of rat kidney. Some properties and kinetic constants.   *Can. J. Biochem.* **52**: 33-41.
Meister, A.   γ-Glutamyl cycle.   *Science* **184**: 587-588.
Orlowski, M., Sessa, G. & Green, J. P.   γ-Glutamyl transpeptidase in brain capillaries: possible site of a blood-brain barrier for amino acids.   *Science* **184**: 66-68.
Takahashi, S., Pollack, J. & Seifter, S.   Purification of γ-glutamyltransferase of rat kidney by affinity chromatography using concanavalin A conjugated with Sepharose 4B.   *Biochim. Biophys. Acta* **371**: 71-75.
Taniguchi, N.   Purification and some properties of γ-glutamyl traspeptidase from azo dye-induced hepatoma.   *J. Biochem.* **75**: 473-480.
Tate, S. S. & Meister, A.   Interaction of γ-glutamyl transpeptidase with amino acids, dipeptides, and derivatives and analogs of glutathione.   *J. Biol. Chem.* **249**: 7593-7602.
Tate, S. S. & Meister, A.   Stimulation of the hydrolytic activity and decrease of the transpeptidase activity of γ-glutamyl transpeptidase by maleate; identity of a rat kidney maleate-stimulated glutaminase and γ-glutamyl transpeptidase.   *Proc. Natl. Acad. Sci. U.S.A.* **71**: 3329-3333.

1975

Binkley, F. & Wiesemann, M. L.   Glutathione and gamma glutamyl transferase in secretory processes.   *Life Sci.* **17**: 1359-1362.
Binkley, F., Wiesemann, M. L., Groth, D. P. & Powell, R. W.   γ-Glutamyl transferase: a secretory enzyme.   *FEBS Lett.* **51**: 168-170.
Curthoys, N. P. & Kuhlenschmidt, T.   Phosphate-independent glutaminase from rat kidney. Partial purification and identity with γ-glutamyltranspeptidase.   *J. Biol. Chem.* **250**: 2099-2105.
DeLap, L. W., Tate, S. S. & Meister, A.   γ-Glutamyl transpeptidase of rat seminal vesicles; effect of orchidectomy and hormone administration on the transpeptidase in relation to its possible role in secretory activity.   *Life Sci.* **16**: 691-704.
Kuhlenschmidt, T. & Curthoys, N. P.   Subcellular localization of rat kidney phosphate independent glutaminase.   *Arch. Biochem. Biophys.* **167**: 519-524.
Rosalki, S. B.   Gamma-glutamyl transpeptidase.   *Adv. Clin. Chem.* **17**: 53-107.
Schulman, J. D., Goodman, S. I., Mace, J. W., Patrick, A. D., Tietze, F. & Butler, E. J.   Glutathionuria: inborn error of metabolism due to tissue deficiency of gamma-glutamyl transpeptidase.   *Biochem. Biophys. Res. Commun.* **65**: 68-74.
Taniguchi, N., Saito, K. & Takakuwa, E.   γ-Glutamyltransferase from azo dye induced hepatoma and fetal rat liver. Similarities in their kinetic

and immunological properties. *Biochim. Biophys. Acta* **391**: 265-271.

Tate, S. S. Interaction of γ-glutamyl transpeptidase with S-acyl derivatives of glutathione. *FEBS Lett.* **54**: 319-322.

Tate, S. S. & Meister, A. Identity of maleate-stimulated glutaminase with γ-glutamyl transpeptidase in rat kidney. *J. Biol. Chem.* **250**: 4619-4627.

Thompson, G. A. & Meister, A. Utilization of L-cysteine by the γ-glutamyl transpeptidase-γ-glutamyl cyclotransferase pathway. *Proc. Natl. Acad. Aci. U.S.A.* **72**: 1985-1988.

1976

Hughey, R. P. & Curthoys, N. P. Comparison of the size and physical properties of γ-glutamyltranspeptidase purified from rat kidney following solubilization with papain or with Triton X-100. *J. Biol. Chem.* **251**: 7863-7870.

Meister, A. & Tate, S. S. Glutathione and related γ-glutamyl compounds. Biosynthesis and utilization. *Annu. Rev. Biochem.* **45**: 559-604.

Miller, S. P., Awasthi, Y. C. & Srivastava, S. K. Studies of human kidney γ-glutamyl transpeptidase. *J. Biol. Chem.* **251**: 2271-2278.

Novogrodsky, A., Tate, S. S. & Meister, A. γ-glutamyl transpeptidase, a lymphoid cell-surface marker: relationship to blastogenesis, differentiation and neoplasia. *Proc. Natl. Acad. Sci. U.S.A.* **73**: 2414-2418.

Tate, S. S. & Meister, A. Subunit structure and isozymic forms of γ-glutamyl transpeptidase. *Proc. Natl. Acad. Sci. U.S.A.* **73**: 2599-2603.

Zelazo, P. & Orlowski, M. γ-Glutamyl transpeptidase of sheep-kidney cortex. Isolation, catalytic properties and dissociation into two polypeptide chains. *Eur. J. Biochem.* **61**: 147-155.

1977

Jung, K. & Liese, W. Influence of temperature on the determination of enzyme activities in human serum γ-glutamyltransferase. *Enzyme* **22**: 213-218.

Kenny, A. J. Proteinases associated with cell membranes. In: *Proteinases in Mammalian Cells and Tissues* (Barrett, A. J. ed.), pp. 393-444 (see pp. 410-417), North-Holland Publishing Co., Amsterdam.

Lu, C. & Steinberger, A. Gamma-glutamyl transpeptidase activity in the developing rat testis. Enzyme localization in isolated cell types. *Biol. Reprod.* **17**: 84-88.

Seymour, C. A. & Peters, T. J. Enzyme activities in human liver biopsies: assay methods and activities of some lysosomal and membrane-bound enzymes in control tissue and serum. *Clin. Sci. Mol. Med.* **52**: 229-239.

Tate, S. S. & Meister, A. Affinity labeling of γ-glutamyl transpeptidase and location of the γ-glutamyl binding site in the light subunit. *Proc. Natl. Acad. Sci. U.S.A.* **74**: 931-935.

Thompson, G. A. & Meister, A. Interrelationships between the binding sites for amino acids, dipeptides and γ-glutamyl donors in γ-glutamyl transpeptidase. *J. Biol. Chem.* **252**: 6792-6798.

1978

Beck, P. R. & King, J. Influence of detergents and organic solvent extraction on human gamma-glutamyltransferase activity. *Enzyme* **23**: 388-394.

Inoue, M., Horiuchi, S. & Morino, Y. Inactivation of γ-glutamyl transpeptidase by phenylmethanesulfonyl fluoride, a specific inactivator of serine enzymes. *Biochem. Biophys. Res. Commun.* **82**: 1183-1188.

Kenny, A. J. & Booth, A. G. Microvilli: their ultrastructure, enzymology and molecular organization. *Essays Biochem.* **14**: 1-44.

Malvoisin, E., Mercier, M. & Roberfroid, M. Gamma glutamyl transferase. Application of a new radiochemical assay to the analysis of its subcellular distribution in the rat liver. *Enzyme* **23**: 373-381.

Persson, A. & Wilson, I. B. A new chromogenic substrate for γ-glutamyl transpeptidase. *Anal. Biochem.* **89**: 408-413.

Smith, R. L. & Heizer, W. D. Gamma-glutamyl transferase of rat and human intestine: greater enhancement of activity by dipeptides than by amino acids or longer peptides. *Biochem. Med.* **19**: 383-394.

Tate, S. S. & Meister, A. Serine-borate complex as a transition-state inhibitor of γ-glutamyl transpeptidase. *Proc. Natl. Acad. Sci. U.S.A.* **75**: 4806-4809.

1979

Boelsterli, U. Gamma-glutamyl transpeptidase (GGT) - an early marker for hepatocarcinogens in rats. *Trends Pharmacol. Sci.* **1**: 47-49.

Booth, A. G., Hubbard, L. M. L. & Kenny, A. J. Proteins of the kidney microvillar membrane. Immunoelectrophoretic analysis of the membrane hydrolases: identification and resolution of the detergent- and proteinase-solubilized forms. *Biochem. J.* **179**: 397-405.

Butler, J. D. & Spielberg, S. P. Influence of phospholipids and their hydrolytic products on γ-glutamyl transpeptidase activity. *J. Biol. Chem.* **254**: 3152-3155.

Colombo, J. P. & Colombo, J. Plasma gamma-glutamyl transpeptidase in heroin addicts. *Clin. Chim. Acta* **95**: 483-486.

Colombo, J. P. & Gigon, P. L. γ-glutamyltranspeptidase (GGTP) and cytochrome P-450 after portacaval shunt in the rat. *Experientia* **35**: 1005-1006.

Curthoys, N. P. & Hughey, R. P. Characterization and physiological function of rat renal γ-glutamyltranspeptidase. *Enzyme* **24**: 383-403.

Das, N. D. & Shichi, H. Gamma-glutamyl transpeptidase of bovine ciliary body: Purification and properties. *Exp. Eye Res.* **29**: 109-121.

Das, N. D. & Shichi, H. Tissue difference in gamma-glutamyl transpeptidase attributed to sialic acid content. *Life Sci.* **25**: 1821-1827.

Echetebu, Z. O. & Moss, D. W. Electrophoretic patterns of γ-glutamyltransferase activity eluted from liver tissue. *Clin. Chim. Acta*

95: 433-441.

Fiala, S., Trout, E., Pragani, B. & Fiala, E. S.   Increased γ-glutamyl transferase activity in human colon cancer. *Lancet* **1**: 1145 only.

Gardell, S. J. & Tate, S. S.   Latent proteinase activity of γ-glutamyl transpeptidase light subunit. *J. Biol. Chem.* **254**: 4942-4945.

Grau, E. M., Marathe, G. V. & Tate, S. S.   Rapid purification of rat kidney brush borders enriched in γ-glutamyl transpeptidase. *FEBS Lett.* **98**: 91-95.

Griffith, O. W. & Meister, A.   Translocation of intracellular glutathione to membrane-bound γ-glutamyl transpeptidase as a discrete step in the γ-glutamyl cycle: Glutathionuria after inhibition of transpeptidase. *Proc. Natl. Acad. Sci. U.S.A.* **76**: 268-272.

Griffith, O. W., Bridges, R. J. & Meister, A.   Transport of γ-glutamyl amino acids: Role of glutathione and γ-glutamyl transpeptidase. *Proc. Natl. Acad. Sci. U.S.A.* **76**: 6319-6322.

Griffith, O. W., Novogrodsky, A. & Meister, A.   Translocation of glutathione from lymphoid cells that have markedly different γ-glutamyl transpeptidase activities. *Proc. Natl. Acad. Sci. U.S.A.* **76**: 2249-2252.

Hamlyn, A. N., Hopper, J. C. & Skillen, A. W.   Assessment of erythrocyte δ-aminolaevulinate dehydratase for outpatient detection of alcoholic liver disease: comparison with γ-glutamyltransferase and casual blood ethanol. *Clin. Chim. Acta* **95**: 453-459.

Hopkins, L. J. & Moss, D. W.   A comparison of reaction conditions for the automated determination of γ-glutamyltransferase activity in serum. *Clin. Chim. Acta* **92**: 443-449.

Hughey, R. P., Coyle, P. J. & Curthoys, N. P.   Comparison of the association and orientation of γ-glutamyltransferase in lecithin vesicles and in native membranes. *J. Biol. Chem.* **254**: 1124-1128.

Huseby, N.-E.   Subcellular localization of γ-glutamyltransferase activity in guinea pig liver. Effect of phenobarbital on the enzyme activity levels. *Clin. Chim. Acta* **94**: 163-171.

Inoue, M., Horiuchi, S. & Morino, Y.   Affinity labeling of rat kidney γ-glutamyl transpeptidase by 6-Diazo-5-oxo-D-norleucine. *Eur. J. Biochem.* **99**: 169-177.

Jaken, S. & Mason, M.   Purification and comparison of several catalytic parameters of the γ-glutamyltranspeptidase of rat mammary adenocarcinoma (13762) and of normal rat mammary gland. *Biochim. Biophys. Acta* **568**: 331-338.

Lisý, V., Stastný, F. & Lodin, Z.   Regional distribution of membrane-bound γ-glutamyl transpeptidase activity in mouse brain: comparison with rabbit brain. *Neurochem. Res.* **4**: 747-753.

Marathe, G. V., Nash, B., Haschemeyer, R. H. & Tate, S. S.   Ultrastructural localization of γ-glutamyl transpeptidase in rat kidney and jejunum. *FEBS Lett.* **107**: 436-440.

Mathieu, M.   La gamma-glutamyl transpeptidase. *Lyon Pharm.* **30**: 5-11.

Minato, S.   A new colorimetric method for the determination of serum enzyme, γ-glutamyl transpeptidase, cystine aminopeptidase, and leucine aminopeptidase. *Clin. Chim. Acta* **92**: 249-255.

Minato, S. Isolation of anthglutin, an inhibitor of γ-glutamyl transpeptidase from Penicillium oxalicum. *Arch. Biochem. Biophys.* **192**: 235-240.

Perantoni, A., Berman, J. J. & Rice, J. M. L-Azaserine toxicity in established cell lines. Correlation with γ-glutamyl transpeptidase activity. *Exp. Cell. Res.* **122**: 55-61.

Puente, J., Varas, M. A., Beckhaus, G. & Sapag-Hagar, M. γ-Glutamyltranspeptidase activity and cyclic AMP levels in rat liver and mammary gland during the lactogenic cycle and in the oestradiol-progesterone pseudo-induced pregnancy. *FEBS Lett.* **99**: 215-218.

Scherberich, J. E., Kleemann, B. & Mondorf, W. Isolation of kidney brush border gamma-glutamyl transpeptidase from urine by specific antibody gel chromatography. *Clin. Chim. Acta* **93**: 35-41.

Sells, M. A., Katyal, S. L., Sell, S., Shinozuka, H. & Lombardi, B. Induction of foci of altered, γ-glutamyltranspeptidase-positive hepatocytes in carcinogen-treated rats fed a choline-deficient diet. *Br. J. Cancer* **40**: 274-283.

Shaw, L. M. & Petersen-Archer, L. Interaction of γ-glutamyltransferase from human tissues with insolubilized lectins. *Clin. Biochem.* **12**: 256-260.

Smith, G. D., Ding, J. L. & Peters, T. J. A sensitive fluorimetric assay for γ-glutamyl transferase. *Anal. Biochem.* **100**: 136-139.

Spielberg, S. P., Butler, J. DeB., MacDermot, K. & Schulman, J. D. Treatment of glutathione synthetase deficient fibroblasts by inhibiting γ-glutamyl transpeptidase activity with serine and borate. *Biochem. Biophys. Res. Commun.* **89**: 504-511.

Tate, S. S. & Orlando, J. Conversion of glutathione to glutathione disulfide, a catalytic function of γ-glutamyl transpeptidase. *J. Biol. Chem.* **254**: 5573-5575.

Thompson, G. A. & Meister, A. Modulation of the hydrolysis, transfer and glutaminase activities of γ-glutamyl transpeptidase by maleate bound at the cysteinylglycine binding site of the enzyme. *J. Biol. Chem.* **254**: 2956-2960.

Thompson, G. A. & Meister, A. Modulation of γ-glutamyl transpeptidase activities by hippurate and related compounds. *J. Biol. Chem.* **255**: 2109-2113.

1980

Allen, L., Meck, R. & Yunis, A. The inhibition of γ-glutamyl transpeptidase from human pancreatic carcinoma cells by (αS,5S)-α-amino-3-chloro-4,5-dihydro-5-isoxazoleacetic acid (AT-125; NSC-163501). *Res. Commun. Chem. Pathol. Pharmacol.* **27**: 175-182.

DeBault, L. E. & Cancilla, P. A. γ-Glutamyl transpeptidase in isolated brain endothelial cells: Induction by glial cells in vitro. *Science* **207**:

653-655.

Elce, J. S. Active-site amino acid residues in γ-glutamyltransferase and the nature of the γ-glutamyl-enzyme bond. *Biochem. J.* **185**: 473-481.

Gardell, S. J. & Tate, S. S. Affinity labeling of γ-glutamyl transpeptidase by glutamine antagonists. Effects on the γ-glutamyl transfer and proteinase activities. *FEBS Lett.* **122**: 171-174.

Ghandour, M. S., Langley, O. K. & Varga, V. Immunohistological localization of γ-glutamyltransferase in cerebellum at light and electron microscope levels. *Neurosci. Lett.* **20**: 125-129.

Griffith, O. W. & Meister, A. Excretion of cysteine and γ-glutamylcysteine moieties in human and experimental animal γ-glutamyl transpeptidase deficiency. *Proc. Natl. Acad. Sci. USA* **77**: 3384-3387.

Griffith, O. W. & Tate, S. S. The apparent glutathione oxidase activity of γ-glutamyl transpeptidase. *J. Biol. Chem.* **255**: 5011-5014.

Hata, K., Abiko, Y. & Takiguchi, H. Species distribution of γ-glutamyl transpeptidase in the parotid gland. *J. Dent. Res.* **59**: 728 only.

Horiuchi, S., Inoue, M. & Morino, Y. Latent active site in rat-kidney γ-glutamyl transpeptidase. The refolding process of the large subunit and characterization of the renatured enzyme. *Eur. J. Biochem.* **105**: 93-102.

Hultberg, B. & Sjögren, U. L-γ-glutamyl transpeptidase activity in normal and leukemic leukocytes. *Acta Haematol.* **63**: 132-135.

Inoue, M., Hayashida, S., Hosomi, F., Horiuchi, S. & Morino, Y. The molecular forms of γ-glutamyl transferase in bile and serum of icteric rats. *Biochim. Biophys. Acta* **615**: 70-78.

Kozak, E. M. & Tate, S. S. Interaction of the antitumor drug, L-(αS,5S)-α-amino-3-chloro-4,5-dihydro-5-isoxazoleacetic acid (AT-125) with renal brush border membranes. Specific labeling of γ-glutamyl transpeptidase. *FEBS Lett.* **122**: 175-178.

Lipsky, M. M., Hinton, D. E., Klaunig, J. E., Goldblatt, P. J. & Trump, B. F. Gamma glutamyl transpeptidase in safrole-induced, presumptive premalignant mouse hepatocytes. *Carcinogenesis* **1**: 151-156.

Lisý, V., Dutton, G. R. & Currie, D. N. Cerebellar specific γ-glutamyltranspeptidase activation by GABA and taurine in postnatal mouse brain. *Life Sci.* **27**: 2615-2620.

Mamelok, R. D., Groth, D. F. & Prusiner, S. B. Separation of membrane-bound γ-glutamyl transpeptidase from brush border transport and enzyme activities. *Biochemistry* **19**: 2367-2373.

Matsuda, Y., Tsuji, A. & Katunuma, N. Studies on the structure of γ-glutamyltranspeptidase. I. Correlation between sialylation and isozymic forms. *J. Biochem.* **87**: 1243-1248.

McIntyre, T. M. & Curthoys, N. P. The interorgan metabolism of glutathione. *Int. J. Biochem.* **12**: 545-551.

Orning, L., Hammarström, S. & Samuelsson, B. Leukotriene D: a slow reacting substance from rat basophilic leukemia cells. *Proc. Natl. Acad. Sci. USA* **77**: 2014-2017.

PetitClerc, C., Schiele, F., Bagrel, D., Mahassen, A. & Siest, G. Kinetic properties of γ-glutamyltransferase from human liver. *Clin. Chem.* **26**: 1688-1693.

Pocius, P. A., Baumrucker, C. R., McNamara, J. P. & Bauman, D. E. γ-Glutamyl transpeptidase in rat mammary tissue. Activity during lactogenesis and regulation by prolactin. *Biochem. J.* **188**: 565-568.

Prusak, E., Siewinski, M. & Szewczuk, A. A new fluorimetric method for the determination of γ-glutamyltransferase activity in blood serum. *Clin. Chim. Acta* **107**: 21-26.

Reyes, E. & Barela, T. D. Isolation and purification of multiple forms of γ-glutamyl transpeptidase from rat brain. *Neurochem. Res.* **5**: 159-170.

Satoh, T., Takenaga, M., Kitagawa, H. & Itoh, S. Microassay of gamma-glutamyl transpeptidase in needle biopsies of human liver. *Res. Comm. Chem. Path. Pharm.* **30**: 151-161.

Sikka, S. C. & Kalra, V. K. γ-Glutamyl transpeptidase-mediated transport of amino acid in lecithin vesicles. *J. Biol. Chem.* **255**: 4399-4402.

States, B. & Segal, S. Levels of gamma-glutamyltranspeptidase in cultured skin fibroblasts from cystinotics and normals. *Life Sci.* **27**: 1985-1990.

Tsuji, A., Matsuda, Y. & Katunuma, N. Characterization of human serum γ-glutamyltranspeptidase. *Clin. Chim. Acta* **104**: 361-366.

Tsuji, A., Matsuda, Y. & Katunuma, N. Purification and characterization of γ-glutamyltranspeptidase from human bile. *Biomed. Res.* **1**: 410-416.

Tsuji, A., Matsuda, Y. & Katunuma, N. Studies on the Structure of γ-Glutamyltranspeptidase. II Location of the Segment Anchoring γ-Glutamyltranspeptidase to the Membrane. *J. Biochem.* **87**: 1567-1571.

Welbourne, T. C., Dass, P. D. & Smith, R. L. Renal glutamine utilization: glycylglycine stimulation of γ-glutamyltransferase. *Can. J. Biochem.* **58**: 614-619.

1981

Allen, L. M., Corrigan, M. V. & Meinking, T. Interaction of AT-125 (αS,5S)-Amino-3-Chloro-4,5-Dihydro-isoxazoleacetic acid, with bovine kidney γ-glutamyl transpeptidase. *Chem. Biol. Interactions* **33**: 361-365.

Allison, R. D. & Meister, A. Evidence that transpeptidation is a significant function of γ-glutamyl transpeptidase. *J. Biol. Chem.* **256**: 2988-2992.

Anjaneyulu, K., Anjaneyulu, R., Sener, A. & Malaisse, W. J. γ-Glutamyltranspeptidase activity in pancreatic islets. *FEBS Lett.* **125**: 57-59.

Busachi, C., Mebis, J., Broeckaert, L. & Desmet, V. Histochemistry of γ-glutamyl transpeptidase in human liver biopsies. *Path. Res. Pract.* **172**: 99-108.

Dass, P. D., Misra, R. P. & Welbourne, T. C. Presence of γ-glutamyltransferase in the renal microvascular compartment. *Can. J. Biochem.* **59**: 383-386.

Ding, J. L., Smith, G. D. & Peters, T. J. Purification and properties of γ-glutamyl transferase from normal rat liver. *Biochim. Biophys. Acta*

657: 334-343.

Ding, J. L., Smith, G. D. & Peters, T. J. Subcellular localization and isolation of γ-glutamyltransferase from rat hepatoma cells. *Biochim. Biophys. Acta* **661**: 191-198.

Griffith, O. W., Bridges, R. J. & Meister, A. Formation of γ-glutamylcysteine *in vivo* is catalyzed by γ-glutamyl transpeptidase. *Proc. Natl. Acad. Sci. USA.* **78**: 2777-2781.

Hada, T., Higashino, K., Yamamoto, H., Okochi, T., Sumikawa, K. & Yamamura, Y. Further investigations on a novel γ-glutamyl transpeptidase in human renal carcinoma. *Clin. Chim. Acta* **112**: 135-140.

Harrison, E. H. & Bowers, W. E. Inhibition of γ-glutamyltranspeptidase by treatment of intact lymphocytes with periodate. *FEBS Lett.* **136**: 289-292.

Hata, K., Hayakawa, M., Abiko, Y. & Takiguchi, H. Purification and properties of γ-glutamyl transpeptidase from bovine parotid gland. *Int. J. Biochem.* **13**: 681-692.

Huseby, N.-E. Separation and characterization of human γ-glutamyltransferase. *Clin. Chim. Acta* **111**: 39-45.

Inoue, M. & Morino, Y. Inactivation of renal γ-glutamyl transferase by 6-diazo-5-oxo-L-norleucylglycine, an inactive precursor of affinity-labeling reagent. *Proc. Natl. Acad. Sci. USA* **78**: 46-49.

Jacobs, J. M., Pretlow, T. P., Fausto, N., Pitts, A. M. & Pretlow, T. G. Separation of two populations of cells with gamma-glutamyl transpeptidase from carcinogen-treated rat liver. *J. Nat. Cancer Inst.* **66**: 967-973.

Nakamura, Y., Kato, H., Suzuki, F. & Nagata, Y. Some properties of γ-glutamyltransferase from hog small intestine. *Biomed. Res.* **2**: 509-516.

Puente, J., Martinez, A. M., Beckhaus, G. & Sapag-Hagar, M. Properties of γ-glutamyltranspeptidase and glutathione levels in rat mammary gland. *Experientia* **37**: 459-462.

Rice, J. M. & Williams, G. M. Histochemical characteristics of spontaneous and chemically induced hepatocellular neoplasms in mice and the development of neoplasms with γ-glutamyl transpeptidase activity during phenobarbital exposure. *Histochem. J.* **13**: 85-99.

Schiele, F., Artur, Y., Bagrel, D., Petitclerc, C. & Siest, G. Measurement of plasma gamma-glutamyltransferase in clinical chemistry: kinetic basis and standardisation propositions. *Clin. Chim. Acta* **112**: 187-195.

Selvaraj, P. , Balasubramanian, K. A. & Hill, P. G. Isolation of gamma-glutamyl transpeptidase from human primary hepatoma and comparison of its kinetic and catalytic properties with the enzyme from normal adult and fetal liver. *Enzyme* **26**: 57-63.

Shine, H. D. & Haber, B. Immunocytochemical localization of γ-glutamyl transpeptidase in the rat CNS. *Brain Res.* **217**: 339-349.

Shine, H. D., Hertz, L., De Vellis, J. & Haber, B. A fluorometric assay for γ-glutamyl transpeptidase: demonstration of enzymatic activity in cultured cells of neural origin. *Neurochem. Res.* **6**: 453-463.

Solberg, H. E., Theordorsen, L. & Strømme, J. H. γ-Glutamyltransferase in human serum: an analysis of kinetic models. *Clin. Chem.* **27**: 303-307.

States, B. & Segal, S. Characteristics of γ-glutamyl-transpeptidase in cultured skin fibroblasts from normals and cystinotics. *Enzyme* **26**: 156-164.

Tarachand, U., Sivabalan, R. & Eapen, J. Enhanced gamma glutamyl transpeptidase activity in rat uterus following deciduoma induction and implantation. *Biochem. Biophys. Res. Commun.* **101**: 1152-1157.

Tate, S. S. & Meister, A. γ-Glutamyltranspeptidase: catalytic, structural and functional aspects. *Mol. Cell. Biochem.* **39**: 357-368.

Vanderlaan, M. & Phares, W. γ-Glutamyltranspeptidase: a tumour cell marker with a pharmacological function. *Histochem. J.* **13**: 865-877.

Viot, M., Thyss, A., Viot, G., Ramaioli, A., Cambon, P., Schneider, M. & Lalanne, C. M. Comparative study of gamma glutamyl transferase, alkaline phosphatase and its α1 isoenzyme as biological indicators of liver metastases. *Clin. Chim. Acta* **115**: 349-358.

Yamamoto, H., Sumikawa, K., Hada, T., Higashino, K. & Yamamura, Y. γ-Glutamyltransferase from human hepatoma tissue in comparison with normal liver enzyme. *Clin. Chim. Acta* **111**: 229-237.

1982

Anderson, M. E., Allison, R. D. & Meister, A. Interconversion of leukotrienes catalyzed by purified γ-glutamyl transpeptidase: concomitant formation of leukotriene D4 and γ-glutamyl amino acids. *Proc. Natl. Acad. Sci. USA* **79**: 1088-1091.

Bernström, K. & Hammarström, S. A novel leukotriene formed by transpeptidation of leukotriene E. *Biochem. Biophys. Res. Commun.* **109**: 800-804.

Daoust, R. The histochemical demonstration of γ-glutamyl transpeptidase activity in different populations of rat liver cells during azo dye carcinogenesis. *J. Histochem. Cytochem.* **30**: 312-316.

Dass, P. D. & Welbourne, T. C. Effect of AT-125 on in situ renal γ-glutamyltransferase activity. *FEBS Lett.* **144**: 21-24.

Ding, J. L., Smith, G. D., Searle, A. & Peters, T. J. In situ determination of the molecular weight of hepatic γ-glutamyl transferase and γ-glutamyl hydrolase activities. *Biochim. Biophys. Acta* **707**: 164-166.

Echetebu, Z. O. & Moss, D. W. Multiple forms of human γ-glutamyltransferase. Modification and possible structures of different molecular weight fractions. *Enzyme* **27**: 9-18.

Echetebu, Z. O. & Moss, D. W. Multiple forms of human γ-glutamyltransferase. Preparation and characterization of different molecular weight fractions. *Enzyme* **27**: 1-8.

Frielle, T., Brunner, J. & Curthoys, N. P. Isolation of the hydrophobic membrane binding domain of rat renal γ-glutamyl transpeptidase selectively labeled with 3-trifluoromethyl-3-(m-[^{125}I]iodophenyl)diazirine. *J. Biol. Chem.* **257**: 14979-14982.

Fujiwara, K., Katyal, S. L. & Lombardi, B. Influence of age, sex and cancer on the activities of γ-glutamyl transpeptidase and of dipeptidyl aminopeptidase IV in rat tissues. *Enzyme* 27: 114-118.

Huseby, N. E. Hydrophilic form of γ-glutamyltransferase: proteolytic formation in liver homogenates and its estimation in serum. *Clin. Chim. Acta* 124: 113-121.

Huseby, N. E. Multiple forms of serum γ-glutamyltransferase. Association of the enzyme with lipoproteins. *Clin. Chim. Acta* 124: 103-112.

Kozak, E. M. & Tate, S. S. Glutathione-degrading enzymes of microvillus membranes. *J. Biol. Chem.* 257: 6322-6327.

Linder, M. & Sudaka, P. [Urinary elimination of multiple forms of γ-glutamyltranspeptidase and aminopeptidase.] *Clin. Chim. Acta* 118: 77-85.

Masuike, M., Ogawa, M., Kosaki, G., Minamiura, N. & Yamamoto, T. Purification and characterization of gamma-glutamyl transpeptidase from human pancreas. *Enzyme* 27: 163-170.

Murray, J. L., Lerner, M. P. & Nordquist, R. E. Elevated γ-glutamyl transpeptidase levels in malignant melanoma. *Cancer* 49: 1439-1443.

Nash, B. & Tate, S. S. Biosynthesis of rat renal γ-glutamyl transpeptidase. Evidence for a common precursor of the two subunits. *J. Biol. Chem.* 257: 585-588.

Orning, L. & Hammerström, S. Kinetics of the conversion of leukotriene C by γ-glutamyl transpeptidase. *Biochem. Biophys. Res. Commun.* 106: 1304-1309.

Satoh, T., Igarashi, T., Hirota, T. & Kitagawa, H. Induction of hepatic γ-glutamyl transpeptidase in rats by repeated administration of aminopyrine. *J. Pharmacol. Exp. Ther.* 221: 795-800.

Selvaraj, P. & Balasubramanian, K. A. Localization of γ-glutamyl transferase on polyacrylamide gels using L-γ-glutamyl-*p*-nitroanilide as substrate. *Clin. Chim. Acta* 121: 291-300.

Shakespeare, P. G., Stokes, S. H. & Futerman, A. H. Heat inactivation of gamma-glutamyltransferase in homogenates of human kidney. *Ann. Clin. Biochem.* 19: 43-46.

Takahashi, S. & Zeydel, M. γ-Glutamyl transpeptidase and glutathione in aging IMR-90 fibroblasts and in differentiating 3T3 L1 preadipocytes. *Arch. Biochem. Biophys.* 214: 260-267.

Takahashi, S., Steinman, H. M. & Ball, D. Purification and characterization of γ-glutamyltransferase from rat pancreas. *Biochim. Biophys. Acta* 707: 66-73.

Welbourne, T. C. & Dass, P. D. Function of renal γ-glutamyltransferase: significance of glutathione and glutamine interactions. *Life Sci.* 30: 793-801.

1983

Abbott, W. A. & Meister, A. Modulation of γ-glutamyl transpeptidase activity by bile acids. *J. Biol. Chem.* 258: 6193-6197.

Felberg, N. T., Shields, J. A., Maquire, J., Piperata, S. & Amsel, J.

Gamma-glutamyl transpeptidase in the prognosis of patients with uveal malignant melanoma. *Am. J. Ophthalmol.* **95**: 467-473.

Frielle, T. & Curthoys, N. P. Characterization of the membrane binding domain of γ-glutamyltranspeptidase by specific labeling techniques. *Biochemistry* **22**: 5709-5714.

Furukawa, M., Higashi, T., Tateishi, N., Ochi, K. & Sakamoto, Y. Purification and properties of bovine liver γ-glutamyltransferase. *J. Biochem.* **93**: 839-846.

Fushiki, T., Iwami, K., Yasumoto, K. & Iwai, K. Evidence for an essential arginyl residue in bovine milk γ-glutamyltransferase. *J. Biochem.* **93**: 795-800.

Gardell, S. J. & Tate, S. S. Effects of bile acids and their glycine conjugates on γ-glutamyl transpeptidase. *J. Biol. Chem.* **258**: 6198-6201.

Kuno, T., Matsuda, Y. & Katunuma, N. The conversion of the precursor form of γ-glutamyltranspeptidase to its subunit form takes place in brush border membranes. *Biochem. Biophys. Res. Commun.* **114**: 889-895.

Matsuda, Y., Tsuji, A. & Katunuma, N. Studies on the structure of γ-glutamyltranspeptidase. III. Evidence that the amino terminus of the heavy subunit is the membrane binding segment. *J. Biochem.* **93**: 1427-1433.

Mojamdar, M., Ichihashi, M. & Mishima, Y. γ-Glutamyl transpeptidase, tyrosinase, and 5-S-cysteinyldopa production in melanoma cells. *J. Invest. Dermatol.* **81**: 119-121.

Sachdev, G. P., Leahy, D. S. & Chace, K. V. Phenobarbital and related compounds as novel inhibitors of γ-glutamyltranspeptidase. *Biochim. Biophys. Acta* **749**: 125-129.

Selvaraj, P. & Balasubramanian, K. A. Comparison of electrophoretic and immunological properties of gamma-glutamyl transpeptidase from human adult liver, fetal liver and primary hepatoma. *Enzyme* **30**: 21-28.

Tsuchida, S. & Sato, K. Purification of detergent-solubilized form and membrane-binding domain of rat γ-glutamyltransferase by immuno-affinity and hydrophobic chromatography. *Biochim. Biophys. Acta* **756**: 341-348.

Viña, J., Puertes, I. R., Montoro, J. B. & Viña, J. R. Effect of specific inhibition of gamma-glutamyl transpeptidase on amino acid uptake by mammary gland of the lactating rat. *FEBS Lett.* **159**: 119-122.

1984

Artur, Y., Wellman-Bednawska, M., Jacquier, A. & Siest, G. Associations between serum gamma-glutamyltransferase and apolipoproteins: relationships with hepatobiliary diseases. *Clin. Chem.* **30**: 1318-1321.

Barouki, R., Finidori, J., Chobert, M.-N., Aggerbeck, M., Laperche, Y. & Hanoune, J. Biosynthesis and processing of γ-glutamyl transpeptidase in hepatoma tissue culture cells. *J. Biol. Chem.* **259**: 7970-7974.

Bussmann, L. E. & Deis, R. P. γ-Glutamyltransferase activity in mammary gland of pregnant rats and its regulation by ovarian hormones, prolactin and placental lactogen. *Biochem. J.* **223**: 275-277.

Crockard, A. D. Cytochemistry of lymphoid cells: a review of findings in the normal and leukaemic state. *Histochem. J.* **16**: 1027-1050.

Finidori, J., Laperche, Y., Haguenauer-Tsapis, R., Barouki, R., Guellaen, G. & Hanoune, J. *In vitro* biosynthesis and membrane insertion of γ-glutamyl transpeptidase. *J. Biol. Chem.* **259**: 4687-4690.

Halsall, S. & Peters, T. J. Effect of chronic ethanol consumption on the cellular and subcellular distribution of γ-glutamyltransferase in rat liver. *Enzyme* **31**: 221-228.

Kuno, T., Matsuda, Y. & Katunuma, N. Characterization of a processing protease that converts the precursor form of γ-glutamyltranspeptidase to its subunits. *Biochem. Int.* **8**: 581-588.

Milnerowicz, H. & Szewczuk, A. Bovine kidney γ-glutamyltransferase. Solubilized forms, biochemical and immunochemical properties. *Enzyme* **32**: 208-217.

Nash, B. & Tate, S. S. *In vitro* translation and processing of rat kidney γ-glutamyl transpeptidase. *J. Biol. Chem.* **259**: 678-685.

Osuji, G. O. An oscillatory mechanism for the γ-glutamyl transpeptidase-mediated translocation of amino acids across the cell membrane. *J. Theor. Biol.* **109**: 1-15.

Pajari, M. Properties of γ-glutamyltransferase in developing rat brain. *Int. J. Dev. Neurosci.* **2**: 197-202.

Rabito, C. A., Kreisberg, J. I. & Wight, D. Alkaline phosphatase and γ-glutamyl transpeptidase as polarization markers during the organization of LLC-PK$_1$ cells into an epithelial membrane. *J. Biol. Chem.* **259**: 574-582.

Selvaraj, P., Rolston, D. D. K. & Balasubrananian, K. A. Separation of hydrophobic and hydrophilic forms of γ-glutamyl transferase from human serum by hydrophobic chromatography on phenyl-Sepharose CL-4B: studies on normal sera and sera of patients with liver disease. *Clin. Chim. Acta* **138**: 141-149.

Smith, G. D., Chakravarty, P. K., Connors, T. A. & Peters, T. J. Synthesis and preliminary characterization of a novel substrate for γ-glutamyl transferase. *Biochem. Pharmacol.* **33**: 527-529.

Valdivieso, M. P., Puente, J. F. & Sapag-Haga, M. Effect of hormones and cyclic AMP on γ-glutamyltranspeptidase activity of rat mammary gland explants. *Experientia* **40**: 710-711.

Wenham, P. R., Horn, D. B. & Smith, A. F. Physical properties of γ-glutamyltransferase in human serum. *Clin. Chim. Acta* **141**: 205-218.

Lysosomal γ-Glutamyl Carboxypeptidase

Summary

EC Number: 3.4.22.12 (formerly 3.4.12.10)

Earlier names: Conjugase, folate conjugase, vitamin Bc conjugase, folic
acid polyglutamate conjugase, pteroyl-oligoglutamyl conjugase, pteroylpoly-γ-glutamate hydrolase (PPH), γ-glutamyl hydrolase, folyl poly-γ-
glutamyl carboxypeptidase, γglutamylglutamate carboxypeptidase,
carboxypeptidase G.

Abbreviation: l-γ-GCP.

Distribution: Intestinal mucosa [1,2], liver [3,4] and kidney [5], where
it generally shows a lysosomal localization [1,4]. Also appears to be
lysosomal in human granulocytes [6] and placenta [7], and to occur in
pancreatic [8] and biliary [9] secretions. Plasma levels of l-γ-GCP are
relatively high in man and rat, but only moderate levels are found in
beef, monkey and guinea pig plasmas. The enzyme is undetectable in goat,
dog and rabbit plasmas [10].

γ-GTP activity has been detected in bacteria, yeast, and avian tissues
such as chicken liver [11] and pancreas [12].

Source: Beef [13] and human [3,14] liver; rat [15] and human [16]
intestinal mucosa. Generally, intestinal mucosa is by far the richest
source [17].

Action: Progressively removes γ-glutamyl residues from pteroylpoly-γ-
glutamyl substrates to yield, as final products, pteroylmonoglutamate
(folic acid) and free glutamic acid. Such an exopeptidase mechanism has
been described for l-γ-GCP from human liver [3] and intestinal mucosa
[16].

In contrast to these reports, the enzyme from rat bile [9] and
intestinal mucosa [15] has been characterized as an "endopeptidase"
because more than one glutamyl residue is removed at a time. However,
this alone does not disqualify the enzye as an exopeptidase (see
Introduction and Comment). A "mixed" hydrolysis pattern has been
described for l-γ-GCP from beef liver [13].

Human liver l-γ-GCP also acts on analogs of pteroyltri-glutamate in which another amino acid (such as glycine, serine, proline, leucine, phenylalanine or aspartic acid) is substituted for the terminal glutamic acid residue, but the rates of hydrolysis are lower. This low degree of specificity is seen only if the terminal residue is in γ linkage. Pteroylglutamyl-γ-Glu-γ-Leu is hydrolyzed at about half the rate of pteroylglutamyl-γ-Glu-γ-Glu, but no hydrolysis occurs if a second leucyl residue is added through the α linkage. The requirement for a C-terminal L-γ-peptide bond is further indicated by the active degradation of poly-L-γ-glutamic acid, whereas no action is seen on poly-D-γ-glutamic acid or poly-L-α-glutamic acid [3,17]. The enzyme is thus a relatively nonspecific γ-carboxypeptidase.

Requirements: pH 4.5 and 37°C in an acetate buffer containing 5 mM 2-mercaptoethanol [9]. Divalent cations such as Zn^{2+}, Ca^{2+}, Mg^{2+} and Mn^{2+} may show activation at 5 mM [6,13].

Substrate, usual: Pteroylglutamyl-γ-Glu-γ-[U-^{14}C]glutamic acid ("pteroyltriglutamate") in conjunction with the charcoal precipitation method [14].

Substrate, special: Pteroyl[U-^{14}C]glutamyl-hexa-γ-glutamate [16], pteroylhexa-γ-Glu-γ-[U-^{14}C]glutamate [14], and shorter polyglutamate chains labeled in intermediate positions as in pteroylglutamyl-γ-[U-^{14}C]Glu-γ-glutamate [14], all of which may be synthesized by a solid-phase procedure [18]. These substrates are used primarily for characterizing the specificity of the enzyme, and for establishing the end point of degradation.

Inhibitors: Thiol-blocking reagents such as p-chloromercuribenzoate, and polyanionic compounds such as dextran sulfate, chondroitin sulfate, heparin and hyaluronic acid [2,13]. DNA, RNA, deoxycholic acid, and polygalacturonic acid have also been cited as inhibitory polyanions [3]. Citrate and phosphate buffers are generally more inhibitory than acetate [3]. Prolonged exposure to chelating agents such as 1,10-phenanthroline may cause inactivation that is reversible with added Zn^{2+} [13].

Molecular properties: Beef liver l-γ-GCP has been characterized as a M_r 108,000 glycoprotein having a possible Zn^{2+} requirement [13]. A range of M_r values has been reported for l-γ-GCP derived from different sources: 45,000 for the human intestinal enzyme [2], 60,000 for the rat pancreatic enzyme [8], and 80,000 for the rat intestinal enzyme, pI 8.2 [15].

Comment: γ-Glutamyl carboxypeptidase (including Entry 18.05) has been included among the omega peptidases because it fails to meet the classical definition of a carboxypeptidase. Although it typically requires an unsubstituted C-terminal α-carboxyl group, it fails to cleave α-peptide linkages [19]. The fact that human liver l-γ-GCP activity releases virtually any γ-linked C-terminal amino acid [19] shows that the binding site specificity of the enzyme does not require the presence of a free C-terminal γ-carboxyl group like that in the pteroylpolyglutamates. On the other hand, its requirement for a free α-COOH, and the fact that such a group occurs at each γ-glutamyl linkage, gives reason to expect that cleavage might occur at other than the terminal γ-peptide bonds.

Although l-γ-GCP derived from some mammalian sources has been characterized as an "endopeptidase" [9,15], on the basis of its ability to remove more than one γ-glutamyl residue at a time, the more typical course of degradation appears to involve the sequential release of glutamyl residues from both pteroylpolyglutamates and poly-γ-glutamates [3,16].

It has been proposed that blood-born γ-GCP, which appears to arise from the liver [10], may serve to degrade folylpolyglutamates released into the blood during red cell death. However, like other lysosomal enzymes that appear in the blood, it probably has no special metabolic role.

Folic acid is ubiquitous in nature, and is commonly found in foods in the form of polyglutamate derivatives. These dietary "conjugates", which consist of one to seven glutamyl residues joined in γ-peptide linkage, subsequently appear in the blood as the unconjugated form of folic acid (pteroylmonoglutamate), largely as the result of their degradation by intestinal l-γ-GCP and microsomal γ-GCP (see Entry 18.05). The pteroylpolyglutamates that occur in liver and other tissues probably serve as storage forms of folic acid that require degradation by l-γ-GCP prior to their release into the blood as the monoglutamate - the only form believed to be capable of entering cells and participating in folate-dependent reactions.

See also Entry 18.05 for additional comment.

References
[1]   Hoffbrand & Peters *Biochim. Biophys. Acta* **192**: 479-485, 1969.
[2]   Reisenauer *et al.* *Science* **198**: 196-197, 1977.
[3]   Baugh & Krumdieck *Ann. N. Y. Acad. Sci.* **186**: 7-28, 1971.
[4]   Silink & Rowe *Biochim. Biophys. Acta* **381**: 28-36, 1975.
[5]   Buehring *et al.* *J. Biol. Chem.* **249**: 1081-1089, 1974.
[6]   Jägerstad & Olsson *Scand. J. Clin. Lab. Invest.* **39**: 343-349, 1979.
[7]   Landon *Int. J. Biochem.* **3**: 387-388, 1972.
[8]   Jägerstad *et al.* *Scand. J. Gastroenterol.* **7**: 593-597, 1972.
[9]   Horne *et al.* *J. Nutr.* **111**: 442-449, 1981.
[10] Lakshmaiah & Ramasastri *Int. J. Vitam. Nutr. Res.* **45**: 183-193, 1975.
[11] Rao & Noronha *Biochim. Biophys. Acta* **481**: 594-607, 1977.
[12] Leichter *et al.* *Proc. Soc. Exp. Biol. Med.* **154**: 98-101, 1977.
[13] Silink *et al.* *J. Biol. Chem.* **250**: 5982-5994, 1975.
[14] Krumdieck & Baugh *Anal. Biochem.* **35**: 123-129, 1970.
[15] Elsenhans *et al.* *J. Biol. Chem.* **259**: 6364-6368, 1984.
[16] Reisenauer & Halsted *Biochim. Biophys. Acta* **659**: 62-69, 1981.

[17] Hoffbrand & Peters *Schweiz. Med. Wochenschr.* **100**: 1954-1960, 1970.
[18] Krumdieck & Baugh *Biochemistry* **8**: 1568-1572, 1969.
[19] Baugh *et al. Biochim. Biophys. Acta* **212**: 116-125, 1970.

Bibliography

1944

Mims, V., Totter, J. R. & Day, P. L. A method for the determination of substances enzymatically convertible to the factor stimulating *Streptococcus lactis* R. *J. Biol. Chem.* **155**: 401-405.

1946

Bird, O. D., Robbins, M., Vandenbelt, J. M. & Pfiffner, J. J. Observations on vitamin B conjugase from hog kidney. *J. Biol. Chem.* **163**: 649-659.

1969

Butterworth, C. E., Jr., Baugh, C. M. & Krumdieck, C. A study of folate absorption and metabolism in man utilizing carbon-14-labeled polyglutamates synthesized by the solid phase method. *J. Clin. Invest.* **48**: 1131-1142.
Hoffbrand, A. V. & Peters, T. J. The subcellular localization of pteroyl polyglutamate hydrolase and folate in guinea pig intestinal mucosa. *Biochim. Biophys. Acta* **192**: 479-485.
Krumdieck, C. L. & Baugh, C. M. The solid-phase synthesis of polyglutamates of folic acid. *Biochemistry* **8**: 1568-1572.
Rosenberg, I. H., Streiff, R. R., Godwin, H. A. & Castle, W. B. Absorption of polyglutamic folate: participation of deconjugating enzymes of the intestinal mucosa. *N. Eng. J. Med.* **280**: 985-988.

1970

Baugh, C. M., Stevens, J. C. & Krumdieck, C. L. Studies on γ-glutamyl carboxypeptidase. I. The solid phase synthesis of analogs of polyglutamates of folic acid and their effects on human liver γ-glutamyl carboxypeptidase. *Biochim. Biophys. Acta* **212**: 116-125.
Bernstein, L. H., Gutstein, S. & Weiner, S. V. Gamma glutamyl carboxypeptidase (conjugase), the folic acid-releasing enzyme of intestinal mucosa. *Am. J. Clin. Nutr.* **23**: 919-925.
Hoffbrand, A. V. & Peters, T. J. Recent advances in knowledge of clinical and biochemical aspects of folate. *Schweiz. Med. Wochenschr.* **100**: 1954-1960.
Krumdieck, C. L. & Baugh, C. M. Radioactive assay of folic acid polyglutamate conjugase(s). *Anal. Biochem.* **35**: 123-129.

1971

Baugh, C. M., Krumdieck, C. L., Baker, H. J. & Butterworth, C. E., Jr. Studies on the absorption and metabolism of folic acid. I. Folate absorption in the dog after exposure of isolated intestinal segments to synthetic pteroylpolyglutamates of various chain lengths. *J. Clin. Invest.* **50**: 2009-2021.

1972

Jägerstad, M., Lindstrand, K. & Westesson, A.-K. Hydrolysis of conjugated folic acid by pancreatic 'conjugase'. *Scand. J. Gastroenterol.* **7**: 593-597.
Landon, M. J. Placental γ-glutamyl carboxypeptidase. *Int. J. Biochem.* **3**: 387-388.

1973

Godwin, H. A. Properties of γ-glutamyl peptidase (folic acid conjugase) from human plasma. *Clin. Res.* **21**: 555 only.

1974

Jägerstad, M., Lindstrand, K., Nordén, A., Westesson, A.-K. & Lindberg, T. The folate conjugase activity of the intestinal mucosa in celiac disease. *Scand. J. Gastroenterol.* **9**: 255-259.

1975

Lakshmaiah, N. & Ramasastri, B. V. Folic acid conjugase from plasma. I. Partial purification and properties. *Int. J. Vitam. Nutr. Res.* **45**: 183-193.
Lakshmaiah, N. & Ramasastri, B. V. Folic acid conjugase from plasma. II. Studies on the source of the enzyme in blood. *Int. J. Vitam. Nutr. Res.* **45**: 194-200.
Lakshmaiah, N. & Ramasastri, B. V. Folic acid conjugase from plasma. III. Use of the enzyme in the estimation of folate activity in foods. *Int. J. Vitam. Nutr. Res.* **45**: 262-272.
Lavoie, A., Tripp, E. & Hoffbrand, A. V. Sephadex-gel filtration and heat stability of human jejunal and serum pteroylpolyglutamate hydrolase (folate conjugase). Evidence for two different forms. *Biochem. Med.* **13**: 1-6.
Silink, M. & Rowe, P. B. The localization of glutamate carboxypeptidase in rat liver lysosomes. *Biochim. Biophys. Acta* **381**: 28-36.
Silink, M., Reddel, R., Bethel, M. & Rowe, P. B. γ-Glutamyl hydrolase (conjugase). Purification and properties of the bovine hepatic enzyme. *J. Biol. Chem.* **250**: 5982-5994.

1976

Eichner, E. R., Loewenstein, J. E. & Cox, E. Effect of ethanol on activity of folate conjugase and on serum binding and cellular uptake of radiolabelled folates. *Biochem. Soc. Trans.* **4**: 908-910.

Halsted, C. H., Reisenauer, A., Back, C. & Gotterer, G. S. In vitro uptake and metabolism of pteroylpolyglutamate by rat small intestine. *J. Nutr.* **106**: 485-492.

Krumdieck, C. L., Boots, L. R., Cornwell, P. E. & Butterworth, C. E., Jr. Cyclic variations in folate composition and pteroylpolyglutamyl hydrolase (conjugase) activity of the rat uterus. *Am. J. Clin. Nutr.* **29**: 288-294.

Shin, Y. S., Chan, C., Vidal, A. J., Brody, T. & Stokstad, E. L. R. Subcellular localization of γ-glutamyl carboxypeptidase and of folates. *Biochim. Biophys. Acta* **444**: 794-801.

1977

Leichter, J., Butterworth, C. E., Jr. & Krumdieck, C. L. Partial purification and some properties of pteroylpolyglutamate hydrolase (conjugase) from chicken pancreas. *Proc. Soc. Exp. Biol. Med.* **154**: 98-101.

McDonald, J. K. & Schwabe, C. Intracellular exopeptidases. In: *Proteinases in Mammalian Cells and Tissues* (Barrett, A. J. ed.), pp. 311-391 (see pp. 344-345), North-Holland Publishing Co., Amsterdam.

Rao, K. N. & Noronha, J. M. Studies on the enzymatic hydrolysis of polyglutamyl folates by chicken liver folyl poly-γ-glutamyl carboxypeptidase. I. Intracellular localization, purification and partial characterization of the enzyme. *Biochim. Biophys. Acta* **481**: 594-607.

Rao, K. N. & Noronha, J. M. Studies on the enzymatic hydrolysis of polyglutamyl folates by chicken liver folyl poly-γ-glutamyl carboxypeptidase. II. Structrual studies. *Biochim. Biophys. Acta* **481**: 608-616.

Reisenauer, A. M., Krumdieck, C. L. & Halsted, C. H. Folate conjugase: two separate activities in human jejunum. *Science* **198**: 196-197.

Thompson, R. W., Leichter, J., Cornwell, P. E. & Krumdieck, C. L. Alterations in the chain length of pteroylpoly-γ-glutamates and in the activity of pteroylpoly-γ-glutamate hydrolase in response to changes in the steady state of one carbon metabolism. *Am. J. Clin. Nutr.* **30**: 1583-1590.

1979

Jägerstad, M. & Olsson, I. Pteroylpolyglutamate hydrolase of human granulocytes. I. Partial purification and kinetic studies. *Scand. J. Clin. Lab. Invest.* **39**: 343-349.

1980

Lakshmaiah, N. & Ramasastri, B. V. Plasma folic acid conjugase. *Methods Enzymol.* **66**: 670-678.

1981

Horne, D. W., Krumdieck, C. L. & Wagner, C. Properties of folic acid γ-glutamyl hydrolase (conjugase) in rat bile and plasma. *J. Nutr.* **111**: 442-449.

Reisenauer, A. M. & Halsted, C. H. Human jejunal brush border folate conjugase. Characteristics and inhibition by salicylazosulfapyridine. *Biochim. Biophys. Acta* **659**: 62-69.

1982

Priest, D. G., Veronee, C. D., Mangum, M., Bednarek, J. M. & Doig, M. T. Comparison of folylpolyglutamate hydrolases of mouse liver, kidney, muscle and brain. *Mol. Cell. Biochem.* **43**: 81-87.

1983

Kesavan, V. & Noronha, J. M. Folate malabsorption in aged rats related to low levels of pancreatic folyl conjugase. *Am. J. Clin. Nutr.* **37**: 262-267.

1984

Elsenhans, B., Ahmad, O. & Rosenberg, I. H. Isolation and characterization of pteroylpolyglutamate hydrolase from rat intestinal mucosa. *J. Biol. Chem.* **259**: 6364-6368.

McGuire, J. J. & Coward, J. K. Pteroylpolyglutamates: biosynthesis, degradation, and function. In: *Folates and Pterins* (Blakley, R. L. & Benkovic, S. J. eds), pp. 135-190, Wiley, New York.

Microsomal γ-Glutamyl Carboxypeptidase

Summary

EC Number: 3.4.22.-

Earlier names: Brush border folate conjugase, membrane-bound folate conjugase.

Abbreviation: m-γ-GCP

Distribution: Membrane-bound on the brush border of human jejunal mucosa [1,2]. Compared to its levels in human biliary and pancreatic secretions, γ-GCP activity is several hundred-fold more concentrated in the jejunal and ileal mucosa [3].

Source: Jejunal mucosa from patients undergoing elective jejunoileal bypass surgery for treatment for obesity.

Action: Catalyzes the sequential release of glutamyl residues from the C-terminus of pteroylpoly-γ-glutamate, leaving pteroylmono-γ-glutamate (folic acid) available for intestinal transport.

Requirements: pH 7.5 Tris-HCl buffer at 37°C, with p-hydroxymercuri-benzoate (0.5 mM) added to inhibit possible contaminating lysosomal γ-GCP [2]. Further, the lysosomal enzyme is virtually inactive above pH 7.0.

Substrate, usual: Pteroyl-γ-Glu-γ-Glu-γ-[U-^{14}C]glutamic acid (K_m 0.16 μM) in conjunction with the charcoal adsorption method [4].

Substrate, special: Pteroyl[U-^{14}C]glutamyl-hexa-γ-glutamate (^{14}C-labeled PteGlu$_7$) synthesized by the solid phase method [5] is especially useful for characterizing the degradative mechanism [2].

Inhibitors: Salicylazosulfapyridine (a drug whose chronic use is associated with reduced serum folate levels) shows competitive inhibition (K_i 0.13 mM).

Molecular properties: The M_r 91,000 enzyme has a pH 6.5 optimum [1].

Comment: Although m-γ-GCP is capable of catalyzing the successive cleavage of γ-glutamyl linkages, the final linkage to the pteridine ring structure is resistant to attack, thus leaving pteroylmonoglutamate (folic acid) available for intestinal absorption. m-γ-GCP is most probably the enzyme that is primarily responsible for folate absorption by the brush border.

A bacterial Zn^{2+}-metalloenzyme known as carboxypeptidase G specifically cleaves the pteroyl glutamic acid linkage to release pteroic acid and free glutamic acid [6]. Carboxypeptidase G also hydrolyzes glutamate from folate analogs such as methotrexate, a drug used for antifolate chemotherapy. Because mammalian tissues lack such an enzyme, it has been possible to administer purified microbial carboxypeptidase G as a "rescue agent" to facilitate the selective treatment of central nervous system tumors [7]. The rationale for this therapeutic approach exploits the fact that, whereas carboxypeptdase G is able to spare most tissues from exposure to high levels of methotrexate, it is unable to cross the blood brain barrier.

Among three drugs whose chronic use is associated with low serum folate levels, namely ethanol, diphenylhydantoin and salicylazosulfapyridine, only the latter was found to inhibit m-γ-GCP. Thus, it appears that folate malabsorption in chronic alcoholism and during diphenylhydantoin therapy is not due to the impaired hydolysis of dietary pteroylpolyglutamates, but rather to the impaired transport of pteroylglutamic acid [2].

Whereas the human intestinal mucosa contains a microsomal γ-GCP in addition to the well known lysosomal enzyme, similar studies conducted with rat intestinal mucosa have revealed only the lysosomal form of the enzyme [8].

See also Entry 18.04 for additional comment.

References
[1] Reisenauer *et al.* *Science* **198**: 196-197, 1977.
[2] Reisenauer & Halsted *Biochim. Biophys. Acta* **659**: 62-69, 1981.
[3] Hoffbrand & Peters *Schweiz. Med. Wochenschr.* **100**: 1954-1960, 1970.
[4] Krumdieck & Baugh *Anal. Biochem.* **35**: 123-129, 1970.
[5] Krumdieck & Baugh *Biochemistry* **8**: 1568-1572, 1969.
[6] Goldman & Levy *Proc. Natl. Acad. Sci. USA* **58**: 1299-1306, 1967.
[7] Abelson *et al.* In: *Chemistry and Biology of Pteridines* (Kisliuk, R. L. & Brown, G. M. eds.), pp. 629-633, Elsevier/North-Holland, New York, 1978.
[8] Elsenhans *et al.* *J. Biol. Chem.* **259**: 6364-6368, 1984.

Bibliography

1977

Reisenauer, A. M., Krumdieck, C. L. & Halsted, C. H. Folate conjugase: two separate ativities in human jejunum. *Science* **198**: 196-197.

1981

Reisenauer, A. M. & Halsted, C. H.   Human jejunal brush border folate conjugase. Characteristics and inhibition by salicylazosulfapyridine. *Biochim. Biophys. Acta* **659**: 62-69.

1984

McGuire, J. J. & Coward, J. K.   Pteroylpolyglutamates: biosynthesis, degradation, and function.   In: *Folates and Pterins* (Blakley, R. L. & Benkovic, S. J. eds), pp. 135-190, Wiley, New York.

Acylaminoacyl Peptidase

Summary

EC Number: 3.4.19.1 (formerly 3.4.14.3)

Earlier names: Acylamino acid-releasing enzyme, formylmethionine-releasing enzyme, N-formylmethionine (fMet) aminopeptidase, N-acetylalanine (AcAla) aminopeptidase, N-acetylaminoacyl-p-nitranilidase, N-acylpeptide hydrolase.

Abbreviation: AcAP.

Distribution: AcAP, which appears to be cytosolic, shows high levels in liver, spleen and reticulocytes; moderate levels in erythrocytes and lung; and low levels in muscle, heart and brain of the rat [1,2]. It is also plentiful in the cytosol of human erythrocytes, but is absent from leukocytes [3].

Source: Rat [2,4], beef [5] and hog [6] liver, human placenta [7], and sheep [8] and human [3] red cells.

Action: Releases acylamino acids from N-acylated peptides, but has no action on N-acylated amino acids or pyroglutamyl peptides [1,5]. Although exceptions have been noted (see below), non-acylated peptides are generally not attacked. N-Acetyl peptides are hydrolyzed 4-, 8-, and 10-times faster than the corresponding formyl, propionyl, and butyryl peptides, respectively [5]. No action occurs on Tos-Arg-OMe, Bz-Arg-OEt, or Ac-Tyr-OEt [4].

A comparison of rates on N-acetylated alanyl peptides (Ac-Ala$_2$ through Ac-Ala$_5$) shows Ac-Ala released fastest from Ac-Ala$_3$ (100%), and slowest from Ac-Ala$_2$ (7%). Rates on the others, including Ac-Ala$_4$-OMe, range from 50-60% [5]. At pH 8.5-9.0, non-acylated Ala$_3$ is acted upon as a consequence of enzymatic misrecognition: the unprotonated terminal alanyl residue is seen as an acylating group, thus giving rise to an apparent dipeptidyl peptidase activity [6].

By comparison with the rate of hydrolysis of Ac-Met-Thr (100), rates for other acetylated peptides were as follows: Ac-Met-Ala 95, Ac-Ala-Tyr 79, Ac-Ser-Tyr 67, Ac-Ser-Thr-NH$_2$ 65, Ac-Leu-Ala-Gly 62, Ac-Ala-Ala 38,

Ac-Met-OMe 38, Ac-Met-Thr-OMe 29, Ac-Phe-Gly-Phe 27, Ac-Gly-Ser 14, Ac-

Met-NH$_2$ 5. No action occurs on Met-Thr, Ac-D-Ala-Ala, Ac-Trp-Ala, Ac-Ala-Leu, or Ac-Gly-Asp-Val-Glu. The rate on formyl-Met-Thr is about 29% of that on Ac-Met-Thr [1].

Requirements: pH 7.5-8.0 in a phosphate buffer at 37°C.

Substrate, usual: N-Acetyl-Met-Thr (K$_m$ 0.8 mM) in conjunction with a ninhydrin colorimetric procedure for determining free threonine [1,2]; N-formyl-Met-Leu (K$_m$ 0.03 mM) [4].

Substrate, special: N-Acetyl-Ala-NPhNO$_2$ (pH 8.1 optimum, K$_m$ 2 mM) [7] and N-formyl-Met-NNap (pH 7.8 optimum, K$_m$ 0.2 mM) [4]. (Amino acid 2-naphthylamides can be formylated according to a published procedure [9].)

Inhibitors: Dip-F (1 mM) and sulfhydryl-blocking reagents (1 mM) such as p-chloromercuribenzoate and HgCl$_2$ show complete inhibition, whereas IAcOH, IAcNH$_2$ and MalNEt have little or not effect [1,4,8]. p-Chloromercuribenzoate and 2,2'-dipyridyl disulfide, which are notably inhibitory, probably have a denaturing effect in view of their atypical, irreversible action [3]. Pms-F (3 mM) is weakly inhibitory [4], whereas sulfhydryl compounds may show slight activation [1]. AcAP is unaffected by EDTA (1 mM) [1,4], and 0.1 mM concentrations of leupeptin, pepstatin, bestatin, amastatin and phosphoramidon [4].

Molecular properties: Reported M_r values for native polymeric AcAP range from 290,000 [4] to 390,000 [1]. The rat liver enzyme (pI 4.25) is reported to contain about five M_r 75,000 subunits [1] or four M_r 72,000 subunits [4]. M_r values of 300,000 and 380,000 have been reported for the human red cell [3] and placental [7] enzymes, respectively. The latter has two major forms (pI 3.9 and 4.5).

   Activity is most stable at pH 6.0-8.0; rapid inactivation occurs below pH 5.0 [3]. Losses of 91% and 16% occur during 2 weeks of storage at -20°C and -80°C, respectively [5]; unstable to freeze-drying [2]. Relatively heat stable, as indicated by the benefits of a heat treatment step (5 min at 60°C) used in purification [2].

Comment: Although some observations [4] make it appear that "N-formylmethionine aminopeptidase" may be a distinct enzyme, it now seems reasonable to attribute this activity to AcAP. This view is supported by the similar molecular properties and inhibitor sensitivities of the enzymes, and especially by recently published explanations for apparent discrepancies in substrate specificity and pH requirements [6]. Additionally, enzymatic misrecognition by AcAP [6] now accounts for the action of AcAP on non-acylated peptides by the sheep red cell [8] and beef liver [5] enzymes. Misrecognition results in the release of N-terminal dipeptides at alkaline pH from non-acylated peptides containing small uncharged terminal amino acids such as glycine or alanine [5]. Caution is required since such action can mistakenly be attributed to a "new" dipeptidyl peptidase, or even to a familiar one such as DPP IV (Entry 13.04).

AcAP was used successfully to initiate the sequence analysis of an octapeptide containing N-acetylserine [2,10]. The lack of action on acetylated native proteins such as albumin and hemoglobin [7], may possibly be overcome by taking advantage of the activity of AcAP in urea (about 50% in 4 M urea [5]).

In view of the finding that 80-90% of the soluble proteins in eukaryotic cells are N-acetylated [11], it seems most probable that AcAP contributes to their catabolism.

References
  [1] Tsunasawa *et al.* *J. Biochem.* **77**: 89-102, 1975.
  [2] Tsunasawa & Narita *Methods Enzymol.* **45**: 552-561, 1976.
  [3] Schönberger & Tschesche *Hoppe-Seyler's Z. Physiol. Chem.* **362**: 865-873, 1981.
  [4] Suda *et al.* *Biochim. Biophys. Acta* **616**: 60-67, 1980.
  [5] Gade & Brown *J. Biol. Chem.* **253**: 5012-5018, 1978.
  [6] Tsunasawa *et al.* *J. Biochem.* **93**: 1217-1220, 1983.
  [7] Unger *et al.* *Eur. J. Biochem.* **97**: 205-211, 1979.
  [8] Witheiler & Wilson *J. Biol. Chem.* **247**: 2217-2221, 1972.
  [9] Sheehan & Yang *J. Am. Chem. Soc.* **80**: 1154-1158, 1958.
  [10] Nakamura *et al.* *Biochem. Biophys. Res. Commun.* **58**: 250-256, 1974.
  [11] Brown & Roberts *J. Biol. Chem.* **251**: 1009-1014, 1976.

Bibliography

1972

Witheiler, J. & Wilson, D. B. The purification and characterization of a novel peptidase from sheep red cells. *J. Biol. Chem.* **247**: 2217-2221.

Yoshida, A. & Lin, M. NH_2-Terminal formylmethionine- and NH_2-terminal methionine-cleaving enzymes in rabbits. *J. Biol. Chem.* **247**: 952-957.

1975

Tsunasawa, S., Narita, K. & Ogata, K. Purification and properties of acylamino acid-releasing enzyme from rat liver. *J. Biochem.* **77**: 89-102.

1976

Tsunasawa, S. & Narita, K. Acylamino acid-releasing enzyme from rat liver. *Methods Enzymol.* **45B**: 552-561.

1978

Gade, W. & Brown, J. L. Purification and partial characterization of α-N-acylpeptide hydrolase from bovine liver. *J. Biol. Chem.* **253**: 5012-5018.

1979

Unger, T., Nagelschmidt, M. & Struck, H. N-Acetylaminoacyl-*p*-nitranilidase from human placenta. Purification and some properties. *Eur. J. Biochem.* **97**: 205-211.

1980

Suda, H., Yamamoto, K., Aoyagi, T. & Umezawa, H. Purification and properties of N-formylmethionine aminopeptidase from rat liver. *Biochim. Biophys. Acta* **616**: 60-67.

1981

Schönberger, O. L. & Tschesche, H. N-Acetylalanine aminopeptidase, a new enzyme from human erythrocytes. *Hoppe-Seyler's Z. Physiol. Chem.* **362**: 865-873.

1983

Tsunasawa, S., Imanaka, T. & Nakazawa, T. Apparent dipeptidyl peptidase activities of acylamino acid-releasing enzymes. *J. Biochem.* **93**: 1217-1220.

Peptidyl Aminoacylamidase

Summary

EC Number: 3.4.19.2 (formerly 3.4.15.2)

Earlier names: Peptidyl glycinamidase, carboxyamidopeptidase, carboxyamidase, peptidyl carboxyamidase, oxytocinase, antidiuretic hormone-inactivating enzyme.

Abbreviation: PAA.

Distribution: A membrane-bound enzyme that is especially plentiful in the renal medulla [1], and possibly the uterus as well. A comparable membrane-bound, non-mammalian enzyme is present in the urinary bladder [2,3] and skin [4] of the toad (*Bufo marinus*), where it appears to be located primarily in the connective tissue rather than the epithelium [3].

Source: Renal plasma membranes from hog kidney medulla [1]. Toad skin is a rich source of the non-mammalian form of the enzyme [4].

Action: Releases aminoacylamide moieties from the C-terminus of polypeptides. Mammalian PAA cleaves the $Lys_8 \diagup Gly_9\text{-}NH_2$ bond in lysine vasopressin to release $Gly\text{-}NH_2$ [1]. The purified toad skin enzyme, which has been more thoroughly characterized, also releases $Gly\text{-}NH_2$ from arginine vasopressin and oxytocin, by cleaving the $Arg_8 \diagup Gly_9\text{-}NH_2$ bond and the $Leu_8 \diagup Gly_9\text{-}NH_2$ bond, respectively [4]. This dual trypsin-chymotrypsin-like activity is also seen on Bz-Arg-OEt and Ac-Tyr-OEt [4].

The toad skin enzyme also shows peptidyl aminoacylamidase activity on substrates such as $Z\text{-}Pro\text{-}Leu\text{-}Gly\text{-}NH_2$, $Z\text{-}Leu\text{-}Gly\text{-}NH_2$, $Z\text{-}Leu\text{-}Phe\text{-}NH_2$, and $Z\text{-}Pro\text{-}Leu\text{-}Lys(Z)\text{-}NH_2$, but not on other peptides such as $Leu\text{-}Gly\text{-}NH_2$, $Leu\text{-}Phe\text{-}NH_2$, $Z\text{-}D\text{-}Leu\text{-}Gly\text{-}NH_2$, $Z\text{-}Pro\text{-}Met\text{-}Gly\text{-}NH_2$, $Z\text{-}Pro\text{-}Glu\text{-}Gly\text{-}NH_2$, $Z\text{-}Pro\text{-}Leu\text{-}Glu\text{-}NH_2$, $Z\text{-}Val\text{-}Ser\text{-}NH_2$, $pGlu\text{-}His\text{-}Pro\text{-}NH_2$ (thyroliberin) and $pGlu\text{-}His\text{-}Trp\text{-}Ser\text{-}Tyr\text{-}Gly\text{-}Leu\text{-}Arg\text{-}Pro\text{-}Gly\text{-}NH_2$ (luliberin) [5].

At pH 5.5, the purified toad PAA exhibits a "serine" carboxypeptidase activity that closely resembles that of lysosomal carboxypeptidase A (cathepsin A, Entry 15.03). Z-Glu-Tyr (a common assay substrate for the lysosomal enzyme) is readily hydrolyzed by PAA, as are several other

Z-protected di-, tri-, and tetra-peptides. Under acidic conditions, a C-terminal attack occurs on substance P that results in the sequential release of Met-NH$_2$ and the penultimate leucyl residue [5].

Requirements: Phosphate buffer at pH 7.5 and 37°C; activity may be enhanced by EDTA [1,4].

Substrate, usual: Oxytocin or vasopressin, with liberated Gly-NH$_2$ determined by reaction with 2,4,6-trinitrobenzenesulfonic acid, the chromatographically isolated trinitrophenylglycinamide being measured spectrophotometrically at 337 nm (ϵ-1.66 x 10^{4}M^{-1}cm^{-1}) [1]. Substrates specifically radiolabeled in the Gly-NH$_2$ moiety may also be used, and the electrophoretically isolated [^{14}C]Gly-NH$_2$ determined with a radiochromatogram scanner [4].

   Bioassays such as the rat pressor assay [1,6] and the uterotonic assay [7] have been used to follow residual levels of vasopressin and oxytocin, respectively.

Substrate, special: Bz-Arg-OEt or Ac-Tyr-OEt may be used to measure the esterase activity of PAA preparations that are free of interfering activities. A spectrophotometric method [8] has been used to follow rates of ester hydrolysis, which are maximal at about pH 8.0 [4].

Inhibitors: Mammalian PAA is reported to be sensitive to Ca^{2+} and Mg^{2+} [1], but little information is available on its sensitivity to common protease inhibitors.

   The toad skin enzyme is inhibited by Dip-F and other serine protease inhibitors including Tos-Lys-CH$_2$Cl, Tos-Phe-CH$_2$Cl, aprotinin, ovomucoid, lima bean trypsin inhibitor, antipain, leupeptin, chymostatin and elastatinal. Pepstatin, bestatin and phosphoramidon are without effect. The reversible inhibition shown by 4-chloromercuribenzoate (50 μM) suggests that a sulfhydryl group is located near the active site of PAA, or possibly is important to its conformation [4].

Molecular properties: The detergent-solubilized enzyme from renal plasma membranes tends to form large aggregates (M_r about 1,500,000) comprised of M_r 442,000 monomers [1]. The toad skin enzyme forms M_r 320,000 aggregates consisting of M_r 100,000 (glycoprotein) monomers that contain two identical M_r 48,000 subunits. Each subunit consists of a heavy chain (M_r 28,000) and light chain (M_r 19,000) covalently linked by disulfide bond(s). The active-site serine is located in the heavy chain [4].

Comment: Peptidyl aminoacylamidase has striking similarities to lysosomal carboxypeptidase A (*l*CPA, Entry 15.03). Both are "serine" proteases that hydrolyze esters at basic pH and release C-terminal amino acids from peptides at acidic pH. Both are relatively heat labile, and unstable to storage above pH 8.0. *l*CPA has been shown to release Met-NH$_2$ from substance P. Lower rates were seen on oxytocin and vasopressin, but only acidic conditions were used [9].

   Early reports described the release of Gly-NH$_2$ from oxytocin by a "soluble" enzyme derived from the uterus [10], kidney [11], brain [12], and liver [13] of the rat. In view of the impure nature of the preparations used, and the recognized peptidyl aminoacylamidase activity of lysosomal carboxypeptidase A, it seems probable that the latter enzyme was responsible for the Gly-NH_2-releasing activity detected.

References

[1] Nardacci et al. Biochim. Biophys. Acta **377**: 146-157, 1975.
[2] Campbell et al. Life Sci. **4**: 2129-2140, 1965.
[3] Simmons & Walter Fed. Proc. Fed. Am. Soc. Exp. Biol. **38**: 946 only, 1979.
[4] Simons & Walter Biochemistry **19**: 39-48, 1980.
[5] Simmons & Walter In: Neurohypophyseal Peptide Hormones and Other Biologically Active Peptides (Schlesinger, D. H. ed.), pp. 151-165, Elsevier/North-Holland, New York, 1981.
[6] Dekanski Br. J. Pharmacol. **7**: 567-572, 1952.
[7] Holton Br. J. Pharmacol. **3**: 328-334, 1948.
[8] Schwert & Takenaka Biochim. Biophys. Acta **16**: 570-575, 1955.
[9] Matsuda J. Biochem. **80**: 659-669, 1976.
[10] Glass et al. Endocrinology **87**: 730-737, 1970.
[11] Koida et al. Endocrinology **88**: 633-643, 1971.
[12] Marks et al. Proc. Soc. Exp. Biol. Med. **142**: 455-460, 1972.
[13] Fruhaufová et al. Coll. Czech. Chem. Commun. **38**: 2793-2798, 1973.

Bibliography

1965

Campbell, B. J., Thysen, B. & Chu, F. S. Peptidase catalyzed hydrolysis of antidiuretic hormone in toad bladder. Life Sci. **4**: 2129-2140.

1969

Glass, J. D., Schwartz, I. L. & Walter, R. Enzymatic inactivation of peptide hormones possessing a C-terminal amide group. Proc. Natl. Acad. Sci. USA **63**: 1426-1430.

1975

Nardacci, N. J., Mukhopadhyay, S. & Campbell, B. J. Partial purification and characterization of the antidiuretic hormone-inactivating enzyme from renal plasma membranes. Biochim. Biophys. Acta **377**: 146-157.

1979

Simmons, W. H. & Walter, R. Degradation of vasopressin by toad urinary bladder peptidases. Fed. Proc. Fed. Am. Soc. Exp. Biol. **38**: 946 only.

1980

Simmons, W. H. & Walter, R. Carboxamidopeptidase: purification and characterization of a neurohypophyseal hormone inactivating peptidase from toad skin. *Biochemistry* **19**: 39-48.

1981

Simmons, W. H. & Walter, R. Enzyme inactivation of oxytocin: properties of carboxamidopeptidase. In: *Neurohypophyseal Peptide Hormones and Other Biologically Active Peptides* (Schlesinger, D. H. ed.), pp. 151-165, Elsevier/North-Holland, New York.

ß-Aspartyl Peptidase

Summary

EC Number: 3.4.13.10 (but see Comment)

Earlier names: β-Aspartyl dipeptidase.

Abbreviation: β-AP.

Distribution: In the rat, the highest levels occur in liver and kidney. Relatively low but still significant activities in brain, lung, skeletal muscle, and heart muscle. The enzyme appears to be freely soluble in the cytosol [1].

Source: Rat liver or kidney [2]. *E. coli* serves as a richer, non-mammalian source of the enzyme [3].

Action: Catalyzes the specific hydrolysis of β-aspartyl linkages at the N-termini of β-aspartyl peptides. The approximate relative rates at pH 7.2 for the rat liver enzyme are: β-Asp-Gly 100, β-Asp-Gly-Gly 95, β-Asp-Met 82, β-Asp-Leu 65, β-Asp-Ser 56, β-Asp-Gly-Ala 55, and β-Asp-Ala 51. α-Asp-Gly, γ-Gly-Leu, and β-Ala-His (carnosine) are not attacked [1].

Requirements: Phosphate buffer at pH 7.5–8.0 and 37°C. Thiol compounds may produce a slight activation.

Substrate, usual: β-Asp-Gly, with liberated glycine determined colorimetrically by the ninhydrin method [4].

Substrate, special: None has been reported, but chromogenic and fluorogenic β-aspartyl arylamide derivatives should be considered.

Inhibitors: Thiol blocking reagents such as 4-chloromercuribenzoate, but not IAcNH$_2$. Activity is unaffected by EDTA.

Molecular properties: Activity is stable to freezing and thawing in 0.1 M sodium phosphate buffer, pH 7.5, but is unstable in water or Tris-HCl buffer. NaCl (10 mM) has a notable enhancing and stabilizing effect. Little else has been reported on the properties of the mammalian enzyme. Bacterial β-AP is reported to be an enzyme of M_r 120,000 [3].

Comment: In the 1984 recommendations of the IUB Nomenclature Committee, β-aspartyl peptidase is classified as a dipeptidase. This would not appear to be justified since its action is not restricted to dipeptides. We classify β-AP as an omega peptidase (EC 3.4.19.-) on the basis of its specific action on N-terminal isopeptide (β-aspartyl) linkages.

References

[1] Dorer *et al. Arch. Biochem. Biophys.* **127**: 490-495, 1968.
[2] Haley *Methods Enzymol.* **19**: 737-741, 1970.
[3] Haley *Methods Enzymol.* **19**: 730-737, 1970.
[4] Matheson & Tattrie *Can. J. Biochem.* **42**: 95-103, 1964.

Bibliography

1968

Dorer, F. E., Haley, E. E. & Buchanan, D. L. The hydrolysis of β-aspartyl peptidase by rat tissue. *Arch. Biochem. Biophys.* **127**: 490-495.

1970

Haley, E. E. β-Aspartyl peptidase from rat liver. *Methods Enzymol.* **19**: 737-741.

INDEX

351